Lecture Notes in Computer Science 2520
Edited by G. Goos, J. Hartmanis, and J. van Leeuwen

Springer
Berlin
Heidelberg
New York
Hong Kong
London
Milan
Paris
Tokyo

Manolis Koubarakis Timos Sellis
Andrew U. Frank Stéphane Grumbach
Ralf Hartmut Güting Christian S. Jensen
Nikos Lorentzos Yannis Manolopoulos
Enrico Nardelli Barbara Pernici
Hans-Jörg Schek Michel Scholl
Babis Theodoulidis Nectaria Tryfona (Eds.)

Spatio-Temporal Databases

The CHOROCHRONOS Approach

Series Editors

Gerhard Goos, Karlsruhe University, Germany
Juris Hartmanis, Cornell University, NY, USA
Jan van Leeuwen, Utrecht University, The Netherlands

Volume Editors

Manolis Koubarakis
Timos Sellis et al.
see page V

Cataloging-in-Publication Data applied for

A catalog record for this book is available from the Library of Congress

Bibliographic information published by Die Deutsche Bibliothek
Die Deutsche Bibliothek lists this publication in the Deutsche Nationalbibliographie; detailed bibliographic data is available in the Internet at <http://dnb.ddb.de>.

CR Subject Classification (1998): H.2, J.1, H.3

ISSN 0302-9743
ISBN 3-540-40552-6 Springer-Verlag Berlin Heidelberg New York

This work is subject to copyright. All rights are reserved, whether the whole or part of the material is concerned, specifically the rights of translation, reprinting, re-use of illustrations, recitation, broadcasting, reproduction on microfilms or in any other way, and storage in data banks. Duplication of this publication or parts thereof is permitted only under the provisions of the German Copyright Law of September 9, 1965, in its current version, and permission for use must always be obtained from Springer-Verlag. Violations are liable for prosecution under the German Copyright Law.

Springer-Verlag Berlin Heidelberg New York
a member of BertelsmannSpringer Science+Business Media GmbH

http://www.springer.de

© Springer-Verlag Berlin Heidelberg 2003
Printed in Germany

Typesetting: Camera-ready by author, data conversion by Christian Grosche, Hamburg
Printed on acid-free paper SPIN: 10873073 06/3142 5 4 3 2 1 0

Volume Editors

Andrew U. Frank
Dept. of Geoinformation
Technical University of Vienna
A-1040 Vienna, Austria
frank@geoinfo.tuwien.ac.at

Stéphane Grumbach
INRIA
Rocquencourt BP 105
78153 Le Chesnay Cedex
France
stephane.grumbach@inria.fr

Ralf Harmut Güting
Praktische Informatik IV
FernUniversität Hagen
58084 Hagen, Germany
Ralf-Hartmut.Gueting
@FernUni-Hagen.de

Christian S. Jensen
Department of Computer Science
Aalborg University
DK-9220 Aalborg Øst
Denmark
csj@cs.auc.dk

Manolis Koubarakis
Dept. of Electronic
and Computer Engineering
Technical University of Crete
University Campus - Kounoupidiana
GR-73100 Chania, Crete, Greece
manolis@intelligence.tuc.gr

Nikos Lorentzos
Informatics Laboratory
Agricultural University of Athens
Iera Odos 75
GR-11855 Athens, Greece
lorentzos@aua.gr

Yannis Manolopoulos
Department of Informatics
Aristotle University
GR-54006 Thessaloniki, Greece
manolopo@csd.auth.gr

Enrico Nardelli
Dipartimento di Informatica
Università degli Studi di L'Aquila
I-67010 L'Aquila
Italy
Istituto di Analisi dei Sistemi
ed Informatica "Antonio Ruberti"
Consiglio Nazionale delle Ricerche
I-00185 Roma
Italy
nardelli@univaq.it

Barbara Pernici
Dip. Elettronica e Informazione
Politecnico di Milano
Piazza Leonardo da Vinci 32
I-20133 Milano
Italy
pernici@elet.polimi.it

Hans-Jörg Schek
ETH Zürich
Institute of Information Systems
ETH Zentrum, IFW C49.1
CH-8092 Zürich
Switzerland
schek@inf.ethz.ch

Michel Scholl
INRIA
Rocquencourt BP 105
78153 Le Chesnay Cedex
France
michel.scholl@inria.fr

Timos Sellis
Dept. of Electrical
and Computer Engineering
National Technical
University of Athens
GR-15773 Zographou
Athens
Greece
timos@dblab.ece.ntua.gr

Babis Theodoulidis
Dept. of Computation
UMIST, P.O. Box 88
Manchester, M60 1QD
United Kingdom
babis@co.umist.ac.uk

Nectaria Tryfona
Department of Computer Science
Aalborg University
DK-9220 Aalborg Øst, Denmark
tryfona@cs.auc.dk

Preface

This book is an introduction and source book for practitioners, graduate students, and researchers interested in the state of the art and practice in spatiotemporal databases. It collects the most important and representative research carried out in the project CHOROCHRONOS and presents it in a unified fashion. CHOROCHRONOS was a Training and Mobility Research Network funded by the European Commission with the objective to study the design, implementation, and application of spatiotemporal database management systems.

This book would never have been possible if it was not for the devoted work of many people. First and foremost, we would like to thank the authors of the nine chapters of this book for their hard work. We would also like to acknowledge the help of Christiane Bernard, our officer from the European Commission, who saw the project to its conclusion, working as hard as we did to make it a thorough success. The constructive comments and feedback of our reviewer Colette Roland (University of Paris-1) are also very much appreciated. Last, but not least, we would like to thank all the students and postdoctoral fellows who were trained during CHOROCHRONOS. We hope the time they spent at CHOROCHRONOS node institutions was rewarding and lots of fun!

March 2003

Timos Sellis
Manolis Koubarakis
Andrew Frank, Vienna
Stéphane Grumbach
Ralf Hartmut Güting
Christian Jensen
Nikos Lorentzos
Yannis Manolopoulos
Enrico Nardelli
Barbara Pernici
Babis Theodoulidis
Nectaria Tryfona
Hans-Jörg Schek
Michel Scholl

Table of Contents

1 Introduction

Manolis Koubarakis[1] and Timos Sellis[2]

[1] Technical University of Crete, Greece
[2] National Technical University of Athens, Greece

1.1 Why Spatio-temporal Databases?

Time and space are ubiquitous aspects of reality. It is hard to imagine a day in our life going by without making use of temporal and spatial information. In today's digital world, there is hardly any information system where time, space or both need not be present as first-class concepts in order to effectively support the targeted application. The reader is invited to consider the information systems of such varied organizations as financial institutions, public administrations, factories, hospitals, environmental agencies, entertainment industries, the military, space agencies, schools or universities, to name a few.

Starting with Aristotle, philosophers, mathematicians and physicists have developed fascinating formal accounts of time and space culminating in Einstein's theory of general relativity [34,35,37,48,11]. In the last three decades, computer scientists took a strong interest in studying time and space from the point of view of various research areas and applications. There has been a lot of interesting work in temporal logic [30,31,18,4,3], computational geometry [33], databases [21,40,46,24,43,16,44,32,38], artificial intelligence [42,12,26,36], [14], geographic information systems [29,49,1], environmental information systems [22], multimedia information systems [17,8,2,45,13] and so on. Researchers in temporal and spatial databases in particular have been very productive, developing important techniques for the representation, and management of spatial information, together with prototype systems. All this work is nicely surveyed in [21,40,46,24,43,16,44,32,38].

Until the beginning of the nineties, it was quite noticeable that temporal and spatial databases were important but separate areas of database research, with their own specialized meetings and conferences [39,47,10,15,41,25]. However, researchers in both areas had felt that there were important connections in the problems addressed by each area, as well as in the techniques and tools used for their solution. Many papers in temporal databases at that time concluded with phrases such as "the ideas in this paper may be extended to spatial databases". Similarly, many papers in spatial databases suggested that techniques developed for spatial databases apply to temporal databases, by restricting attention to one dimension only. But with the exception of pioneering papers such as [20,9,50], little systematic interaction and synergy between these two areas had occurred. The project CHOROCHRONOS, initiated by the European Commission in August 1996, aimed to achieve exactly this kind of interaction and synergy. Two years earlier, in North America, the National Center for Geographic Informa-

T. Sellis et al. (Eds.): Spatio-temporal Databases, LNCS 2520, pp. 1–8, 2003.
© Springer-Verlag Berlin Heidelberg 2003

tion and Analysis had started the initiative "Time in Geographic Space", with somewhat similar goals.[1]

1.2 CHOROCHRONOS

CHOROCHRONOS was established as a European Commission funded Training and Mobility of Researchers Network,[2] with the objective of studying the design, implementation, and application of spatio-temporal database management systems (STDBMS). The participants of the network were the following institutions: National Technical University of Athens (Timos Sellis), Aalborg University (Christian S. Jensen), FernUniversität Hagen (Ralf Güting), Universita Degli Studi di L'Aquila (Enrico Nardelli), UMIST (Manolis Koubarakis and Babis Theodoulidis), Politecnico di Milano (Barbara Pernici), INRIA (Stéphane Grumbach and Michel Scholl), Aristotle University of Thessaloniki (Yannis Manolopoulos), Agricultural University of Athens (Nikos Lorentzos), Technical University of Vienna (Andrew Frank), and ETH Zürich (Hans-Jörg Schek). All these institutions had established research groups in spatial and temporal database systems, most of which had so far been working exclusively on spatial or temporal databases. CHOROCHRONOS enabled them to collaborate closely and to integrate their findings in their respective areas. The project was coordinated by Timos Sellis.

CHOROCHRONOS had the following main objectives:

- To stimulate research in the areas of spatial and temporal databases, paying particular attention to the many real-life problems that require spatio-temporal concepts that go *beyond* traditional research in spatial and temporal databases.
- To allow researchers working on spatial and temporal databases to improve their understanding of each other's work, to integrate their results, and to avoid duplication of work. The joint design and partial implementation of one or many STDBMS prototypes would be a desirable outcome of this collaboration.
- To allow researchers working on temporal and spatial databases to cooperate with researchers from other disciplines (environmental sciences, multimedia, transportation etc.) that are faced with spatial and temporal information and that would benefit from spatio-temporal database technology.

To achieve these objectives, CHOROCHRONOS pursued an extensive research program, covering issues related to the ontology, structure, and represen-

[1] For more information, see `http://www.spatial.maine.edu/~max/I10TechReport.html`.

[2] The objective of Training and Mobility of Researchers Networks funded under the ESPRIT programme of the European Commission was to pull together a significant amount of expertise in an area of interest, to push the frontiers of research in this area, and at the same time facilitate the training and mobility of young researchers in Europe.

tation of space and time; data models and query languages for STDBMS; graphical user interfaces for spatio-temporal information; query processing algorithms, storage structures and indexing techniques; and architectures and implementation techniques for STDBMSs.

1.3 Contributions

Put briefly, a spatio-temporal database embodies spatial, temporal, and spatio-temporal database concepts, and it captures simultaneously spatial and temporal aspects of data. All the individual spatial and temporal concepts (e.g., rectangle or time interval) must be considered. However, attention focuses on the area where the two classes of concepts intersect, which is challenging, as it represents inherently spatio-temporal concepts (e.g., velocity or acceleration) observed in objects that may be changing their spatial characteristics continuously (e.g., moving objects). Simply extending a spatial data model with a temporal dimension or vice versa will result in a temporal data model that may capture spatial data, or in a spatial data model that may capture time-referenced sequences of spatial data. However, this simple aggregation of space and time is inadequate for general spatio-temporal data management.

Research in CHOROCHRONOS concentrated on the following six tasks:

- *Ontology, structure, and representation of space and time.* This involved the study of temporal and spatial ontologies, including their interrelations and their utility in STDBMS. As explained above, particular emphasis was put on concepts that are clearly spatio-temporal and cannot be obtained by a simple aggregation of temporal and spatial dimensions.
- *Models and languages for STDBMS.* The focus here was on three topics: (i) the study of languages for spatio-temporal relations, (ii) the development of models and query languages for spatio-temporal databases, and (iii) the provision of techniques for designing spatio-temporal databases. This work built on previous proposals and covered relational, object-oriented and constraint databases.
- *Graphical user interfaces for spatio-temporal information.* Research in this area had two goals: (i) to extend graphical interfaces for temporal and spatial databases, and (ii) to develop better visual interfaces for specific applications (e.g., VRML for time-evolving spaces).
- *Storage structures, indexing techniques, and query processing algorithms for spatio-temporal databases.* Techniques for the efficient evaluation of queries were the focus of this area. These techniques included high-level algebraic optimizations, new indexing techniques, and low-level strategies for page/object management.
- *Architectures for STDBMS.* The study of alternative architectures and implementation techniques for STDBMS was of high interest. As a result, several prototype STDBMS were developed in the project to demonstrate the innovations of more theoretical work.

- *Applications of STDBMS.* Finally, applying STDBMS in realistic problem settings guided our research throughout the project. We were particularly interested in applying our concepts and techniques not only to traditional target applications (e.g., GIS), but also to more advanced ones where the connection to a standard database approach was not obvious.

In summary, CHOROCHRONOS was a very exciting project. New ideas came out and collaboration among the specific teams has brought a lot of new and interesting issues to work on. The participating teams were very enthusiastic about this collaboration, and a lot of high-quality work was produced. Many joint papers in major conferences and journals were published, and various prototype STDBMS and related systems were developed. Several dissemination activities were held including an Intensive Workshop in Austria [19], a seminar at Schloss Dagstuhl [23] and an International Workshop in Edinburgh [7]. In all these events there was lively participation of researchers from other disciplines faced with temporal and spatial information. Last but not least, the project enabled many talented young researchers (pre-docs and post-docs) to remain in Europe and contribute to the research area of spatio-temporal databases.

1.4 Organization of the Book

The present book collects important and representative research carried out in CHOROCHRONOS and presents it in a unified fashion. The various chapters have been written in a tutorial way making the research contributions of the project accessible to a new generation of researchers interested in spatio-temporal information.

After this introductory chapter, Chapter 2 sets the stage for our contributions to the field of spatio-temporal databases, with a far-reaching study of the role of ontologies in spatio-temporal databases and information systems. Ontology and the related term "semantics" have recently gained increased attention in the database community particularly with the emphasis on ontology-based information integration [27] and the Semantic Web efforts [6,5]. Compared with traditional DBMS, STDBMS must be prepared to make stronger ontological commitments to capture the rich meaning of space and time. Such an ontology is necessarily more involved and the connection to the application area stronger. Moreover, the designer of a database application has to reconcile the ontological concepts from the application area, with the ontology built into the STDBMS. Optimally, an STDBMS involves in its built-in ontology a minimal commitment on how space and time is structured and is thus most open for application specific refinements. Exploring the minimal set of ontological commitment is the goal of Chapter 2.

Chapter 3 concentrates on conceptual database models for spatio-temporal information systems, and presents two related frameworks: the Spatio-Temporal Entity Relationship Model and the Extended Spatio-Temporal Unified Modeling Language. The emphasis in this chapter is on expressing application requirements in high-level conceptual notations, so that designs become understandable to application stakeholders and users, and can be translated easily into the logical spatio-temporal models to be developed in Chapter 4.

Chapters 4 and 5 present CHOROCHRONOS' main contributions in terms of logical models for STDBMS. These chapters develop data models and query languages to deal with geometries changing over time. In contrast to most of the earlier work on this subject, these models and languages are capable of handling continuously changing geometries, or moving objects. These chapters focus on two basic abstractions called moving point and moving region. A moving point can represent an entity for which only the position in space is relevant. A moving region captures moving as well as growing or shrinking regions.

Chapter 4 takes a data type oriented approach. The idea is to view moving points and moving regions as three- or higher-dimensional entities whose structure and behavior is captured by modeling them as abstract data types. These data types can then be integrated as attribute types into relational, object-oriented, or other DBMS data models; they can be implemented as extension packages for suitable extensible DBMSs. Chapter 5 takes a different approach and explores extensions of the constraint database model [28] for the representation of geometries changing. The benefits of this approach is that the main concepts of the relational model are kept intact, and at the same time, one has powerful constructs to express infinite or indefinite phenomena.

Chapter 6 discusses index structures and query evaluation algorithms for STDBMS. The discussion in this chapter is far-reaching going from simple extensions of well-understood index structures for temporal and spatial databases, to very innovative recent proposals especially designed for moving object databases.

Chapter 7 is devoted to architectural and implementation aspects of spatio-temporal database management systems. This chapter gives a general introduction to architectures and commercial approaches to extending databases by spatio-temporal features. Then several prototype systems developed by CHOROCHRONOS researchers are discussed together with their intended applications.

Chapter 8 is devoted to the study of how the techniques developed in earlier chapters of this book can be put to use in applications that cannot be characterized as traditional database applications, but where there is a strong spatio-temporal emphasis. The application chosen is that of composing interactive multimedia presentations where a strong spatio-temporal connection is evident.

Chapter 9 is the epilogue to this book. The readers are challenged with the discussion of three important application areas (mobile and wireless computing, data warehousing and mining, and the Semantic Web) and the role that ideas from spatio-temporal databases can play in these.

References

1. N. Adam and A. Gangopadhyay. *Database Issues in Geographic Information Systems.* Kluwer Academic Publishers, 1997.
2. P. Apers, H. Blanken, and M. Houtsma, editors. *Multimedia Databases in Perspective.* Springer, 1997.

3. H. Barringer, M. Fisher, D. Gabbay, and G. Gough, editors. *Advances in Temporal Logic*, volume 16 of *Applied Logic Series*. Kluwer Academic Publishers, 1999.
4. H. Barringer, M. Fisher, D. Gabbay, R. Owens, and M. Reynolds, editors. *The Imperative Future: Principles of Executable Temporal Logic*. Research Studies Press, 1996.
5. T. Berners-Lee and M. Fischetti. *Weaving the Web: The original design and ultimate destiny of the World Wide Web, by its inventor*. Harper, 1999.
6. T. Berners-Lee, J. Hendler, and O. Lassila. The Semantic Web. Scientific American, May 2001.
7. M.H. Böhlen, C.S. Jensen, and M.O. Scholl, editors. *Spatio-Temporal Database Management. International Workshop STDBM'99. Edinburgh, Scotland, September 1999, Proceedings*, volume 1678 of *Lecture Notes in Computer Science*. Springer, 1999.
8. S.-K. Chang and E. Jungert. *Symbolic Projection for Image Information Retrieval and Spatial Reasoning*. Academic Press, 1996.
9. T. Cheng and S. Gadia. A Pattern-Matching Language for Spatiotemporal Databases. In *Proceedings of the ACM Conference on Information and Knowledge Management*, pages 288–295, 1994.
10. J. Clifford and A. Tuzhilin, editors. *Recent Advances in Temporal Databases (Proceedings of the International Workshop on Temporal Databases, Zürich, Switzerland, September 1995)*, Workshops in Computing. Springer, 1995.
11. H.D. and J. Stachel, editors. *Einstein and the history of general relativity*. Boston, 1989.
12. E. Davis. *Representations of Commonsense Knowledge*. Morgan Kaufmann, 1990.
13. A. Del Bimbo. *Visual Information Retrieval*. Morgan Kaufmann Publishers, 1999.
14. B. Donald, K. Lynch, and D. Rus, editors. *Algorithmic and Computational Robotics: New Directions*. A.K. Peters, 2001.
15. M.J. Egenhofer and J.R. Herring, editors. *Advances in Spatial Databases, 4th International Symposium, SSD'95, Portland, Maine, USA, August 6–9, 1995, Proceedings*, volume 951 of *Lecture Notes in Computer Science*. Springer, 1995.
16. O. Etzion, S. Jajodia, and S.M. Sripada, editors. *Temporal Databases: Research and Practice*, volume 1399 of *Lecture Notes in Computer Science*. Springer, 1998.
17. C. Faloutsos. *Searching Multimedia Databases by Content*. Kluwer Academic Publishers, 1999.
18. M. Fisher and R. Owens, editors. *Executable Modal and Temporal Logics*, volume 897 of *Lecture Notes in Artificial Intelligence*. Springer-Verlag, 1995.
19. A. Frank and S. Winter, editors. *First Chorochronos Intensive Workshop on Spatio-Temporal Database Systems*, November 1997.
20. S. Gadia and S. Nair. Temporal Databases: A Prelude to Parametric Data. In Tansel, A., et al., editor, *Temporal Databases: Theory, Design, and Implementation*, chapter 2. Benjamin/Cummings Pub. Co., 1993.
21. O. Günther. *Efficient Structures for Geometric Data Management*, volume 337 of *Lecture Notes in Computer Science*. Springer, 1988.
22. O. Günther. *Environmental Information Systems*. Springer, 1998.
23. O. Günther, T. Sellis, and B. Theodoulidis, editors. Dagstuhl Seminar on Integrating Spatial and Temporal Databases, November 22–27, 1998. Schloss Dagstuhl, Germany. See `http://dblab.chungbuk.ac.kr/~kwnam/dagseminar.html`.
24. R.H. Güting. An introduction to spatial database systems. *VLDB Journal*, 3(4):357–399, 1994.

25. R.H. Güting, D. Papadias, and F.H. Lochovsky, editors. *Advances in Spatial Databases, 6th International Symposium, SSD'99, Hong Kong, China, July 20–23, 1999, Proceedings*, volume 1651 of *Lecture Notes in Computer Science.* Springer, 1999.
26. D. Hernandez. *Qualitative Representation of Spatial Knowledge*, volume 804 of *Lecture Notes in Artificial Inteligence.* Springer, 1994.
27. R.J.B. Jr., B. Bohrer, R. S. Brice, A. Cichocki, J. Fowler, A. Helal, V. Kashyap, T. Ksiezyk, G. Martin, M. H. Nodine, M. Rashid, M. Rusinkiewicz, R. Shea, C. Unnikrishnan, A. Unruh, and D. Woelk. InfoSleuth: Semantic Integration of Information in Open and Dynamic Environments (Experience Paper). In *Proceedings of ACM SIGMOD International Conference on Management of Data*, pages 195–206, 1997.
28. P. Kanellakis, G. Kuper, and P. Revesz. Constraint Query Languages. *Journal of Computer and System Sciences*, 51:26–52, 1995.
29. R. Laurini and D. Thompson. *Fundamentals of Spatial Information Systems.* Academic Press, 1992.
30. Z. Manna and A. Pnueli. *The Temporal Logic of Reactive and Concurrent Systems: Specification.* Springer Verlag, 1991.
31. Z. Manna and A. Pnueli. *The Temporal Logic of Reactive and Concurrent Systems: Safety.* Springer Verlag, 1995.
32. Y. Manolopoulos, Y. Theodoridis, and V.J. Tsotras, editors. *Advanced Database Indexing.* Kluwer International Series on Advances in Database Systems. Kluwer Academic Publishers, 1999.
33. F. Preparata and M. Shamos. *Computational Geometry.* Springer Verlag, 1985.
34. A. Prior. *Time and Modality.* Clarendon Press, 1957.
35. A. Prior. *Past, Present, and Future.* Clarendon Press, 1967.
36. R. Reiter. *Knowledge in Action: Logical Foundations for Specifying and Implementing Dynamical Systems.* MIT Press, 2001.
37. N. Rescher and A. Urquhart. *Temporal Logic.* Springer Verlag, 1971.
38. P. Rigaux, M. Scholl, and A. Voisard. *Spatial Databases: With Application to GIS.* Morgan Kaufmann Publishers, 2001.
39. C. Rolland, F. Bodart, and M. Lèonard, editors. *Proceedings of the IFIP TC 8/WG 8.1 Working Conference on Temporal Aspects in Information Systems, Sophia-Antipolis, France, 13–15 May, 1987.* North-Holland/Elsevier, 1988.
40. H. Samet. *The Design and Analysis of Spatial Data Structures.* Addison-Wesley, 1990.
41. M. Scholl and A. Voisard, editors. *Advances in Spatial Databases, 5th International Symposium, SSD'97, Berlin, Germany, July 15–18, 1997, Proceedings*, volume 1262 of *Lecture Notes in Computer Science.* Springer, 1997.
42. Y. Shoham. *Reasoning About Change: Time and Causation from the Standpoint of Artificial Intelligence.* MIT Press, 1987.
43. R. Snodgrass, editor. *The TSQL2 Temporal Query Language.* Kluwer Academic Publishers, 1995.
44. R.T. Snodgrass. *Developing Time-Oriented Database Applications in SQL.* Morgan-Kaufmann, 1999.
45. V. Subrahmanian. *Principles of Multimedia Database Systems.* Morgan Kaufmann Publishers, 1998.
46. A. Tansel, J. Clifford, S. Gadia, S. Jajodia, A. Segev, and S. R., editors. *Temporal Databases: Theory, Design and Implementation.* Database Systems and Applications Series. Benjamin/Cummings Pub. Co., 1993.

47. ARPA/NSF International Workshop on an Infrastructure for Temporal Databases, Arlington, Texas, June 1993.
48. J. van Benthem. *The Logic of Time.* D. Reidel Publishing Company, 1983.
49. M. Worboys. *GIS: A Computing Perspective.* Taylor and Francis, 1995.
50. M.F. Worboys. A Unified Model for Spatial and Temporal Information. *The Computer Journal*, 37(1):26–34, 1994.

2 Ontology for Spatio-temporal Databases

Andrew U. Frank

Technical University of Vienna, Austria

2.1 Introduction

Ontology and the related term "semantics" have recently found increased attention in database discussions. Early discussions of ontology issues important for databases [126,78] were lost in a sea of papers on technical, mostly performance issues, despite the fact that textbooks as early as [134] discussed briefly the relationship between information system and real world. This is different today; interest in semantics has increased, and this will be more so in the future given the current interest in the Semantic Web [11,10] (see Chapter 9 of the book for related discussion).

Information systems and their implementation as databases rest on ontological commitments. Decisions about the type system used, how identifiers are managed, and so on, are derived from a specific view of the world to which the database relates, in other words from a specific ontology. The ontologies of standard database models make very limited assumptions and therefore the data model is widely applicable. Spatio-temporal databases must make stronger commitments to capture the meaning of space and time. Such an ontology is necessarily more involved and the connection to the application area stronger. The designer of a database application has to reconcile the ontological concepts from the application area with the ontology built into the database. Optimally, a spatio-temporal database involves in its built-in ontology a minimal commitment on how space and time is structured and is thus most open for application specific refinements. Exploring the minimal set of ontological commitment is the goal of this chapter.

The ontology built into a DBMS can be insufficient or it can be too restraining. It is insufficient if the ontological categories necessary for numerous applications are not available and must be reconstructed for each application anew; the resulting incompatibilities will be very costly to correct later [81]. It is too restraining if the ontology commits those who apply it to assumptions which do not hold for novel applications. Spatio-temporal databases are typically constructed to integrate the knowledge of many agents and face the problem of heterogeneous environments, a point already raised by Wiederhold et al. [190, Chapter 22]. Current databases do not allow us to model joint beliefs of groups of agents which do not correspond to similar beliefs of other groups of agents; for example, Reuter works with groups of scientists, who manage terabytes of reports of results from experiments in cellular biology, where the validity of the results and their interpretation are debated among the groups. Current ontological investigations related to databases and information systems have been

T. Sellis et al. (Eds.): Spatio-temporal Databases, LNCS 2520, pp. 9–77, 2003.
© Springer-Verlag Berlin Heidelberg 2003

extended into the spatial domain [36,37,62,63,65,160,188], but their extension into the spatio-temporal domain [44,97,98,120,139] has proved more difficult than expected [86,182]. An overview of Time Ontology for computer science was published by [170]; Montanari and Pernici discuss the different proposals for temporal reasoning [190, Chapter 21].

I will investigate the questions which arise when information systems are built for purposes involving the representation of real space and time. Examples from the domain of Geographic Information Systems demonstrate the issues. Geographic Information Systems are especially suited for our purposes because they model real-world situations including their spatial and temporal aspects. Their application area is very broad and extends from the administrative and legal rules governing land ownership and registration [54] to systems built for environmental purposes [111] and for research into global change [145]. The situation is not substantially different for other spatio-temporal systems, like systems for motor traffic monitoring or tracking airplanes. Spatio-temporal databases are often built from data from many different sources, which is notoriously difficult [103,199]. Data to be integrated differ in their semantics and representation, and a meaningful combination requires bridging the gap created by ontological assumptions as well as translations between the representations once their meaning is in the same context. But even for databases where all data are from the same source, the gap between the ontology of the data collectors and the ontological assumptions of the designer of the GIS software and later the users must be bridged.

I propose a multi-tier ontology, where different rules apply to each tier (Table 2.1). The approach used here is empirical and starts with the observation of physical properties for specific locations and instants. Objects are formed as areas of uniform properties which endure through time as identical. Cultural conventions link names to objects and construct objects of "social reality" [12,172], which are meaningful within a set of culture-dependent rules. For example, the legal system of a country gives a meaning to concepts like "parcel" and "ownership". But the corresponding objects do not have physical existence; they are social artifacts. Agents – human beings or organizations which behave like persons with respect to the aspects considered here – make all observations. Agents derive decisions about actions from the knowledge they have acquired. An agent's knowledge evolves over time and spatio-temporal databases must therefore document the temporal evolution of an agent's knowledge. The historical state of an agent's knowledge must be considered to make a fair assessment of an agent's actions.

The proposed tiers are ordered from data for which data collections from multiple sources are more likely to agree, to data for which disagreement is more likely; they help with the integration of data from different sources to understand the processes which result in agreement or disagreement between data. Debates on the length of a year are limited to scientific discussions on the 12th decimal, the measured height of mountain tops may differ between countries by a few meters; but debates about the location of boundary lines occur occasionally, the limits of areas with economic problems are debated in parliaments and

Table 2.1. The five tiers of ontology

Ontological Tier 0: Physical Reality:
- the existence of a single physical reality,
- determined properties for every point in time and space,
- space and time as fundamental dimensions of this reality.

Ontological Tier 1: Observable Reality:
- properties are observable now at a point in space,
- real observations are incomplete, imprecise and approximate.

Ontological Tier 2: Object World:
- objects are defined by uniform properties for regions in space and time,
- objects continue in time.

Ontological Tier 3: Social Reality:
- social processes construct external names,
- social rules create facts and relationships between them,
- social facts are valid within the social context only.

Ontological Tier 4: Cognitive Agents:
- agents use their knowledge to derive other facts and make decisions,
- knowledge is acquired gradually and lags behind reality,
- reconstruction of previous states of the knowledgebase is required in legal and administrative processes.

the judgment on desirable areas for vacations is mostly a question of personal preference. This leads to separation of physical reality, object reality and socially constructed reality in different tiers of an ontology.

A multi-tier ontology allows to integrate different philosophical stances, from an extreme realist or positivist view to the current post-modern positions. The multiple tiers recognize that various approaches contribute to our understanding of certain aspects of the world around us and take the philosophically unusual position that none is universal [164].

The goal of this chapter is to investigate what the minimal ontological commitments for spatio-temporal databases are. To this end, the concept of ontology in the context of database design is clarified first and then an "observation-based", empirically justified minimal ontology is designed. The ontology is designed to facilitate a computational model. The approach owes much to the efforts in formal ontology by Guarino [107,175,178,183] and related researchers [71,70,201]. It connects, however, their findings with the concept of "social reality" introduced by Searle [172], which gives a foundation to most of the semantics of administrative data processing. Multi-agent theory [75,203] provides a framework to justify the two-time perspectives used in temporal databases [187].

2.1.1 Ontology to Drive Information System Design

Guarino [109] and Egenhofer [76] promote the term "ontology driven information system design" and experiments abound to formalize ontology description languages. This is the continuation of the observation that application programs incorporate various properties of the objects handled, properties which are coded many times in the application code – and not always consistently. Database schemata, originally used only for the structuring of storage of data, were soon discovered to be useful for the generation of reports (e.g. the report generation language of CODASYL [43]) and later also for the automatic generation of data. These application-independent properties of the object represented in a program can be described as ontology. If such ontological properties of the objects are concentrated in one place and if they can be used by various programs, simplification of the software development process can be achieved and applications may even gain in usability as more consistency in the operations is achieved [81]. There is today substantial – even commercial – interest in shareable ontologies (witness all the recent activity on the Semantic Web [11,10]), and there are companies which construct and sell ontologies (CYC [53] or Ontek [151]). Several international standardization bodies, from ISO (ISO/TC211) [122] to OMG [150] and OGC [149], standardize spatial and temporal aspects of ontology.

2.1.2 Ontological Problems of Geographic Information Systems and Other Spatio-temporal Information Systems

Geographic Information Systems have to reflect truthfully the state of the world; information systems which do not provide reliable and correct information are useless. This correspondence between reality and information system will be used throughout this chapter to define ontology and to explain its role in the design of information systems. The high cost of collecting and maintaining spatial data has led to more demand for data sharing: data should be collected and maintained once and used by many [103,199]. This forces differences in the ontological commitments into the open:

- the continuous nature of reality compared to the discrete approximation of space and time in a database (see Chapter 4 in this book);
- the fact that the world changes continuously and the database lags behind;
- differentiation between "valid time" and "transaction time" in temporal databases is an ontological differentiation [186];
- the closed world assumption [162], which is conveniently assumed in databases, is not valid for spatio-temporal databases [81,180];
- the interoperability problem, which is the inability of comparable systems to co-operate [26];
- the internal hierarchical structures for space, time and categories [95], and the consequent stratification in the ontological categories [25];
- the difficulty to combine solutions developed for different applications (composability in linguistic terminology [123]);
- the difficulty of describing data and the quality of data: the so-called metadata discussion [38,144,167];
- the differences in classification found across cultural boundaries [32,113,143].

2.1.3 Structure of the Chapter

The chapter discusses the meaning of the notion of ontology in Section 2.2. The following Section 2.3 introduces typical application domains for spatio-temporal databases, which are used as examples in later sections. Section 2.4 discusses the fundamental aspects of information systems and shows how they relate to ontology. The next Section 2.5 gives an overview of the five tiers of ontology. Section 2.6 discusses the languages which could be used to describe an ontology. The major part of the chapter is formed by Sections 2.7 to 2.11, which treat tiers in detail. A summary Section 2.12 lists the ontological commitments encountered. Conclusions and future work are discussed in the final Section 2.13.

2.2 The Notion of Ontology

Ontology describes what is; it is "the metaphysical study of the nature of being and existence" [74]. In a naïve view, there should be only one ontology, as there is only one world. In practice, we observe different conceptualizations of the world by different people. Ontology, especially if the term is used in its plural form, describes a conceptualization of the world and is closely related to software engineering activities like conceptual analysis, domain modeling, etc. [108].

2.2.1 Classical View

The notion of ontology is borrowed in Computer Science from philosophy. The Greek philosophers, especially Aristotle in his Metaphysics, inquired what the properties of the world and the objects in it are and how we perceive them [179,183]. Philosophy uses the term ontology to describe that which is (ontos, Greek, to be; ontology, therefore: the science of what is) and in this sense, it is used as a synonym for "metaphysics". Ontology is often used in contradistinction to epistemology, which is "the field of philosophy, which deals with the nature and source of knowledge" [148]. Quoted after [107] epistemology in brief is a "theory of knowledge". It is difficult for us living in the world to separate the description of the world from our knowledge of the world and how it is expressed in language. Strictly speaking, if there is only one reality, there must also be only one ontology, and the human views or conceptualizations of this world are not ontologies in the strict sense. We need another term for the "theory what people believe the world is like"; one could call it "projected ontologies" or "epistemological ontologies" [155].

Ontologies are modeled after scientific theories (or the naïve counterparts thereof), especially physics [116] and geography [66], and they generalize the rules found there. Recently, philosophical ontologists have begun to study practical problems from law, engineering and commerce [179] and they have started to identify the limits of ontologies based on empirical observations of physical objects.

Augustine introduced the related notion "universe of discourse". As far as a rational discussion is concerned, the notions of ontology or universe of discourse are related to the concept of a closed system and its boundary. In this sense computer science has borrowed the term ontology to list what is considered within a discussion or an information system [116]; database specialists talk about database schema, often with the same meaning. Davis [55] separates three levels in the analysis of a microworld: a level of definition of a domain model, a level of formalization with types and axioms, and an implementation level (Figure 2.1). In this chapter, this approach is extended for spatio-temporal databases which must support multiple applications and therefore multiple domain models which fit in a single generic model.

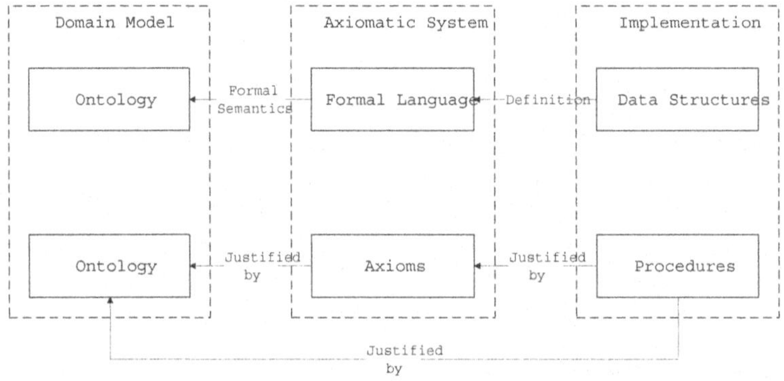

Fig. 2.1. The modeling of a single application after [55, p.7]

2.2.2 Social Reality

Ontology describes what is independent of an observer, what exists for every observer. The ontology describes the common reality. The discussion in the past has concentrated on physical reality: on physical objects in the real world. The applicability of this ontology for most of the information systems in administration is doubtful. Much of what is collected in databases are facts not about physical reality (e.g., the position of a building in coordinate space), but rather about human agreements (contracts), about classification according to some culturally fixed rules (e.g., who is an adult) and about social arrangements (ownership rights). These are not physical but nevertheless very "real". Many important aspects of our daily life fall in this category: neither money nor marriage or companies are physically existing and can be touched. They are related to physical objects and have specific relations between them but their existence is not physical in nature.

The physical ontology can only describe the physical part of reality; things like money, ownership, social status, etc., are not real in the same way as is light, physical objects, etc. Multiple observers may see the same objects (e.g., pieces of paper) but may not agree on their value, because for some observers and contexts some of the pieces of paper represent money but not for other observers in other contexts. Sociologists have pointed out that part of what seems real to us is constructed by society [12], and Searle [172] has provided a succinct analysis of the kind of reality behind money, property of land, marriages, etc., which is extremely helpful to avoid some of the confusing tangles of ontological discussion.

Agents create social reality; they can be single individuals (persons) or aggregations of persons in agencies or organizations. Some agents – in general, the agencies of the state – have the power to make other agents observe the same rules regarding the objects they create. The standard example is money, which physically is nothing else than printed paper; given that all members of a group treat money in the same way, it functions to facilitate commerce, despite the fact that the fiction that one can exchange paper money into gold has been dropped a long time ago.

Social reality, like language symbols, is meaningful only in a context. The law of a country creates a (local) context in which institutional reality as a part of social reality is defined. We will find it helpful to use here the example of the land-registration process to demonstrate how social reality is created, because it demonstrates spatial and temporal aspects [1,16,79,146]. Ownership of land requires spatial delimitation, i.e., boundaries, and is, in some countries, created by registration in some sort of information system (cadastre, land registry).

Much of social interaction is based on understanding and speculating what other agents think. Social rules of fairness dictate that agents are not responsible for not knowing facts which they had no possibility to learn. To judge the effects of actions, one must therefore be able to reconstruct what the agents have known at a specific time. In other circumstances posting some facts publicly is crucial to their being established socially. For example, ownership of land must usually be registered in a public register to be enforceable against others; many western movies describe the race between two gold diggers in order to register their claims first.

2.3 Application Domains

Ontologies are influenced by the examples the designer uses. Classical ontology studies, from Aristotle onwards, are based on material objects, preferably solid bodies, the human body or animals, as well as the actions and events in which such objects are engaged. Hayes has studied the ontology of liquids and found it to be very complicated [115]. Different domains have different ontological foundations.

Three quite different application domains are used as examples here to assure that the ontological base for spatio-temporal databases does not include commitments which will exclude the application to other domains:

- a table-top situation, with solid and liquid material objects, as they are customarily found on a dinner table; these objects are moved around by humans;
- a city environment, where persons or cars move between buildings and along streets (similar to examples found later, for example, in Chapter 4).
- a geographic situation, with plots of rural land, forest, roads and rivers, where people and animals can move unrestricted across the land (somewhat related to the ski resort example of Chapter 5 in this book).

The different examples should demonstrate the breadth of the realm of applications requiring spatio-temporal databases and the differences in their conceptualization of reality. I have recommended that particular ontologies be developed for specific application areas [81]. For example, an ontology for farming is highly desirable to connect the rules for data collection, calculation of agricultural subsidies in the European Union and the integration of the resulting database for policy [82]. Bernasconi has documented an ontology for the sewer systems of a commune [13]. Further we urgently need concrete ontologies for land registration to build the base software usable in several countries with different legislation [16,146]. Last but not least, an ontology of traffic, private and public, would be very useful in the exchange of data between different traffic guidance systems, transportation schedule services and car navigation aids. The foundation ontology incorporated in the spatio-temporal database must be open to allow each of these ontologies, indeed it must be possible to integrate more than one of them.

2.3.1 Table-Top Situation

A well-researched abstraction of the situation on a dinner table was one of the first examples of a computer science ontology: the blocks world [8]. It consists of solid blocks, which can be stacked on top of each other. This has served as a fruitful example to discuss the meaning of ontologies [106] and to discuss the formal definition of the semantics of spatial relations [92,120]. More complex is an environment which includes liquids in bottles or cups [115]. Liquids do not have a fixed form, but fill the holes in other objects (only specific kinds of holes can be used to contain liquids [36]). Liquids can be poured and mixed, but it is generally impossible to separate two pieces of a liquid once they are merged [139,140].

The objects on a table are under control of a person manipulating them (Figure 2.2). Possession of an object by a person may signal legal ownership and we see here a close connection between physical possession and ownership.

One can see that objects are conceived in such a way that important invariants are maintained. The regular laws of conservation of matter apply and material properties, for example, color, specific weight, remain invariant under a large number of operations. Solid objects on a table maintain their size, volume and form. More complex ontologies apply for cooking, where less invariants are maintained: neither form, color, or volume, nor weight is preserved.

Fig. 2.2. Table-top situation with solid objects and liquids

2.3.2 Cityscape

A city contains buildings and streets (Figure 2.3). We can understand the buildings as containers which are further subdivided into rooms. Persons can be in these rooms. Doors between the rooms allow people to move between the rooms or leave the buildings. Streets are formed by the empty space available for movement. Streets and plazas can, again, be seen as containers, but for navigation in a city, a linear conception of a street as a path between doors is a more effective conceptualization. For most purposes, the details of the movement of a person in a street is irrelevant, important is only that the person follows the street from intersection to intersection.

But not only a container and a linear model of space are applicable, we find also an areal one: Considering the rainfall on buildings, the amount of rain running off a roof is proportional to the area. The runoff then follows the streets and in modern cities disappears in the sewer network (again a linear, graph like, structure) [31].

Buildings, streets, plants, etc., do not move from their location and the processes of creation take much time. Persons, cars and other vehicles move among them rapidly; their movement is restricted to certain pathways.

This example shows how different tasks lead to different conceptualizations of space: the same cityscape is seen in terms of volumes, areas and lines. But even within a single type of geometry, for example, the linear network structure of a street network, different levels of detail are used, depending on the specific task: planning a trip uses a less detailed representation of the street network than the description of a path to take, where every intersection must be mentioned. Finally driving in lanes and changing between lanes is yet a third level of detail in a street graph [194]. A hierarchy of containers is also useful to navigate

Fig. 2.3. City situation with buildings, streets, and people

in a city environment and to produce maps at different levels of cartographic generalization for this purpose [193].

Physical possession is not sufficient to indicate ownership of land. Legal institutions, often called land tenure, are necessary to transfer and publicize ownership and other rights in land. The registry of deeds or a registry of title are maintaining public knowledge about these rights.

2.3.3 Geographic Landscape

The first object of the geographic world is the surface of the world and its form (Figure 2.4). The landscape is seen as an undulated surface (a two-dimensional geometrical object) embedded in three-dimensional space. The geological processes create this surface, most importantly through erosion caused by water flow. The general importance of water and water flow for our lives leads to the concept of height measured as potential with respect to a reference potential assumed as "sea level".

Water flows under the force of gravity over surfaces and forms rivers at the bottom of valleys. Streams form a linear network and watersheds form a functionally defined subdivision of space – for every point along a street network there exists a corresponding watershed (namely all the area from which water flows to this point) [90,152,153].

Couclelis has pointed out the contradiction between objects and fields: "people manipulate objects but cultivate fields" [48]. The surface of the earth is divided into parcels, which are manipulated like objects, bought and sold like books or shoes. Fences divide the fields and streets link fields to populated places.

Remote sensing allows observation of large areas and permits the classification of actual land use. Areas of uniform use, for example, forest area, do not

Fig. 2.4. Landscape with hills and valleys

automatically correspond to the areas of land ownership. The maps in planning offices show the intended use of some area, but this does not automatically correspond to actual use.

All objects in the geographic world change and move, but some move much faster than others. Most geographic processes are so much slower than the majority of human activities that geography seems to be the "stable" backdrop against which other processes are played out. Mountains and rivers do not move, people move between them. Considering a geological scale, mountains rise and are eroded, rivers change their courses; changes in land use are relatively rapid and woods can appear or disappear within a few decades. Movements of geographic objects are qualitatively different from the movement of persons along a street or across a field [91], like the airplanes in the examples of Chapter 4 of this book.

Man-made objects in the landscape are sharply delimited, but most natural objects do not have sharp boundaries. Various methods have been discussed – from fuzzy logic to qualitative reasoning – to deal with objects with undetermined boundaries, from forests to geographic regions like "the North Sea" [28,27].

2.4 Model of Information Systems

Information systems are advanced forms of symbol manipulation, but this point of view is not sufficient to understand the relation between reality and the information the system provides. Information systems are used to make mental experiments when real experiments are undesirable, too expensive, etc. They are useful and valuable only if the information they represent corresponds to the state of the real world. This correspondence between reality and information system is used to define formally the meaning of ontology in a model.

2.4.1 Information Systems as Vehicles of Exchange between Multiple Agents

The simplest situation in which ontological issues become important requires at least two cognitive subjects, both of which consider reality possibly in different ways, and both of which communicate about this reality. The cognitive subjects here will be called "agents" to stress that we include single persons as well as multi-person organizations, for example, state agencies, companies, etc. A practical example is the collection of street information by one agent, which is then provided to another agent to help him find his way; this is done, for example, by national mapping agencies, which collect topographic information and distribute this information to the public in the form of maps; but it is also encountered when somebody informs a friend how to find his way home [85].

Fig. 2.5. An agent producing a map and another agent using a map for navigation [85]

The basic situation is sketched in Figure 2.5: a person observes the world and builds a database of his observation ("beliefs" in the terminology of [55]). He gives this description of the world, which is a small database, to another person, who uses it to find his way to a goal. The data in the database are only useful to this other person if his planned path is effective and brings him to the desired location. For this it is necessary that the information gained from querying the database is the same information as that the agent would gain if he would inspect the world directly. For example the length between the two street intersections must be predictable from the database with sufficient precision to select the shortest path.

Abstracting from the particulars of Figure 2.5 we arrive at Figure 2.6, which shows reality and the model of reality in the information system and the operations which enable persons to interact with the information system and reality: R stands for the reality, D for the realm of the data representation, f for the mapping between the data and reality. An operation g_0 carried out on the data must have the same effect as the corresponding operation g_n carried out in reality; mathematically a homomorphism must exist between real world and data [101,106,118]:

$$f(g_0(d_i)) = g_n(f(d_i)) \tag{2.1}$$

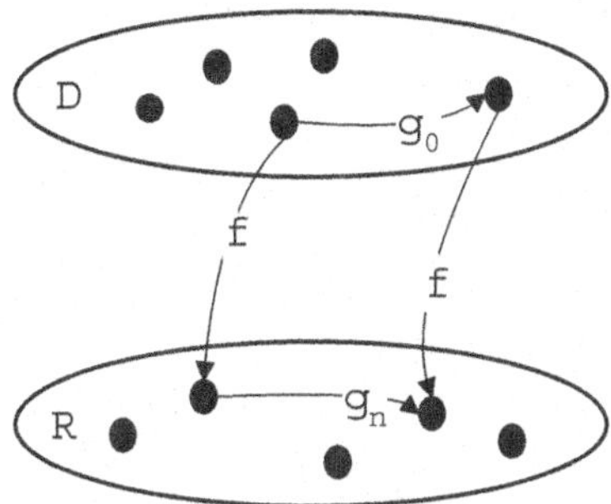

Fig. 2.6. Homomorphism [78, p.18]

2.4.2 Correctness of Information System Related to Observations

The correctness of an information system is expressed as a homomorphism between the information system and some portion of reality. The mapping between data representation and reality is based on observations. A generic observe operation links reality to a data value. Observation provides a homomorphism for all operations defined in the information system: the programmed method which returns the shortest path between two nodes in a navigation system and the shortest path traveled in the city must correspond (otherwise the information system does not properly inform about the portion of reality it pretends to model).
For all objects in r in R and all op in R and op' in D

$$observe\, op(r) = op'(observe\, r) \tag{2.2}$$

The approach suggested here is related to the correspondence theory of truth introduced by Aristotle and reformulated by Tarski [191]. It goes, however, beyond the regular correspondence between concepts and reality, but links all concepts to observations of reality and operations to actions applied to the objects in reality (Figure 2.7). The objects 1 and 2 are observed, operation d applied to objects 1 and 2 results in object 3 – for example, the shortest path between two locations 1 and 2 requires turn 3 at location 1. Carrying out the action "turn

3" completes the loop from the observation of the world to acting on the world and observing the results of the action:

$$\begin{aligned} & act(op'(observe(r_1), observe(r_2))) \\ = & act(op'(d_1, d_2)) \\ = & op(act(d_1), act(d_2)) \\ = & op(r_1, r_2) \end{aligned} \tag{2.3}$$

with $act = observe^{-1}$ and $act(op'(r)) = op(act\, r)$ (corresponding to Eq. 2.2).

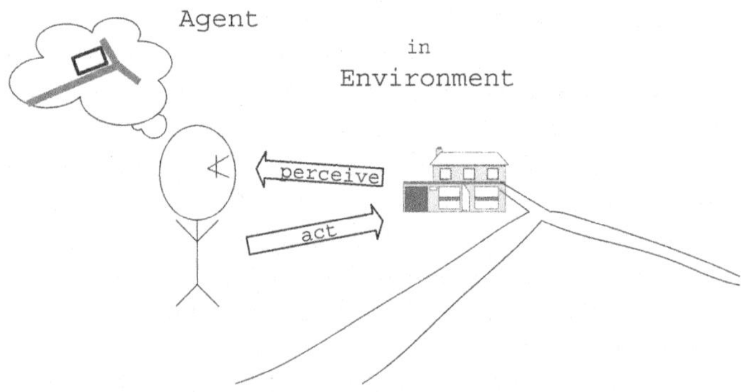

Fig. 2.7. Observation and action form a closed loop

Observations link reality to data and actions link data to reality. In this closed loop, the connection between the observation operation and the ontology applied by the observing agent can be compared to the action and the ontology it implies.

2.4.3 Semantics for Terms in Information Systems

The meaning of the symbols in an information system are linked by conventions with the objects in reality at the level of individual instances or tokens; Saussure already has pointed out that words are only meaningful in the context of a language [168]. Individuals in reality correspond to entities in the database. Database books as early as 1978 [134] described perception and codification conventions, which connect the real world in which humans live to the information system. The database schema lists the types of objects, usually describing them with common nouns. Software engineering tools equally rely on natural language and a common understanding of words [23,47].

It is, however, well known that natural language terms have multiple meanings, and even common terms, like "road width", may have multiple, slightly differing interpretations in different contexts [40]. Defining the natural language terms using other natural language terms leads to infinite recursion. Some linguists try to identify a small number of base words from which all others can be defined; it is claimed that a list of about 100 base words occurring in all natural languages is sufficient for this [204]. Easily accessible is the Wordnet project, which defines words by sets of synonyms and covers currently over 150000 English words [74].

The specification of semantics is a deep problem, which makes it difficult or impossible to link different databases [103,199] to produce comprehensive databases covering a large area. It has proven very difficult to construct databases covering the European Union, due to differences in the interpretation of terms. Take the simple concept of dividing a population in minors and adults; European countries use different age thresholds for adulthood and therefore to establish the number of European adults is a questionable project. However, to make statistics covering each age group, defined by numerical age, is much less error prone, as these concepts are more likely used uniformly. The customary approach for database integration is based on the comparison of some static properties of the database schemata, but the important decisions are left to the team of database designers to establish links where possible [57].

2.4.4 Grounding of Semantics in Physical Operations

Cognitive linguists, in particular Lakoff and Johnson, suggest that the meaning of words is related to the bodily experience of humans with the world [124,130]. The possible base interactions of humans in the world are simple and limited, and their meaning is captured in so-called "image schemata". An incomplete list is given in Table 2.2 and the close connection to the list prepared by Wierzbicka [204] is striking, despite the completely different approaches followed.

Table 2.2. Image schemata (after [124])

Container	Balance	Full-Empty	Iteration	Compulsion
Blockage	Counterforce	Process	Surface	Restraint Removal
Enablement	Attraction	Matching	Part-Whole	Mass-Count
Path	Link	Collection	Contact	Center-Periphery
Cycle	Splitting	Merging	Object	Scale
Near-Far	Superimposition			

The specifications of the image schemata relate to operations humans can perform and their results. The semantics of closely related terms are described in a cluster. For example, the container image schema is described by operations of put-in, take-out, the effects of which can be observed by testing whether something is in the container or not (Table 2.3). The definition of the meaning of symbols in the information system must be such that an isomorphism between information system and reality obtains [92]. This overcomes most of the classical problems with the definition of words, at least for the meaning of words used to describe the physical, spatial and temporal world. The complex, abstract concepts humans are capable of are combined from these base concepts by transformation (Lakoff calls them metaphorical mapping in [132,130], Fauconnier and Turner use the term "blend" [73,197]). Goguen has shown how to formalize such blends [101].

Table 2.3. Algebra Container $s\ a$ with Eq for a

Operations:

$$\begin{aligned} empty &:: s\,a \\ isEmpty &:: s\,a \rightarrow Bool \\ size &:: s\,a \rightarrow Int \\ put &:: a \rightarrow s\,a \rightarrow s\,a \\ isIn &:: a \rightarrow s\,a \rightarrow Bool \end{aligned}$$

Axioms:

$$\begin{aligned} isEmpty(empty) &= True \\ isEmpty(put(a,s)) &= False \\ size(empty) &= 0 \\ size(put(a,s)) &= 1 + size(s) \\ isIn(a, empty) &= False \\ isIn(a, put(b,s)) &= \text{ if } a = b \text{ then } True \text{ else } isIn(a,s) \end{aligned}$$

2.5 The Five Tiers of the Ontology

2.5.1 Physical Reality Seen as an Ontology of a Four-Dimensional Field

The physical laws which describe the behavior of the macroscopic world can be expressed as differential equations, which describe the interaction of a number of properties in space – seen as forming a continuum. For each point in space and time a number of properties can be observed: color, the forces acting at that point, the material and its properties, like mass, melting temperature at that point, etc. Movement of objects can be described as changes in these properties; even the movement of solid objects can be described as the result of the cohesive forces in the body maintaining its shape. The description of reality by differential equations (e.g., the description of forces on a plate under a load p) is widely used in mechanical and civil engineering, geology, etc. Models of mechanisms or building structures are described and their reaction under various applied forces is analyzed. This view is also quite natural for most studies under the heading of "global systems" [145].

A field model can be observed at every point in space and time for different properties:

$$f(x, y, z, t) = a \tag{2.4}$$

Abstracting from the temporal effects, a snapshot of the world can be described by the formula which Goodchild called "geographic reality" [102].

$$f(x, y, z) = a \tag{2.5}$$

The processes occurring in this physical reality have spatial and temporal extensions: some are purely local and happen very fast; others are very slow and affect very large regions. The processes of objects moving on the tabletop are fast (m/sec) and the spatial extent is small (m); movement of persons in cities is again fast (m/sec) and the movements of the buildings very slow (mm/annum); geological processes are very slow (mm/annum) and affect large areas (1000 km^2). One can thus associate different processes with different frequencies in space and time [95]. Each science is concerned with processes in a specific spectrum of space and time, which interact strongly; other processes, not included in this science, appear then to be either so slow or so fast that they can be considered constant. Space and time together form a four-dimensional space in which other properties are organized. Giving space and time a special treatment results in simpler formulations of the physical laws that are of particular interest to humans. For example, the mechanics of solid bodies, e.g., the movement of objects on a tabletop, is explainable by Newtonian mechanical laws, which relate phenomena which are easily observable for humans in a simple form ($s = vt$, etc.). Other sciences, e.g., astrophysics, prefer other coordinate systems in which mass is included.

However, the assumption that the formula $a = f(x, y, z, t)$ describes a regular function which yields only a single value, is equivalent to the assumption that there is only one single space-time world and excludes "parallel universes" as parts of reality.

2.5.2 Observation of Physical Reality

Agents can – with their senses or with technical instruments – observe the physical reality at the current time, the "now". Results of observations are measurement values on some measurement scale [189], which may be quantitative or qualitative. Such observations are assumed in later chapters of this book to describe, for example, the movement of airplanes. Observation with a technical measurement system comes very close to an objective, human- independent observation of reality. A subset of the phenomena in reality is objectively observed. Many technical systems allow the synchronous observations of an extent of space at the same time, for example, remote sensing of geographic space from satellite (Figure 2.8). Typically a regular grid is used and the properties observed are energy reflected in some bands of wavelength (the visible spectrum plus some part of infrared).

The same kind of observation as sampling in a regular grid can be used in any other situation, the blocks world on my table as well as the city, including moving objects. They can be sampled and described as a raster. Such observations are mainly used for robots, where TV cameras which sample the field in a regular

Fig. 2.8. Remote sensing image

grid are used to construct "vision" systems to guide the robot's in manipulating objects on a table or moving through buildings [129].

2.5.3 Operations and Ontology of Individuals

Our cognitive system is so effective because, from the array of sensed values, it forms individuals, which are usually called objects, and reasons with them. Thinking of tables and books and people is much more effective than seeing the world as consisting of data values for sets of regularly subdivided cells (i.e., three-dimensional cells, often called voxels). It is economical to store properties of objects and not deal with individual raster cells. As John McCarthy and Patrick Hayes have pointed out:

> ...suppose a pair of Martians observe the situation in a room. One Martian analyzes it as a collection of interacting people as we do, but the second Martian groups all the heads together into one subautomaton and all the bodies into another. ...How is the first Martian to convince the second that his representation is to be preferred? ...he would argue that the interaction between the head and the body of the same person is closer than the interaction between the different heads. ... when the meeting is over, the heads will stop interacting with each other but will continue to interact with their respective bodies. [137, p.33]

Our experience in interacting with the world has taught us that the appropriate subdivision of continuous reality is that into specific types of individuals. Instead of reasoning with arrays of connected cells (as is done, for example, in computer simulations of strain analysis or oil spill movements), we select the shorter and more direct reasoning with individuals: The elements on the tabletop are divided into objects at the boundaries where cohesion between cells is low; a spoon consists of all the material which moves with the object when I pick it up and move it to a different location. This is obviously more effective than individual efforts to reason about the content of each cell. Animals and most plants form individuals in a natural way.

The cognitive system is very fast in identifying objects with respect to typical interactions. We see things as chairs or cups if they are presented in situations where sitting or drinking are of potential interest (under other circumstances, the same physical objects may be seen as a box or a vase). The detection of "affordances" of objects is immediate and not conscious. The identification of affordances implies a breakup of the world into objects: the objects are what we can interact with [100]. Cognitive science has demonstrated that even infants at the early age of three months have a tendency to group what they observe in terms of objects and to reason in terms of objects. It has been shown that animals do the same. Most of the efforts of our cognitive system to structure the world into objects are unconscious and so it is not possible for us to scrutinize them. There are a number of well-known effects where a raster image is interpreted in one or the other way; for example, Figure 2.9 can be seen as cube or a corner, but not both at once. In Figure 2.10 the decision what is foreground and what is background is arbitrary, but we can alternatively see the two faces or the vase, not both at the same time.

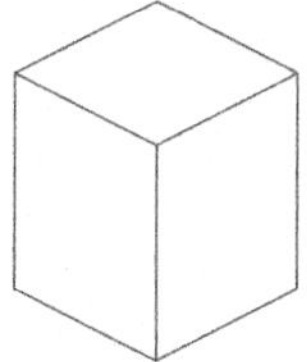

Fig. 2.9. A cube or a corner?

Efforts to explain the categorization of phenomena in terms of common nouns based on a fixed set of properties, as initiated by Aristotle, occasionally lead to contradictions. Dogs are often defined as "barking", "having four legs, etc."; but from such a set of attributes it does not follow that my neighbor's dog, which lost a leg in an accident, is no longer a dog. Modern linguistics assumes generally that prototype effects make some exemplars better examples for a class than others. A robin is a better example for a bird than a penguin or an ostrich [165,166]. Linguistic analysis suggests that the ways objects are structured are closely related to operations one can perform with them, and empirical data support this [74,123].

Fig. 2.10. Two faces or a vase?

Humans have a limited set of interactions with the environment – the senses to perceive it and some operations like walking, picking up, etc. – and these operations are common to all humans. Therefore the object structure, at least at the level of direct interaction, is common to all humans, and it provides the foundation on which to build the semantics of common terms [131]. In general, the way individual objects and object types are formed varies with the context, but is not arbitrarily. Different from the viewpoint of physics, humans experience space and time not just as different dimensions of a continuum. Time is experienced by all biological systems as a vector and processes are not reversible – energy is used and dissipated, and entropy increases by the laws of thermodynamics [49]. All observation of the world is limited to the observation at the time "now". "Now" is not only a difficult philosophical problem [77] but also a tricky problem for temporal query languages.

Human experience of time contrasts with space, which is isotropic: it has the same properties in all directions. Humans experience the direction of gravity as most salient "up-down" axis, which leaves the plane orthogonal to gravity as space which is experienced isotropically – what is in front of me is behind me if I turn around [198]. Objects can move, nearly without effort, in this plane and these movements are reversible. The geometry of the object – especially the distance between two points on the object or angles – remains the same, independent of movement.

Points in space seem natural, despite the fact that they are abstractions, which cannot be materialized. Similarly, time-points, called instants, are important to mark boundaries between intervals [97]. Spatial objects have boundaries, which are lines and surfaces which bound volumes. The objects of the tabletop are modeled as solid volumes, most of them with fixed form (except for liquids and similar). Their surfaces can touch, but the volumes cannot overlap. Euler has described the rules for the manipulation of polyhedrons, so-called Euler operators, for merging and splitting of solid objects and these rules were used to construct an ontology for Computer Aided Design systems [58]. The movement of solid objects can be represented as a translation of the center of gravity and a rotation around this point.

An agent – and its database – may abstract space in one of several ways, for example, a regular raster of observations or an object concept, or may use them in combination. The linkage between these views poses difficult theoretical problems, which are addressed in the spatial reasoning community. The question "what is special about spatial?" has been asked by several authors, but no generally satisfactory answer has been given [59,141].

2.5.4 Social Ontology

Human beings are social animals; language allows us to communicate and to achieve high levels of social organization and division of labor. These social institutions are stable, evolve slowly and are not strongly observer dependent. Conventionally fixed names for objects, but also much more complex arrangements which are partially modeled according to biological properties, for example, the kin system, or property rights derived from physical possession, can be refined and elaborated to the complex legal system of today's society.

Names. The common names in language are clearly the result of a social process: words as names for individuals. This gives identifiers for objects, which are different from predicates to select an individual based on some unique set of properties. Nevertheless, socially agreed identifiers seem to be part of the individual, because they exist outside of the observing agent. Pointing out that "chien", "Hund" and "cane" are equally good words to describe what in English is called a dog should make it clear that none of these names is more natural than any other. Examples for proper names and similar identifiers reach from names for persons and cities to license plates for cars; there are also short-lived names created, like "my fork", during a dinner.

Institutions. Social systems construct rules for their internal organization [12], for example, laws, rules of conduct and manners, ethics, etc. Such rules are not only procedural ("thou shalt not kill"), but often create new conceptual objects (e.g., marriage in contradistinction to cohabitation without social status, adult person as a legal definition and not a biological criterion, etc.). Institutions are extremely important in our daily life and appear to us as real (who would deny the reality of companies, for example, Microsoft Corporation).

Much of what administration and therefore administrative databases deal with are facts of law - the classification of reality in terms of the categories of the law. The ontology of these objects is defined by the legal system and is only loosely related to the ontology of physical objects; for example, legal parcels behave in some ways similar to liquids: one can merge them but it is not possible to recreate the exact same parcels again (without the agreement of the mortgage holders) [139,140].

2.5.5 Ontology of Cognitive Agents

Cognitive agents – persons and organizations – have incomplete and partial knowledge of reality, but they use this knowledge to deduce other facts and make decisions based on such deductions. Agents are aware of the limitations of the knowledge of other agents; social games, social interaction and business are to a very large degree based on the reciprocal limitations of knowledge. Game theory explores rules for behavior under conditions of incomplete knowledge [6,56,147]. The knowledge of a person or an organization increases over time, but the knowledge necessarily lags behind the changes in reality. Decisions are made based on this "not quite" up-to-date knowledge. Social fairness dictates that the actions of agents are judged not with respect to perfect knowledge available later, but with respect to the incomplete knowledge the agent had or should have had if he had shown due diligence. Sometimes the law protects persons who have no knowledge of certain facts. The popular saying is "Hindsight is 20/20" or "afterwards, everybody is wiser". A fundamental aspect of modern administration is the concept of an audit: administrative acts must be open to inspection to be able to assess whether they were performed according to the rules and regulations or not. Audits must be based on the knowledge available to the agent, not on the facts discovered later. For audits it must therefore be possible to reconstruct the knowledge which an agent, for example, in a public administration, had at a certain time.

2.6 The Language to Describe the Ontology

Some formal language is necessary for the description of an ontology. Database schema, for example, are described in the Data Description Language [3]. The description of ontologies using logic or the use of data description languages resting on the relational data model to describe the schema of a database is (barely) sufficient to capture the meaning of the terms for a snapshot, an a-temporal database. Numerous practical experiences show that describing spatial data types with these means is very difficult [87,94,169], and the problems encountered when integrating data from different sources are so far not resolved [57]. The description of ontologies for spatial-temporal databases is even more demanding. In this section we propose to use algebras, which are not restricted to static relations, but permit to describe objects and operations in the same context and thus better capture temporal aspects. Practical proposals for the description of conceptual models for spatio-temporal applications follow later in this book (Chapter 3).

The language used to describe an ontology should have the following properties:

- formal, independent of subjective interpretation, i.e., it must be described as an algebra with (abstract) types, operations and axioms fixing the behavior up to isomorphism [67];
- declarative and independent of implementation;

- typed, to avoid the difficult logical tangles of untyped languages: Russell's antinomy with sets containing themselves and Gödel's undecidability problem are not existing in a typed universe;
- automated methods to check the consistency of ontologies must exist [88]; it is not humanly possible, to write substantive formal systems without error [52], quoted after [84,156].
- executable, at least as a prototype: it is very difficult to assess if a given formal description captures the correct intuition about the world; however, human beings are very good in judging if a model is a correct description of their experience if one can execute it [88].

Many are tempted to invent a new language to describe ontological models [133], but this is not necessary. Ontologies are traditionally investigated using logic, primarily first-order predicate calculus, where the variables range over the individuals (instances or tuples in the database jargon) [35]. This approach is very useful to construct rules to capture the foundation classes for reusable ontologies [109]. The differentiation between an extensional and intensional interpretation is important, i.e., possible world semantics [127] must be considered. Ontologies for spatio-temporal systems can be formulated in temporal logics, or in situation calculus introduced by McCarthy, and extended to a useful formalization for actions by Reiter [163]. Mereology [173] and mereotopology [177] extend ontological studies to the spatial domain. The formalism of simple logic formulae is easy to understand, but when the numerous technical restrictions are added to deal with time and space, the resulting discussions are very difficult to follow. Further, the use of logic is very often leading to formalizations which are not constructive and thus not directly translatable into implementations. For example, the widely referenced RCC calculus uses non-constructive axioms [161] and is therefore "not suitable for direct implementation in a reasoning system" [24, p.2].

An alternative with equally good mathematical pedigree is algebra. Here technically, by "algebra" we understand universal algebras (specifically heterogeneous or multi-sorted algebra) as introduced by Birkhoff [14,15,136]. An algebra is a triple, namely a set of carriers, a set of operations with signatures and a set of axioms which define the operations [136]. An algebra describes some abstract behavior of a set of objects, called the carrier, which is not further specified. Heterogeneous algebras allow multiple carriers for their objects, which correspond roughly to the notion of type in computer science [34]. There may exist several realizations for an algebra, often called models or implementations (for example, in Chapter 4 of this book), which cannot be separated with the methods included in the algebra. One says that an algebra defines objects and their behavior up to an isomorphism; all models of the algebra are isomorphic, they show the same behavior with respect to the observations possible within the algebra. Technically, the world and the information system are then models for the abstract behavior described algebraically. The definition of structure up to isomorphism is exactly what is desirable for an ontology used for the design of information systems: the ontology should describe the behavior of reality and information system equally. The rules observed in reality and the rules used in the information system must be structurally the same (they cannot be the same rules, as the former apply to physical objects and the latter to the data objects representing these in the database). As an example, we use here the familiar natural numbers N (Equations 2.6-2.11).

The axioms for natural numbers are as given by Peano [138]:

$$1 \in N \tag{2.6}$$

$$\forall m \in N \exists m' \in N (m' \text{ is the successor of } m.) \tag{2.7}$$

$$\forall m \in N\, m' \neq 1 (1 \text{ has no predecessor}). \tag{2.8}$$

$$\forall m, n \in N (m' = n' \rightarrow m = n) \tag{2.9}$$

$$K \subseteq N \rightarrow K = N \text{ provided that the following conditions hold:}$$

$$1 \in K$$

$$k \in K \rightarrow k' \in K. \tag{2.10}$$

Addition:

$$m, k \in N,\ m + 1 \stackrel{df}{=} m', \exists (m + k) \rightarrow m + k' = (m + k)' \tag{2.11}$$

For natural numbers there are many different realizations – as Arabic numbers, Roman numerals, apples, sheep in a flock or binary numbers (Figure 2.11). In all cases, the rules for addition hold the same way. $VII + II = IX$ is the same as $7 + 2 = 9$.

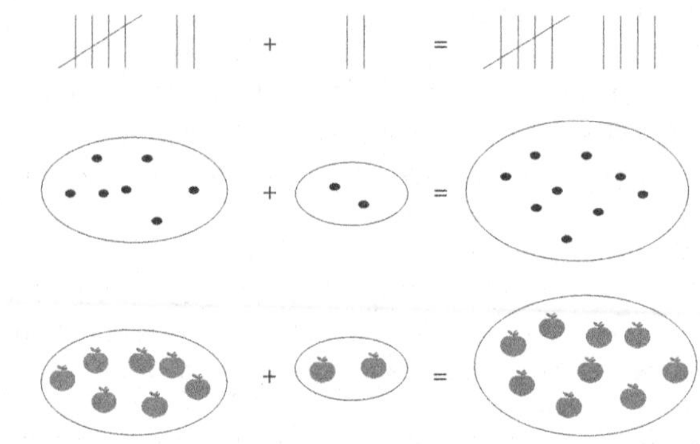

Fig. 2.11. Different realizations of the natural numbers

The essence of an algebra, completely abstracting from the representation, is captured in category theory [5,9,157]. This can be applied to query languages [117] or to algebraic specifications [68,135].

2.6.1 Tools to Implement Ontologies

The algebraic approach for the definition of ontologies for spatio-temporal databases can use modern functional programming languages as tools for formalization and to build executable models. Universal algebra and the development of functional programming languages rest on the same mathematical foundations: category theory. In category theory properties of operations are discussed in complete abstraction from the application of the operation on objects [5].

Functional programming languages can be implemented today close to mathematical concepts and serve as tools to apply the corresponding theories to practical problems. Haskell is the result of a decade of experimentation and is the unification of different proposals. It is a standardized and widely used language [154,156,192]. It is a non-strict (lazy) purely functional language with classes. In Haskell algebras are described as classes (abstract data types) and simple models constructed for testing. In order to create executable models, care must be taken to include only constructive axioms; other axioms can be included as tests for the model and exclude non-intended interpretations. The following code gives the example of Peano's algebra in Haskell with the executed example `t5 = 2 + 3 = 5`.

```
class PeanoNum n where
    suc :: n -> n
    eq :: n -> n -> Bool
    plus :: n -> n -> n

data Nat = One | Suc Nat

instance PeanoNum Nat where
    suc = Suc
    eq One One = True
    eq (Suc m) (Suc n) = eq m n
    eq _ _ = False
    plus m One = Suc m
    plus m (Suc n) = Suc (plus m n)

t5 = plus (Suc (One)) (Suc ( Suc (One)))
result:          Suc (Suc (Suc (Suc One)))
```

In a functional language, everything is a function which yields a value. Haskell is a second-order language, which allows variables which are functions. Most current programming languages are first-order languages and permit variables only to range over constant values, e.g., natural numbers, floating point numbers, etc.; in a second-order language, a variable can also be a function, e.g., *cos* or *absoluteValue*. Research in database formalization has previously identified a need for second-order languages [112].

Haskell is a strongly typed language [33] extending the Hindley-Millner type inference systems further. In a typed language, every value has a type and operations are applicable only to values of the correct type; the type system used here assumes type inference, i.e., the type checker infers types for expressions which are not explicitly typed from the types of inputs and outputs. Haskell has classes and allows parametric polymorphic application of operations defined in classes to all elements for which the class has been instantiated. This covers what is usually called "multiple inheritance" in a consistent and rigorous framework, in which the algebra, an abstract data type, is parameterized and the instantiations separately describe how an operation is applied to a specific representation (data type). In this framework, operations which apply to an element in a data structure are then polymorphically extended to apply to the data type; this is called "lifting".

Second-order formalizations are extremely useful to deal with spatio-temporal data types. A data type *movingPoint* can be seen as a function, sometimes called a "fluent", which for every point in time (i.e., instant) yields a point in space. In a second-order language, such functions are properly typed (they have a type described as $movingPoint :: Instant \rightarrow Location$). Operations can be applied to such "function types", for example, two movements can be added; in a polymorphic language like Haskell, the operation "+" can be lifted to extend to this new data type *movingPoint* and thus it becomes possible to add two movements simply with the operation "+". The result is defined as vector-addition for each instant.

As an example, we show how from time-varying values "moving points" are constructed in Haskell. We assume floating-point numbers with the operations "+", "–", "*", square and square root, which are implemented for a data type Float[1].

```
class Number a where
   (+), (-), (*)   :: a -> a -> a
   sqr, sqrt :: a -> a
   sqr a = a * a
```

A type "Moving T" for any type T is used to represent a family of time variable types; for this parameterized type the operations of the class *Number* are implemented as synchronous: "+" applied to two moving values produces a moving value, which is for each time point the sum of the values at this time point. This lifting of the operations from values to time-varying values permits to operate with time-varying values as simply as we operate with constant values.

```
type Moving v = Time -> v
instance Number v => Number (Moving v) where
   (+) a b = \t ->  (a t) + (b t)
   (-) a b = \t ->  (a t) - (b t)
   (*) a b = \t ->  (a t) * (b t)
   sqrt a = \t ->  sqrt (a t)
```

Points are defined with operations to combine the two coordinate values and the projection operations x and y and the operator to calculate the distance between

[1] In Haskell, classes define abstract algebras and instances give the implementation for a specific datatype. Implementation can be parametrized and the parameters restricted to instances of certain classes – here used for the operations on Points. Function application is written without parentheses e.g., `f(x)` is written as `f x`. The lambda construction is written as `\x` → `ax`, defining a function `f (x) = a (x)`.

two points. We also lift the operations "+" and "–" to apply to points as the regular vector addition and subtraction.

```
class Number s => Points p s where
   x, y :: p s -> s
   xy :: s -> s -> p s
   --
   dist :: p s -> p s -> s
   dist a b = sqrt (sqr ((x a) - (x b)) +
                sqr ((y a) - (y b)))

data Point f  = Point f f

instance Number v => Points Point v where
   x (Point x1 y1) = x1
   y (Point x1 y1) = y1
   xy x1 y1 = Point x1 y1
instance Number v =>  (Point v) where
   (+) a b = xy (x a + x b) (y a + y b)
   (-) a b = xy (x a - x b) (y a - y b)
```

Moving points are created as points with moving values as coordinates and for such moving points, the code necessary to calculate distances between points is derived automatically. We can define two points *np*1,*np*2 and calculate the distance between these two moving points as a moving value; *movingDist_1_2* is the distance as a function (which could be passed as an argument to a function to find its minimum or maximum value) and *dist_at_1* is the value of the distance function for time 1.0 – this is all the code necessary to execute!

```
np1, np2 :: Point (Moving Float)
np1 = xy (\t -> 4.0 + 0.5 * t) (\t -> 4.0 - 0.5 * t)
np2 = xy (\t -> 0.0 + 1.0 * t) (\t -> 0.0 - 1.0 * t)
movingDist_1_2 = dist np1 np2
dist_at_1 =  movingDist_1_2 1.0
```

2.6.2 Multi-agent Systems and Formalization of Database Ontologies

The framework in which spatio-temporal databases must be discussed, shown in Figure 2.5, is quite similar to the logical framework used for the discussion of multi-agent systems [75,203]. Indeed, the multi-agent system provides the theoretical framework, in which ontological discussions can be grounded following the idea expressed in Figure 2.12 (and an attempt to create such a view is already present in [109]).

A database represents beliefs (in the terminology of [55]) some agent has collected about the world. Agents can, by definition, have only a partial and

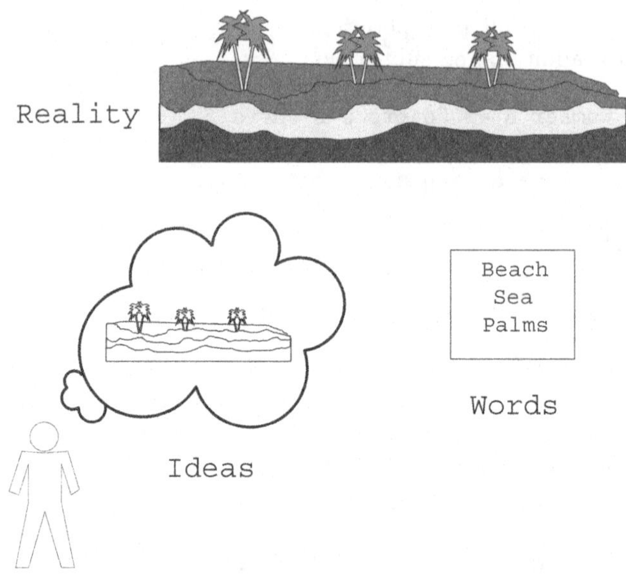

Fig. 2.12. Agents in the world

approximate knowledge of the world (the observations of Tier 2). The knowledge stored in the database is likely incomplete, incorrect and approximate.

An agent in a multi-agent framework can be a single activity [142], a single living entity with some cognitive abilities (an animal or a human being) or a larger organizational unit, for example, an agency, a ministry or a company. For building ontologies, single persons and organizations (from small companies or research groups to whole nations) are considered as agents.

In a multi-agent system, a formal discussion of the correspondence between the simulated reality and the simulated database content is possible. In most discussions of ontology the non-formalized reality is contrasted with the formalized model in an informal discussion. Here, we posit a formally constructed fragment of reality, which is then connected with the model. This allows us to construct formal models of an ontology in which the theoretical issues can be discussed and the necessary theories constructed. Such systems are not directly useful as information systems, but useful to design ontologies and to demonstrate how different domain ontologies can be integrated. For example, the situation described in Subsection 2.4.1 and Figure 2.5 has been translated to an executable model [85].

2.7 Ontological Tier 0: Ontology of the Physical Reality

Ontology, in a naïve view, should describe what is. In this section, the necessary minimal assumptions about physical reality are described. We assume that

- physical reality exists,
- it has determined properties at any point in space and time,
- different types of properties exist.

This gives a minimal ontology of points in space and time. Property values at each point in space and time are quantitative or qualitative values.

This section is necessarily very brief because there is little we know objectively about the physical reality. Even from millions of empirical observations no deductive knowledge about the world ever follows. It is often overlooked that all we know about the world is based on observation; a database represents the beliefs of some agents about the world, never the physical world directly. This section talks about the assumption necessary that the inductions in the following sections, which have led to the theoretical sciences, are possible.

2.7.1 Properties

Properties of the world combine a point in time and space with a value expressed on a continuous scale, represented by real numbers. Some properties can be observed and the property values are transformed by the observation process in measurement values.

There are different properties at every point of reality. The properties at a point in time and space are determined, i.e., multiple observations will always start with the same property value (but may yield different results, due to imperfections in the observation process, see Section 2.8.1). The same properties have at different points in time and space different values. Typing rules avoids nonsensical operations (Figure 2.13).

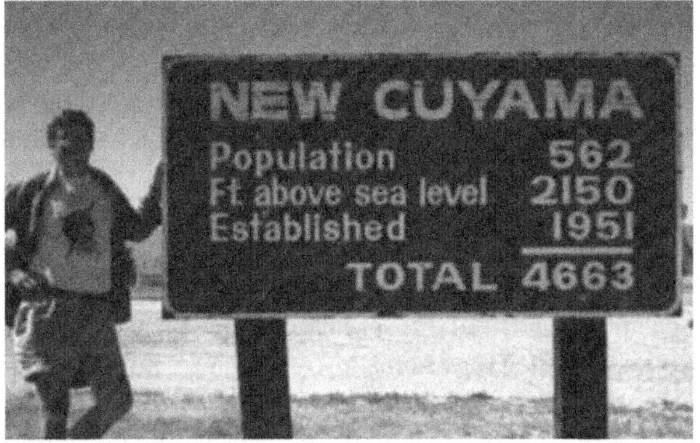

Fig. 2.13. Billboard found in New Cuyama, California

2.7.2 Physical Space-Time Field

For the abstract (non-constructive) level of ontology, points in space and time are described with coordinate values of real numbers to establish continuous time and space. This is the classical model of space and time of physics, where both real-world space and time is mapped to an n-cube with real number axes. This physical abstract field model is described with a function. An observation at a given point in space and time yields a single value. The temporal and spatial coordinates are formally equivalent. There is no special treatment of natural constants – they are understood as varying in time and space, and most of the "natural constants" are indeed varying in time or space (see Section 2.5.2). The world is implemented as a function.

$$\begin{aligned} reality :: world \rightarrow property \rightarrow spacePoint \rightarrow \\ timePoint \rightarrow value \end{aligned} \tag{2.12}$$

This formal model of reality as a function (see 2.12) expresses the ontological assumption of a single reality as observable by a (single-valued) function reality (comparable to $f(x, y, z, t) = a$); if we had allowed multi-valued functions here, then we would permit "parallel" universes, as made famous by Asimov in Living Space [4]. With such a single-valued function the situations shown in Figure 2.2, Figure 2.3, and Figure 2.4 can be represented.

2.8 Ontological Tier 1: Our Limited Knowledge of the World through Observations of Reality

Observations of reality are necessarily limited: we can only know a (very small) subset of reality, with very limited precision. We can only observe at specific locations and at specific times, and human observers are restricted to observations of the properties for the moment "now". Continuous observations are actually rapid samples at discrete points. Measurements are observed with unavoidable error and are expressed with limited resolution; sometimes this is all subsumed under the notion "discretization" [102]. These limitations are modeled in Tier 1.

2.8.1 Observations

Observations translate the value of a property at a specific point in time and space into a measurement value. Observations are realized as physical processes which translate the intensity of some property into an observation value, expressed on some measurement scale; observations are always made at the present time ("now"):

$$\begin{aligned} observation :: world \rightarrow observationTypes \rightarrow \\ location \rightarrow value \end{aligned} \tag{2.13}$$

The domain of this function is composed of:

- the world observed;
- the types of observation agents (humans, with or without technical means);
- the location on the earth; it is finite space, but unbounded, empty space is isotropic.

The range of the function consists of values on some measurement scale.

Measurement Values. Measurement values describe the result of the observation process. Values are formal, i.e., mathematical objects, introduced here to capture the outcome of the observation. Values are defined as algebras, and are typically described as derived algebras using some base (given) fundamental algebra. For example, integers are used to construct values of type `Money`. The algebras for values must connect the values to the outside world. All values have operations to translate from and to a human readable form.

For example, the natural numbers are defined by the axiom system given by Peano (see Equations 2.6-2.11). From natural numbers rational and real numbers and other number systems can be defined similarly; for example, rational numbers can be defined as pairs of integers.

Stevens, in a landmark article [189], has shown the fundamental properties of the measurement scales. He listed four measurement scales, namely the

- Nominal scale: only the equality between values can be tested (example: names of persons);
- Ordinal scale: values are ordered (example: grades in school, rank in a race);
- Interval scale: differences between values are meaningful (example: temperature in degree Celsius, height above sea level);
- Ratio scale: ratios between values can be computed and an absolute zero exists (example: temperature in degree Kelvin, population counts, money in a bank account).

These measurement scales correspond to algebras; we often find the roughly corresponding algebras of equality ($=, \neq$), order ($<,>,\geq, \leq$), integral ($+, -$) and fractional ($+,-, *, /$). Other measurement scales exist but are not as prominent or well researched [41,79]. The nominal and the ordinal scale are often called qualitative, especially when the number of different values is small. For example, the size of a garment can be expressed on an ordinal, qualitative scale with the values "small", "medium", "large", "extra large".

Observation Error. All observations are imperfect realizations and imply error. This is in the limit a fundamental consequence of Heisenberg's uncertainty principle, but most practical observations are far removed in precision from the fundamental limits. Measurements better than 1 part in a million are generally difficult (i.e., distance measurements with an error of 1 mm per kilometer are very demanding, few centimeters per kilometer are standard performance of surveyors today) and the best observations are for time intervals, where 10^{15} is achieved, but the theoretical limit would be 10^{23}. Parts of the error of real observations are the result of random effects and can be modeled statistically. Surveyors report measured coordinates often with the associated standard deviation, which represents – with some reasonable assumption – an interval with approximately 60% chance to contain the true value. Error propagates through the computation. The Gaussian law of error propagation approximates the propagation of random and non-correlated error; it says that the error propagates with the first derivation of the function of interest. Given a value $a = f(b, c)$ and random errors for b and c estimated as e_b and e_c (standard deviations), then, following Gauss, the error on a is:

$$e_a = \sqrt{\frac{df}{db} e_b^2 + \frac{df}{dc} e_c^2} \tag{2.14}$$

A number system can be extended in such a way that every value is associated with an error estimation. Numeric operations on values are lifted to calculate not only the result but also the estimated error in the result using Gauss's formula [83].

```
data Efloat = EF Float Float
instance Num Efloat where
    (EF v' s') + (EF v2 s2) =
       EF (v' + v2) (sqrt (sqr s' + sqr s2))
```

Resolution and Finite Approximation. The results of observations are expressed as finite approximation to real numbers (often called floating point numbers). Geographers often use the term "resolution" to describe the smallest discernible difference between two intensities, not necessarily one unit of the last decimal. For example, distance measurements are often read out to mm, but the error (one standard deviation) is much larger: 1 cm + 1mm /km.

Constructing software for geometrical calculation using the finite approximations available in computers is difficult [94]. Some solutions have been developed recently [105,119,169].

2.8.2 Measurement Units

Measurements describe the quantity or intensity of some properties at a given point in comparison with the intensity at some other, standard, point or standard situation. Well known is the former meter standard, defined as the distance between two marks on a physical object manufactured from precious metal and kept in Paris (it is superseded today by a new definition, which links to a physical process that can be reproduced at any location). The temperature of melting ice is used as the reference point for the Celsius scale [69].

Observation systems are calibrated by comparing their results with the standard. They are expressed as a quantity times a unit e.g., 3 m, 517 days or 21

Table 2.4. SI units

length:	meter	m
mass:	kilogram	kg
time:	second	s
electric current:	ampere	A
thermodynamic temperature:	kelvin	K
amount of substance:	mole	mol
luminous intensity:	candela	cd

degrees Celsius. The Système International d'Unites (SI) is founded on seven SI base units for seven base quantities assumed to be mutually independent (Table 2.4). Before, people used the cgs-system (centimeter-gram-second). For example, the unit of gravity in the cgs-system was Gal, named after Galilei (1 Gal = 1 cm/s^2), but newer books refer to the SI standard (m/s^2).

For the same kind of observation different units are used, most important the metric units and the Anglo-Saxon units (which come in Imperial and U.S. variants). The fundamental physical dimensions (length, mass, etc.) are easily converted, but practically errors occur often. Most spectacular was the recent loss of a probe to land on Mars due to a lack of conversion between metric and Anglo-Saxon measurement units of length. Practically, conversions are a problem, not for the different measurement units, but for the differences in the observation methods, which result in somewhat different properties observed, even when expressed on the same physical scale (for example, noise level measured in dB). Conversion can be achieved, as the two observation methods can be applied at the same location and time, and the results can be compared. From sufficiently well comparable observations a conversion formula can be deduced.

2.8.3 Classification of Values

In daily life and in most applications, the results of observations are expressed on qualitative scales, i.e., scales with only a few values. People are classified in small, medium, tall and very tall; days are hot or cold; etc. These classifications capture sufficient information for the task at hand [96]. For example, to meet a person at an airport, a description using adjectives like "gray, tall, bespectacled, 50ish male" is usually sufficient to identify the person.

Classification translates observations from a larger set of values to a smaller set of values (discretization can be seen as a special case of classification). The age of a person is mapped to the set child, adult with the rule

```
classify :: valueType_1 -> valueType_2
classify age = if age > 20 then adult else child
```

Roads for maps are classified from the width (measured in m with two decimals) to road classes first order, second order, and third order:

```
classify roadwidth = if width > 5.50 then firstOrder else
   if width > 4.0 then secondOrder
   else thirdOrder
```

These two examples clearly demonstrate the difficulties involved in classification and the use of classified data. Seldom do the classifications for two different tasks correspond: How to compute the number of adults in Europe? There exists no uniform concept of adulthood, as age limits vary. A road map which shows a road changing in classification at the border (Figure 2.14) allows two interpretations: either the road changes its width at the border or the classification scheme for roads is different in the two countries.

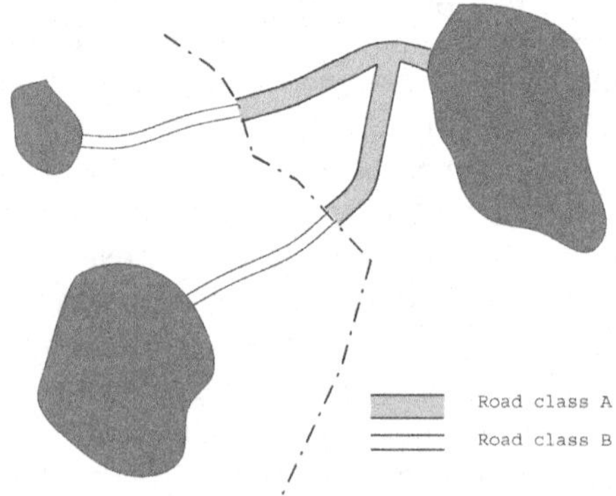

Fig. 2.14. Apparent changes of road classification across border

2.8.4 Special Observations: Points in Space and Time

Descartes discovered that calculation with coordinate values can simulate geometric constructions. Surveyors and engineers make extensive use of analytical geometry. The algebra of vector space, constructed from the field of real numbers with operations for scalar multiplication, vector addition and multiplication, is extremely convenient. Orthogonality of vectors, etc., can be tested and areas of figures calculated with simple arithmetic operations.

These computations with length measurements and the positions of points are expressed as distances from a conveniently selected point of origin. No physical process can measure absolute time or points in space; only relative measurements are possible. Distance measures (in space or time) are always relative to other points and no direct coordinate measurements are possible. Relative measures are tacitly converted into coordinates assuming a conventional origin. These computations make statistical descriptions of errors in the coordinate values difficult.

Buyong describes a method to represent locations by relative measurements and uses coordinates only for computation [29,30]. A database would contain as

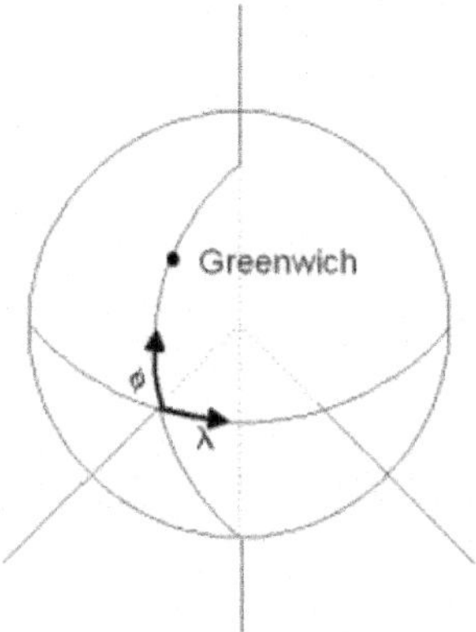

Fig. 2.15. Geographic coordinate systems

original determination of location the relative measurements and new measurements can be added freely. Coordinates are calculated from measurements using adjustment calculation [128], either when needed and immediately discarded, or stored and recalculated after each addition of new measurements. In the traditional coordinate-based system, location information deteriorates over time with the integration of new point locations and occasionally a complete new survey of all locations must be made (typically every 30 to 50 years). In a measurement-based system, the quality of the determination of location improves with the addition of new measurements.

Continuous Time: Instants and Time Measurement Values. We measure time with respect to a time scale with a conventionally selected origin. The customary Gregorian calendar has the origin related to the birth of Jesus Christ (and other religions select origins related to the history of their religion). The conventional time measurement scale has some interesting particularities; for example, there is no year 0, after year 1 BC follows 1 AD; the way the duration is commercially calculated between two dates depends if you borrow or lend, etc. [80].

Continuous Space: Points and Coordinates. Every country has selected a prominent point as an origin for the national grid of coordinate values. Two global systems are used: the well known system of geographic longitude and latitude, where distances are measured by angles and the origin is the intersection of the meridian of the astronomical observatory in Greenwich (exactly the optical center of the old passage instrument) with the equator (Figure 2.15); whereas the modern system uses three orthogonal axis, situated in the center of the earth mass.

For each coordinate system, origin point and direction of the axes must be fixed. In practice, the origin of the coordinate system is of no relevance as the

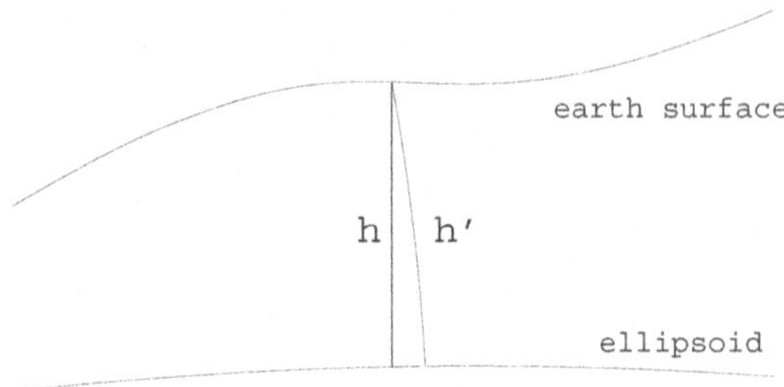

Fig. 2.16. Differences in height measurements

coordinate system is realized by the totality of all points for which coordinate values are determined, and which can be used to measure distances and directions to new points.

Most surveying systems separate the position in the plane from the height and introduce a separate zero point for the height measurement. The conventional origin at sea level has no precise definition and leads to confusing differences between the heights of points expressed in different national systems; well known is the apparent height difference of several meters between Germany and Belgium. Height can be measured as the potential from an assumed zero potential surface or can be measured as a distance from this same surface (Figure 2.16); precise height measurement systems are not a completely resolved topic in geodesy.

2.8.5 Approximate Location

Points (in time and space) are mathematically without extension. Real observations identify small extents which are all mapped to the same value. The dissertation of Bittner [17] showed how operations with such approximations are possible (the recent paper by Smith and Brogaard exploits some similar ideas [181]).

Bittner and Stell describe the location of spatial objects within sets of regions of space (cells) that form regional partitions (Figures 2.17 and 2.18). The location of spatial objects is characterized by sets of relationships to partition cells. Figure 2.17 shows the approximation of a non-regular shaped region, r, with respect to a raster-shaped regional partition. The raster-shaped partition simplifies the example; in general arbitrary partitions of 2-D space are possible. Bittner and Stell distinguish three relations between the region, r, and a partition cell, $g \in G$: (1) Full-overlap, i.e., r contains or is equal to g (Cell K in Figure 2.17). (2) Partial-overlap, i.e., r and g share parts, but do not fully overlap. (All partition cells except K,A,I, and M). (3) Non-overlap, i.e., r and g do not even overlap partially. (Cells A, I, and M). There are coarser and finer distinctions possible (see [19] for details). Formally, the approximate location of the region r within the partition G as a mapping of signature $\alpha_r : G \rightarrow \Omega$. The mapping ($\alpha_r$) returns for every partition element $g \in G$ the relation between r and g, i.e., fo for full-overlap, po for partial-overlap, or no for non-overlap.

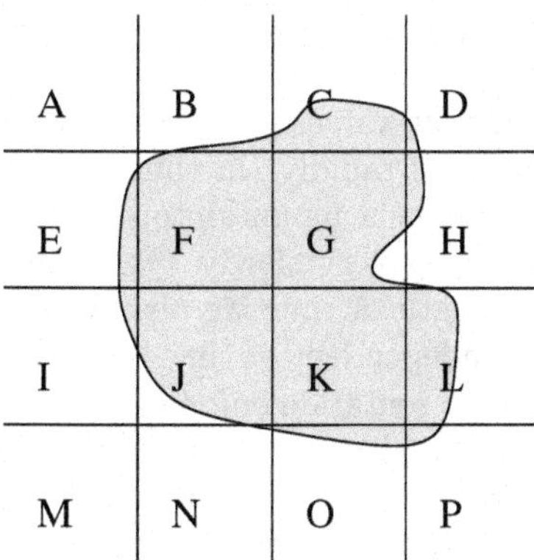

Fig. 2.17. A region approximated by a regular partition

Fig. 2.18. The position of National Parks with respect to some western states of the USA

Approximations represent sets of regions, i.e., all those regions that are represented by the same approximation mapping. Bittner and Stell [19] defined union and intersection operations on approximation mappings, such that their outcome constrains the possible outcome of union and intersections operations between the approximated regions. Moreover, they derived sets of possible relations between spatial regions, given their approximations [20]. Bittner [18] showed how approximate spatial reasoning can be applied to the temporal domain.

Consider, for example, the approximations of a water reservoir and a pollution area with respect to an underlying soil classification partition. Assume both approximations are based on independent observations in different moments of time. Bittner and Stell [19] derive which relations can hold between the water reservoir and the area of pollution from the knowledge about their approximate location. This can be used, for example, to determine if it is possible that both objects overlap or it is certain that they do (or do not) overlap.

2.8.6 Discretization and Sampling

An observation yields different values, depending where and when we observe reality. Some processes change rapidly (in time or in space), others vary very slowly: Observing the height of a mountaintop is (nearly) independent of the time as the value changes only very slowly. Observing the position of a car is highly dependent on the length of time we observe the car. Measuring gravity gives very similar results, independent of time or location.

The limitation of the representation enforces some careful optimization when building models of parts of reality – the model is necessarily limited and an approximation. Information systems concentrate on some processes in reality, e.g., the movement of cars on highways, development of cities or the weather. Each of these processes (or complex of processes) has a certain "scale". Space and time can be treated equally: some properties change quickly if we move in space (e.g., elevation) and others change very slowly (e.g. geology). One can speak of a temporal and spatial frequency of events [95]. For example, geological processes typically have a resolution of 10 to 100 meter in space and thousand to millions of years in time. Movement of cars on highways has a spatial resolution of several m in space and seconds to minutes in time.

If we observe a process, then our observations must have a minimum density to avoid misleading "aliasing". A process can be described with a frequency of change (in space and time). The sampling theorem states that observations must be made with at least twice the frequency of the highest frequency in the process of interest. To avoid erroneous observations, so-called aliasing, frequencies higher than half the sampling frequency must be filtered out before the signal is sampled (Figure 2.19).

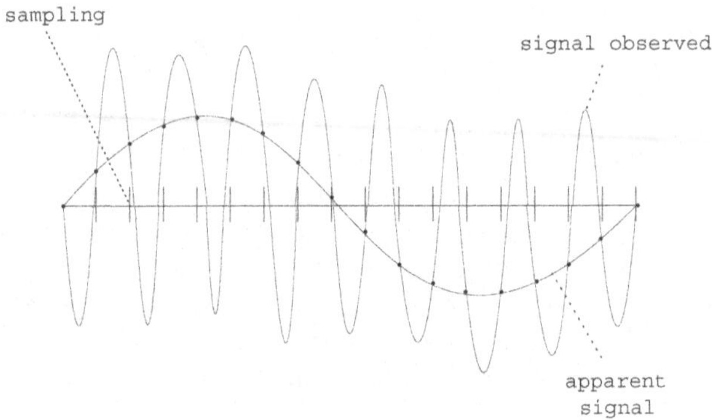

Fig. 2.19. A low frequency signal emerges from a high frequency signal not correctly sampled

The practically very important method of data collection by remote sensing (Figure 2.8) takes the average value over the area of the pixel and thus applies at the same time a filter which eliminates too high (spatial) frequencies. If point sampling is employed, then the high frequencies present in the signal can lead to significant errors in area estimates; this is particularly dangerous in areas with regular structures, e.g., the road network in the U.S. Midwest.

2.8.7 Virtual Datasets: Validity of Values

It is convenient to organize large numbers of observations in such a way that a comprehensive model of reality emerges. It is possible to link datasets available on different computers internally and to provide an interface which gives the same functionality as the real observation of the world (Eq. 2.12):

$$ObsValue :: dataStore \to location \to time \to observationType \to value \quad (2.15)$$

Vckovski has labeled such datasets as "virtual" and discussed their properties [200]. Unlike the observation function of reality in Section 2.8.1, the observations of virtual datasets are partial - not for every combination of input values a result is available; often the value is unknown. Values resulting from simulated observations from a dataset and not direct observation of the world must be qualified:

- Bounded knowledge in time and space: the data collected always covers only some limited time span and some area. For positions and times outside this area, the result is unknown.
- Values for points other than those observed must be interpolated. The selection of appropriate interpolation methods is a difficult problem [51].
- Values were measured with limited precision. The user must be made aware of the error resulting from measurement and the error resulting from interpolation. The effects of classification or different observation methods must be tracked and values converted as far as possible.

Observation operations to the information system are partial: requests for values for points in time or space, before the first or after the last value available (or outside of the spatial limit), must not return a regular value. Computation with total functions is simpler. Wrapping the result value in a data type *Maybe* allows us to transform the partial to a total function and lift operations to this data type (*Maybe* is a monad [202], but this is not important here).

```
data Maybe a = Just a | Nothing
obsValue :: dataStore -> location -> time ->
     observationType -> Maybe value
class Number a => Number (Maybe a) where
    (Just a) + (Just b) = Just (a + b)
    _ + _ = Nothing
```

For virtual data sets, time series, coverages [41] or raster images, the regular operations can be applied ("lifted"). It is possible to add two time series, to apply mathematical formulae to values of corresponding cells, etc. Tomlin has shown how this "map algebra" can be extended with operations useful for planning [196,195].

2.9 Ontological Tier 2: Representation – World of Individual Objects

Humans have the ability to see objects – and they use very different criteria in the way they carve reality into objects: objects are formed according to the current needs, i.e., the task a human agent tries to complete at a given instance. An ordinary tree stomp can be seen as a table, a seat or platform to stand on. The human interactions with the environment, especially with solid bodies as included in the table-top environment, are ubiquitous and probably prototypical for our understanding of objects. But there is also the large class of geographic objects, which are typically unmovable and not physical themselves, but made up from other physical objects (a road or a forest is an area of space, not a physical object). Humans "see" objects with respect to their interaction with them. Gibson called these potentials for interaction "affordance" [100].

Physical objects are extremely real for the human cognition. But instead of following the philosophical tradition, which assumes a preexisting understanding of objects, a very pragmatic approach is followed: objects are defined by uniform properties. The properties which must be uniform for an object are related to the possible ways of interaction with an object. Depending on the property, which is uniform, very different types of objects are formed and these objects then follow different ontological rules (which we call lifestyles, see Section 2.9.5). The properties, which are fixed to determine uniformity, can be used to define a topological, morphological or functional unity [109].

Objects preserve invariants in time and are therefore a method of human cognition to reduce the complexity of the world, by grouping areas of uniform properties with respect to potential interactions. The most salient example are solid bodies, which preserve form, volume, material, weight, color, etc. There are transformations from point observations to observations of objects, typically integrating specific properties (e.g., specific weight) over the volume of the object. Representations for objects are best selected to respect the invariants; the geometric form of an object is best expressed in a coordinate system fixed with the object and a vector which indicates the location of the object and an angle of rotation; from this coordinates in an exterior system can be deduced.

The approach selected here – namely to define objects simply as areas of uniform observable properties – avoids the difficulties philosophers have found in the foundation classes of ontologies, where the most fundamental classes of entity, object, etc., are defined. Guarino and his colleagues have compared several ontologies and have found conflicts and hidden assumptions, which made comparison between the ontologies and transfer of knowledge integrated in one ontology to another ontology difficult [110].

The object concept used here is restricted to "physical objects" (which contains as a subset the "material objects" but is much larger), i.e. things which exist in the physical world and can be observed by observation methods. This is not the most general concept of an object, as it is often used in philosophy or software engineering; specifically, constructions like abstract ideas, social constructions, etc., are not included and will be discussed in the next sections.

2.9.1 Objects Are Defined by Uniform Properties

Objects are defined as spatio-temporal regions of some uniform property. The uniformity can be in the material type, in what moves jointly, etc. Table-top objects are typically delimited by what forms a solid body, but other properties are often used, for example, color, texture, etc., is typically used to identify objects on photographs, including remote sensing images. More complex properties, like "same DNA" for a living animal body, are also possible.

Examples for objects on the tabletop are the cup, knife, piece of bread, etc. In the cityscape, objects are buildings, persons or cars. In the landscape, forests, lakes, mountains and roads are all objects with boundaries of varying degrees of sharpness [28]. In order for things to have uniform properties, the properties must be classified and small variations in reality, or by the errors in the observation process, eliminated. The classifications can be applied to values which are the result of some computations, combining multiple values; for example, to detect areas which are connected, one can observe direction and speed of movement (the same result can be achieved with a static analysis of the resistance to stress and strain, which indicates where a collection of material bodies will separate). Ultimately, the classification results in a binary result – a point in space or time is part of or is not part of an object.

2.9.2 Geometry of Objects

The geometry of objects results from a classification of some property values. Delimiting areas of uniform value have some desirable geometric properties. They form a partition, i.e., they are jointly exhaustive and pairwise disjoint (see Figure 2.20).

Spatial Objects Have a Geometry. Spatial objects have boundaries. In many cases, specific observation systems are organized to find the boundary positions directly in the terrain and not from the point-wise observation of the environment. Surveyors go out and measure the boundary of the forest by detailed observation in the field and then measure the location of the boundary.

If the classification is applied to a collection of raster values, then the discretization effects of the observation step shows. The smallest object which is certainly detectable must have an area of at least four times the cell size (follows from the sampling theorem, see Section 2.8.6).

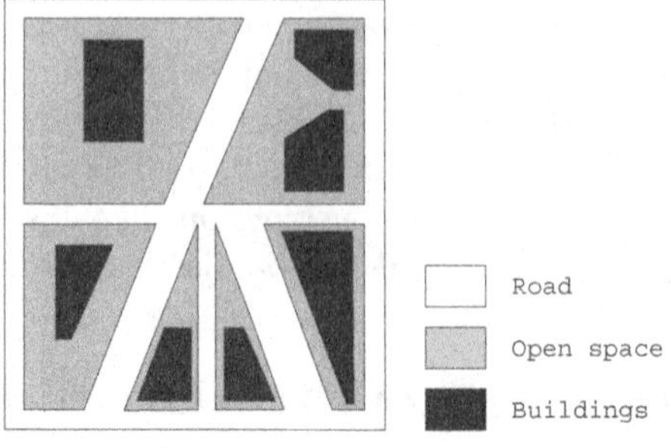

Fig. 2.20. Classification of a part of the remote sensing image from Figure 2.8

Geometric Objects Relate to Other Objects. For spatial objects, the location of the centroid and the boundary of the objects can be deduced:

```
getObjBoundary :: env -> spatobj -> boundary
centroid :: env -> spatobj -> point
```

Both the boundary returned and the centroid point are objects and have properties. For example, the position of a point can be asked, the length of the boundary or the area delimited can be found, and a complementary set of geometric operations for point, line and area are provided in most Geographic Information Systems. The Open GIS Consortium works towards standardization in conjunction with the ISO standard for extension of the query language SQL.

```
position :: point -> env -> coordinate
area:: boundary -> env -> areaValue
length :: boundary -> env -> lengthValue
```

Boundaries can be approximated by a sequence of straight lines and the corners of a boundary are a set of points:

```
corners:: boundary -> env -> [points]
```

The object-object relations are necessary to document that the boundary of parcel A is as well the boundary for the parcel B (Figure 2.21). An inverse function to *getObjBoundary* retrieves for a boundary line the two areas which are bounded by this line:

```
getBoundedArea :: boundary -> env -> [object]
```

The object-object relations are also necessary to model containment of objects with different granularity. One object may be part of a larger object and the larger object may contain a number of smaller objects.

```
contains :: obj -> env -> [objs]
contained:: obj -> env -> obj
```

Containment relations may form hierarchies, for example, for the political subdivision of a continent in countries, regions, provinces, communes, etc. (for example, the European NUTS subdivision forms a hierarchy of partitions, where each higher level forms a refinement of the previous one). Timpf has investigated how such containment hierarchies are formed, how they relate to other hierarchies (for example, functional), and how they are used for cartographic generalization [193].

Objects Resulting from Classification form Topological Complexes. The boundaries of a set of objects coming from a single classification form a complex [99]; each boundary bounds an areal object on each side. All the objects form a partition, i.e., they jointly exhaust the space and are mutually disjoint (often described as JEPD, jointly exhaustive, pairwise disjoint). In the extreme of a binary rule forming a single object, the object formed and the background together are JEPD.

Given a classification to determine what is a "uniform" object one can ask for a list of all objects within an area.

For geometric objects in a complex, operations can can return the boundary of the object or the co-boundary. For a cell or simplicial complex (Figure 2.21), these operations form algebras with well-defined properties. The boundary of an area are the lines, the boundary of a line are the two bounding points. The co-boundary is the converse operation; the co-boundary of the line are the two areas bounded by this line, and the co-boundary of a point are the set of lines starting or ending in this point.

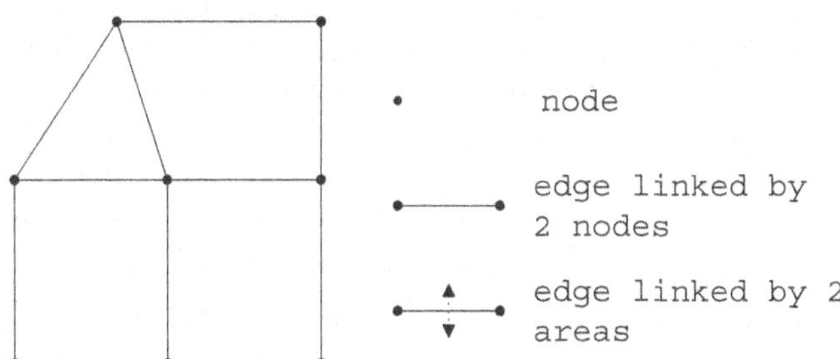

Fig. 2.21. Simplicial complex of nodes, edges and areas with boundary and co-boundary relations

A spatial database for which objects have these or comparable operations is often called a topological GIS or a topological data structure. These properties were identified very early in the history of GIS and used for checking for errors in data freshly input [46].

The topological relations between objects are usually described by the set of topological relations, which were proposed by Egenhofer in his dissertation

[64]. They are a generalization of Allen's relations for temporal intervals for two (and higher) dimensional regions. They are similar (but differently defined) to the relations the RCC calculus proposes [45].

2.9.3 Properties of Objects

Some properties of the objects are related to the problem of identifying regions of uniform properties. Other properties of objects can be related to the properties of the observable (and sometimes non observable) physical properties of the field ontology [61]. For example, the weight of an object is the integral over the weight of its material. The effort necessary to move along a path is related to the accumulated height differences along the path.

Operations to integrate observations for a geometry can be defined as an extension of point observations: integrate along a path, integrate over an area for some property (if the property is time-variant, the movement in time must be provided). From such integrations, properties of objects follow: the length or the area of a geometry by integrating over the constant function 1, but it can also be used to determine the height difference along a path, the annual rainfall over an area, etc.

2.9.4 Geographic Objects Are not Solid Bodies

The classical concept of object is a generalization from the physical objects on the tabletop; such material objects are exclusive: where one object is, no other object can be. This is correct only for solid body objects and not the case for other physically observable objects: in most applications, more than one classification is possible [181]. In the city environment the classification can be based on a pedestrian viewpoint or a legal-ownership viewpoint: a pedestrian is interested in the areas which are uniformly "not obstructed", whereas a bank is interested in seeing what areas have an uniform ownership.

This gives more than one object at a single location. Similar differences in the classification of land for planning purposes can be observed: classifications based on natural habitat, car traffic, pedestrian traffic, residential constructions, all are examples of objects of different types which overlap and coexist (Figure 2.22).

This is a fundamental problem for any object ontology: the division of the world into objects is not unique and depends on the observer and his intentions. A special case is given if a classification is finer than another [93] (Figure 2.23). There, the objects form a lattice [138].

2.9.5 Objects Endure in Time

Objects, especially physical objects, endure in time. Physical objects have individual properties, which differentiate them from other objects which look similar, even if these differences are often not relevant and not noticed by human observers. Objects in reality maintain their identity from begin to end – even a grain

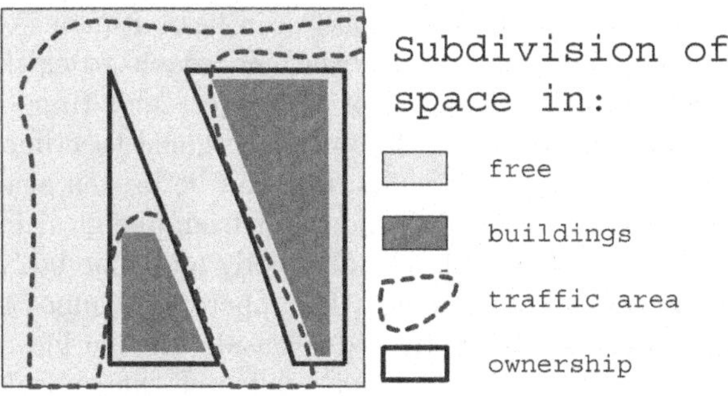

Fig. 2.22. Subdivision of space in building objects and ownership objects which overlap

Fig. 2.23. Three different classifications for urban land use (with 4, 7, and 24 classes)

of salt has an identity, which is lost when it is dissolved in the soup. Objects are "worms in four-dimensional space". Worboys and Pigot explored the mathematics of such constructions [158,205], referring back to classical geographic conceptions of space-time diagrams [114].

The concept of physically observable objects is a generalization of material objects, from which we know from experience that they endure in time: the piece of bread on my table now (Figure 2.2) will remain a piece of bread even five seconds later. Many of its properties remain the same; they are invariant with respect to short intervals of time.

The stable identity of objects is modeled in a database with a (stable) identifier, which replaces the combination of numerous properties which make each individual physical object different from all others. Identifiers are nominal values; they support only the operation *equality* (for performance reasons, it is useful to make internal use of lexigraphical order in the identifiers, but this must not be construed as meaningful).

Objects can be seen as functions from an identifier, an observation type and time to a value (Formula 2.16). Objects are formed such that many important properties remain invariant, primarily with respect to advancing time but also with respect to other operations.

$$observation :: id \rightarrow time \rightarrow obs \rightarrow value \tag{2.16}$$

Object Lifestyles. The changes of objects can be continuous or catastrophic: an object can move or it can be destroyed and ceases to exist. Catastrophic changes are affecting the "existence" of the object and these changes follow different rules: The solids on the desktop can be glued together such that two objects become one and later this connection can be broken again and the two original objects reemerge. If we pour the water from one glass into the wine in the other glass, the two liquid objects water body and wine body cease to exist and a new "water-wine-body" emerges. This operation cannot be undone; the two original liquids cannot be restituted. Considering the life of an object in time, we observe that different objects have different "life styles". Solids can be glued together and reemerge, but the liquids mixed cannot be separated again.

Definition of Lifestyles. In the real world objects are perceived as having their life – a span of time bounding the existence of objects as separable identifiable entities. In a spatio-temporal database objects are modeled in the same manner, i.e., a span of time bounding the *existence of object identifiers in the database* between the two fundamental events: creation and destruction. The way objects emerge and later change the modus of their existence differs for different categories of objects. For example, a liquid object while flowing into a container and overflowing it, gives rise to two new liquid objects: the first one remains in the container, and the second one spills on the ground [115]. This behavior is completely different from the blocks world: solid objects on a table can be piled one upon the other, but their identities are preserved.

Lifestyles are sets of special, identity changing operations applicable to object identifiers of different kinds of objects (Figure 2.24). These special operations form a finite set. Combined together they describe a large number of object categories. Beside the inevitable creation of an object identifier, possible destruction, the concept of temporary loss of identity for an object, has been introduced with operations suspend and resume with the same meaning as kill and reincarnate in [42]. An object may change its identifier keeping track of its predecessor through evolution, modeled as a composition of a creation and a deletion. Complex lifestyles are composed from simpler ones. Thus, of aggregation the parts are suspended, whereas the melting of objects is described as fusion (parts are destroyed). The fundamental difference is that the inverse of the former process (segregation) is reversible while the inverse of the latter (fission) is not: the contents of a glass of water and a glass of wine poured into a carafe cannot be restituted. Examples of lifestyles described in [139] range from physical reality like simple movable objects (stones, blocks), immovables (man-made buildings and bona fide objects like valleys and mountains), living beings, containers, liquids, to abstract concepts like ownership rights, marriages, and partnerships.

The concept of lifestyles allows the designer of a spatio-temporal database to use the same classes of operations for apparently different kinds of objects. In modeling abstract objects, one can benefit from already achieved models of the

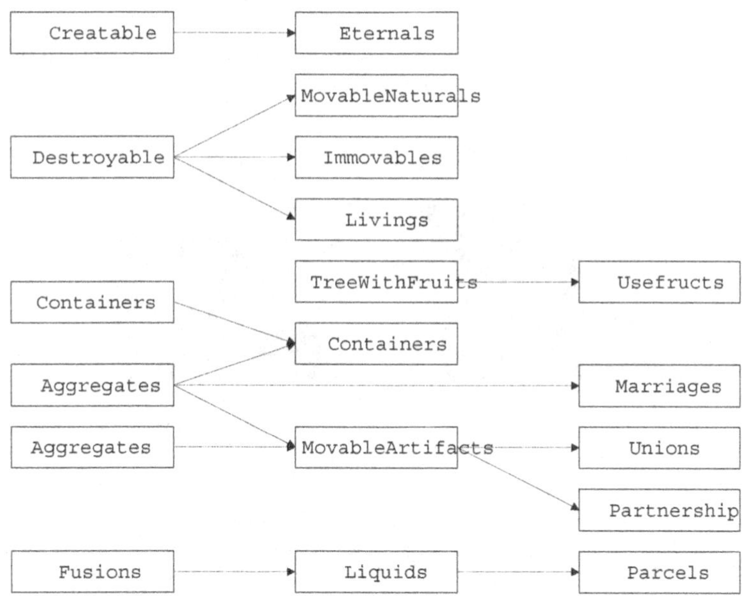

Fig. 2.24. The subsumption-graph of lifestyles

simpler physical realm. For example, legal cadastral parcels follow the "lifestyle" of liquids: two parcels which are merged are not re-emerging after a split [89].

Lifestyles describe the change in object identifiers: an object exists in the database or does not exist (or it is suspended). A possible extension is the change in object types: a liquid object can evaporate changing its appearance. Melting of solid objects into a liquid state is an important transition as well. The change of identity in objects produces many side effects: topological relations, for example, are of the greatest importance in spatial domain. Thus, the investigation relates lifestyles and the change in topology of emerging objects [120].

Moving and Changing Objects. Objects have permanence in time and can move their position or change shape. The observation of movement or change of shape for table-top objects which are under our permanent scrutiny is easy. Solid objects move but maintain their shape, other objects on the table may change form as the result of actions.

The recognition of moving and changing objects in geographic space is more difficult. What can be stated about the sand dunes of Figure 2.25, where we have observations which are half a year apart? One might conclude that the sand dune X in spring 1999 is the effect of merging the two dunes A and B from fall 1998, but this is not necessarily the correct interpretation. The question is generalized to the problem: Given two observations of snapshot spatial objects i_1 at t_1 and i_2 at t_2; construct a time varying object o, such that o at t_1 is i_1 and o at t_2 is i_2.

In this case, we are justified to label the two observed non-temporal objects with the same identifier and consider them as projections of the single object. The detailed rules depend on the particulars of the application.

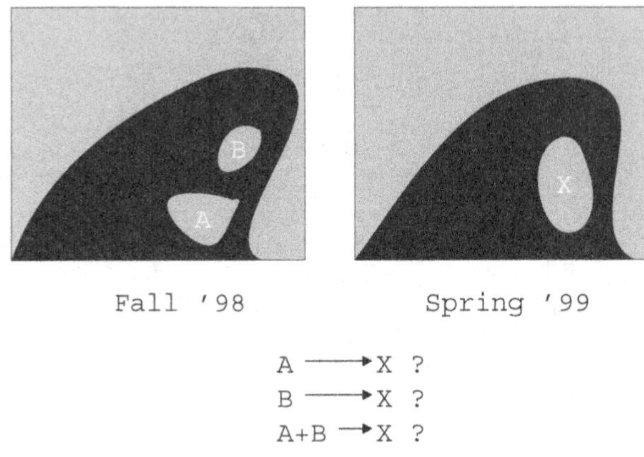

Fig. 2.25. Wandering sand dunes

Time-Varying Geometry. If we allow for time-varying values as the result of determination of object property, then the geometric operations can all be applied to moving objects. For moving objects, they return a time-varying value for the location of the centroid, for the length of the boundary, etc. The operation applicable to fixed values can be lifted to apply to time-varying values but care must be taken for functions like minimum or maximum (see Chapter 4 in this book). For example, the distance between two moving objects is a changing floating-point value (see Section 2.6.1).

2.9.6 Temporal, but A-Spatial Objects

The same principles of identification of area with uniform properties which were using a fixed time point can be inversed to identify temporal regions with uniform properties for a fixed location. To form objects as regions from a snapshot (fixed time) is more common, but sequences in time with uniform properties can be considered as "temporal objects" as well [174]. Examples would be my summer vacation period, an employees sickness leave or the presidency of Bill Clinton. Such temporal objects have a start and an end and one can ask for their duration, they are time intervals:

```
ObjStart :: id -> time
ObjEnd :: id -> time
```

Allen defined topological relations between time intervals [2].

2.10 Ontological Tier 3: Socially Constructed Reality

In the last section objects were constructed with respect to the human interaction with the world. This physical reality includes a large number of the things we interact regularly with, from fruits and other small objects to land, lakes and weather.

Unfortunately, most of what an administrative spatio-temporal database contains are not the physical properties of the world, but the legal and administrative classification of the world, classified and named within the context of social, especially institutional, rules. In the city, building lots, street names, and building zones are administrative facts; in the landscape, county boundaries, right of way and areas of nature parks are administratively constructed facts. These areas created by administrative rules are further simplifications of the complexity of reality to the restricted view of the law. These administrative constructs are valid only within a legal context.

2.10.1 Social Reality Is Real within a Context

A concentration on the ontology of physical things leaves out a very large part of the reality humans perceive. Naming things and using the names to communicate with others is one of the most important cultural achievements. Society consists of a complex web of relationships between people and things, which need to be defined and named. Elements of social reality thus named appear as real as physical reality to us. There is a strange belief – manifest in witchcraft – that an object has a direct and natural link to its name; pointing out that "Hund", nor "chien" or "cane" describes the same species as the English "dog", reveals the contextual nature of the naming conventions. Social constructs, from marriage to ownership, all appear as real to us as physical forces or electricity, but are meaningful only in the social context.

Social reality includes all the objects and relations which are created by social interactions. Human beings are social animals and social interaction is extremely important. The reason to separate physical reality, object reality and socially constructed reality is the potential for differences in observations: within errors of observations, the results of observations of the same point in time and space should be the same. The construction of objects can be based on the uniformity of various properties, and thus objects may be formed differently – for example, the definition of forest can be based on various criteria and thus leads to different extensions of a "forest" (indeed one should speak of different kinds of forest: legal forest, land-use forest, forest as physical presence of trees, etc.); differences for object formation can be traced back to different methods in classification if enough care is applied to the domain-specific interests and procedures.

For socially constructed reality, agreement between different agents from different contexts in the construction is not to be expected. Objects are named with different names in different languages and only naïve persons assume that there are exact translations between terms. Not even countries using the "same" language, apply the same terms with corresponding meaning; well known is the motto "England and the United States are separated by a common language" based on various examples of differences in vocabulary; the same applies to Germany, Switzerland and Austria. Each country, specifically each cultural system, creates its own "conceptualization" of the cultural organization. The results are quite different conceptual systems, and one must not expect that the same concept in different cultures will have the same meaning. There is a European attempt to extend the WordNet dictionary with five European languages to make it multi-lingual.

The names and the concepts are only meaningful within the defining social context. They are not binding outside of this context. What is quite easy to accept with regard to different languages is more difficult to understand with respect to smaller cultural communities: public agencies, administrations, etc.; each creates its own vocabulary and logical organization of the part of reality and cultural institution it is concerned with. It is surprising to see how different the terminology and the concepts of law in Austria, Switzerland and Germany are; neither do the terms correspond, nor do they have the (exact) same meaning. What Austrians call "Kataster", a map and a list of the parcels, is the "Liegenschaftskarte" in Germany. Even smaller communities create their own terminology: the laws for urban planning are in the competence of the Bundesland (federal state) in Austria; therefore there are nine laws, each creating its own set of terms which have meaning within this set of rules. Terms in one Bundesland do not correspond to the same or to other terms used in another Bundesland. Nobody assumes that the different branches in the administration of a town relate the same concept to the word "building"; the prototypical case, a single family dwelling, may be included everywhere, but the treatment of special cases – very small utility constructions, underground constructions, etc. – will vary. Using the concept of radial category [166], one can say that agencies create radial categories, which partially overlap. This makes the construction of databases, or the integration of databases from different origin, very difficult. The smallest common denominators must be found by human specialists; attempts of automatic database schema integration at best provide helpful tools [57].

2.10.2 Names

Objects have names – especially persons – and these names are perceived as "real" properties of the things. The proponents of remote sensing images of the environment always need to be reminded that remote sensing cannot see the names of towns [126], nor the boundaries of countries. Sometimes physical phenomena indicate where boundaries are, sometimes not; towns can be seen, but not their names, because all these elements are not part of the physical reality (which we analyzed in the previous two sections).

It is culturally assumed that names of things are stable. It is improper to change one's name (except for women when they marry in most of western Europe) or use multiple names (only criminals and artists do this). Everybody has to have a name (including "the artist formerly known as Prince"). Culture assumes for many important things (but not all) that there exists functions

```
getname :: obj -> name
findObj :: name -> env -> obj
```

Names are clearly a social construct in the sense of Searle [172]. Names can come in many forms: as strings of characters, as numbers (e.g., the names of the days of a month) or as arbitrary strings (social security numbers, license plates of cars, serial numbers, etc.). Names are always on a nominal scale – only comparison for equality is a relevant operation – and often a lexicographic ordering is exploited for searching (e.g., in telephone directories). Some names, especially surnames, are structured in such a way that they hint to relationships between people: Peter Smith maybe the father of Paul Smith (or his brother, or completely unrelated).

Many uses of names rely on a small context, in which the name is likely unique. The best example is the use of Christian names to identify people, there are thousands of "Rudi" living in Vienna, but within the context of my department, "Rudi" is unique (not so for "Martin"). Usually the context of a situation is sufficient to disambiguate a statement and identify the person. One should not be tempted to think that the usual combination of Christian name and family name is the person: there are three persons with the same name "Martin Staudinger" listed in the Vienna phone directory!

2.10.3 Institutional Reality

Much of what seems very real is, at a second glance, far from real. Neither status, honor or marriage, nor ownership are physically real. A large number of the constructions of social reality are related to institutions, especially the legal system. We concentrate here on legal concepts, as they are the most important for the construction of spatio-temporal databases, for example, about land ownership and the planning of the use of space.

Administration and law has a need to simplify the infinitely complex world to general rules which can be applied generally and uniformly. The complex judgment if a child is mature enough to act as an adult person is replaced by a summary rule which links the age of the person to its classification as a minor, not capable of making legally binding decisions, or an adult. Such rules are important for an efficient functioning of our modern world, where we deal with a large number of strangers and regulate our interactions based on few, typically quickly observable, properties: instructions given by a person in a police uniform are followed when we drive a car, but the same signs made by a non-uniformed person will go mostly unobserved.

Searle observed that some speech acts are not descriptive of reality like "the forest is green" which can be true or not depending on the color of the forest, but are constitutive – they create the described fact. The most famous example is certainly "I declare you husband and wife", which, if spoken by a duly authorized person and after the proper interrogations, creates the fact "marriage"

[171]. Often institutions associate specific treatments – fixed in rules and laws – with such constitutive acts. Incorporation of a company, marriage or submitting a letter of resignation constitute legal facts; these legal facts have well-defined consequences which are evident, when the constitutive act is made. The institutions typically keep registries of these constitutive facts, a registry of deeds is an example, or provide a document as evidence of the fact, for example, a driver's license or a marriage certificate. Confusing are "birth certificates", where the certificate does not constitute the fact that somebody was born – this is an ontological problem of Tier 3 – but constitutes the legal acceptance that birth was given at a specific location and time, which has consequences like conferring nationality – for example, a birth certificate from an U.S. registry is sufficient for entry into the USA.

Searle in his theory of institutional facts starts with the observation that paper money is nothing else than printed paper, but that this special kind of printed paper has a particular function within the context of a society. He sees that "special printed paper" serves as "money" in the context of a national economy. In the theory provided by Searle to explain institutional facts, the formula "x serves as y in the context of z" is very important, but not likely to cover all aspects of social reality [184]. This "x counts as y" assigns to the physical object x (from ontological tier 3) a specific function y. The meaning of the function y and the rule that x counts as y are both part of the context, for example, the legal institution. The function y, for example, "ownership", is then defined in the context of the legal system: ownership links a person to a piece of land, the owner of a piece of land can sell this land or can use it to secure a debt, etc. The meaning of ownership is fully defined within the legal system of a country. The German Grundgesetz says "ownership is guaranteed within the limits of the law...", clearly pointing out the social and legal context in which the term must be understood. On the other hand, some Reform Country has defined new institutions, avoiding the term "ownership" for land; in the opinion of experts, if a piece of land can be owned, sold, inherited and mortgaged, there is no substantial difference to "ownership" (in the meaning of the context of European or American law), independent of the word that this country uses [121].

Important for the application of spatio-temporal databases to land registration is the separation between the physical properties of things in the world, for example, boundary markers, buildings, streets and rivers, and the legal facts. Competent surveyors can measure the positions of boundary markers. There should not be cause for debate about the result. Similarly, the reconstruction of a boundary using the documented measurements in the registry is a (mostly) physical process and not dependent on a legal context.

Smith has separated fiat and bona fide boundaries [176]. Bona fide boundaries exist in reality: they may be natural boundaries, like those enjoyed by an island, watersheds, or clearly monumented artifact; the physical reality constitutes the boundary, the registry only points out that these physical elements are the boundary and may contain measurements or other observation values, which can be used to reconstruct the boundary. For fiat boundaries, the registry gives the exact location in terms of observation, and competent surveyors are required to indicate the location of the boundary in the real world. In this case, the registry constitutes the boundary and its location. Practically, this difference is important when the location of a boundary in the registry and the boundary in reality do not correspond – which one is the ruling one? In most countries, for bona fide boundaries, reality wins; for fiat boundaries, the registry wins.

Confusion in databases of institutional facts may arise from an incomplete separation what are recordings of constitutional facts, which cannot be wrong by definition, and which are facts based on observation of physical reality, which can, obviously, be incorrect descriptions of reality. One can demonstrate that a value does not describe a real property correctly – by inspection of the appropriate place; one cannot demonstrate that a constitutive registration is wrong. However, one can prove that the process that leads to its constitution was not following the prescribed rules and therefore the registration should be void.

2.11 Ontological Tier 4: Modeling Cognitive Agents

Agents acquire and construct knowledge about the world – the physical and the social world – in which they exist. The knowledge they construct is not necessarily and automatically corresponding to reality. Cognitive agents use the accumulated knowledge to derive new knowledge from the accumulated knowledge and make decisions using derived knowledge about actions.

The cognitive system of human beings is very similar to the aggregated cognitive behavior of organizations: They collectively acquire knowledge, which is subject to similar effects which result in only partial correspondence between reality and the knowledge accumulated. The treatment here, which deals mostly with the effects and does not concentrate on the processes and the influences on processes which lead to non-conformance of accumulated knowledge, need therefore not differentiate between single cognitive agents – mostly humans, but to some degree also animals – and organizations seen as cognitive agents.

2.11.1 Logical Deduction

Cognitive agents are capable of logical deduction. From the knowledge accumulated other facts are deduced and used to guide the actions of the agent. Logical deduction can be very simple; for example, a database lookup to check if a person is a client of a bank or to find out how many years a student is already enrolled in the university, which is a simple calculation starting with the year of his first enrollment. More complex deductions are checks if a student can graduate, which must consider a number of requirements.

The rules used for deduction are built into application programs and database query languages. The latter usually follow the axioms pointed out by Reiter [162]:

- the domain closure assumption
- the unique name assumption
- the closed world assumption

These axioms are closed to assumptions built into legal rules and regulations and administrative customs – they are not universally applicable. For example, for spatial databases, the closed world assumption is usually not valid: from the absence of knowledge that a tree or a building exist on a parcel one must not conclude that there are no trees or buildings on the parcel; it is possible that a tree has grown since the last observation or that a building has been erected without informing the authorities (Figure 2.26). The details about an owner of a parcel do not demonstrate that this person is still alive. Spatial information systems require more complex reasoning than ordinary administrative processing [60].

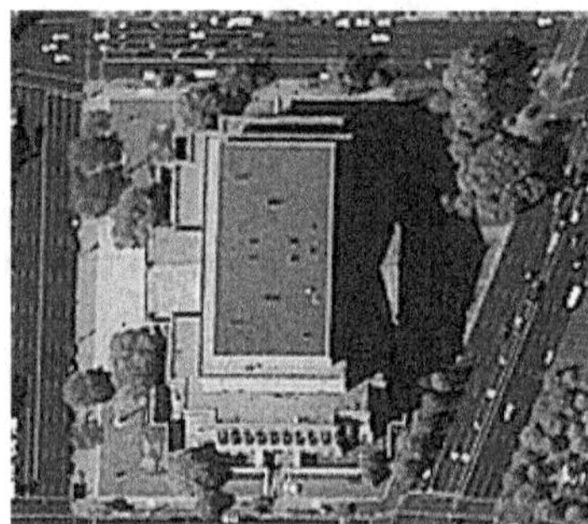

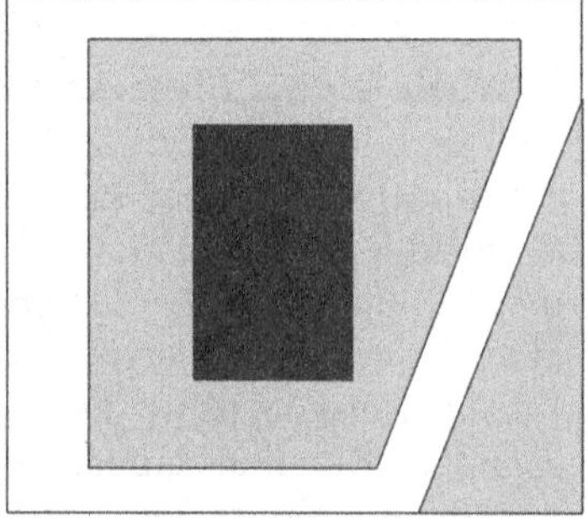

The map does not show all: missing trees, footpaths

Fig. 2.26. The map does not show all!

2.11.2 Two Time Perspectives

A cognitive agent must separate two time perspectives: There is the time in which the world evolves: trees sprout, buildings are constructed, and people die at certain points in time. Second, there is the time at which these facts are entered into the database: Trees are observed and entered into the database, buildings are surveyed and shown on maps, death records are filed; these acts of "knowledge acquisition" occur at a certain point in time – measured along the same time line, but different instants: for each event, two instants are relevant: when it occurred and when knowledge was acquired.

From an agent's perspective, the database (or his own collection of knowledge) is a time-varying value – changing in discrete steps at each transaction.

Therefore, the semantics of a temporal database can be understood as a function from time to a snapshot database, "as of time t" queries can be expressed as a-temporal queries to the snapshot database valid at t.

$$database \rightarrow time \rightarrow snapshot\, database \tag{2.17}$$

There are a number of difficulties arising from the combination of a time-varying collection of facts with a deduction system:

The result of a deduction from a snapshot database is – sound deduction rules assumed – a single result. The result becomes time varying for a time-varying database: depending what data are collected, a deduction yields a different result: the request to withdraw $500 from my account is denied today (balance only $150), but after I received a $1000 reimbursement from a company, the same request to withdraw $500 is granted.

If the results of deductions are stored, then a fact acquired later can make the result invalid and the stored result of the deduction must be identified and corrected (monotonicity of the logic; usually not given). This is usually described as "belief maintenance problem", when agents deduce beliefs from their knowledge and knowledge added later requires a revision of these beliefs. From simple observation in an environment one may deduce that green fruits are unripe and not edible, and red fruits are ripe and sweet; this empirical rule must be revised after tasting of ripe, green figs or grapes, and a more sophisticated rule for ripeness must replace the simple one [104].

Social fairness often leads to rules where not the date of a fact, but when an agent learned about it, is important. The social system does not punish honest "not-knowing" if the agent has made all reasonable efforts to discover the facts. For example, the knowledge of the law is assumed, but only after it has been officially published. A case can be brought before a court within a certain deadline and the deadline is counted not from the offending act, but from the time the plaintiff has learned (or could have learned if diligent) of the act.

In other instances, legal rules depend on having knowledge or not; in European cadastral law, the "bona fide" buyer is protected, even if he buys from a non-owner. In all such cases, it is crucial how much an agent knew, when a fact became true and when an agent learned about it. The registry of deeds or the cadastre is one of the legal domains where most detailed rules were developed over the centuries [80]. Entering a legal fact in a registry often creates the assumption that all parties concerned do know about the fact, because they can access the information if interested.

2.11.3 Sources of Knowledge

Agents acquire much knowledge through communication with other agents and not from direct observations. This knowledge is "hearsay" in legal terminology [21] and considered of much lower reliability than what is directly and immediately observed by an agent. Human beings are extremely well equipped to keep information from different sources separated and maintain a mental link from the information to the source. Reuter has pointed out that databases are not prepared to keep track of collections of facts which form areas of consistency, but are not overall consistent [206].

2.12 Ontological Commitments Necessary for a Spatio-temporal Database

Using a spatio-temporal database implies the acceptance of the ontological commitments built into it. The previous sections discussed these commitments in detail. The conclusions are summarized here:

2.12.1 Existence of a Single Reality

We assume – as generally is the case in the positivist philosophy of science and engineering – that there is a single reality in which we live, and which we gain knowledge of. This assumption results form empirical evidence that I and everybody else live and interact with the same objects, following the same physical laws. Effects of the actions of another person can be seen and others can see the effects of my actions.

2.12.2 Values for Properties Can Be Observed

Our knowledge of the world is through the observation of values of observable properties. Models may link observable values to assumed properties, which are not directly observable. Observations are linked to the point in time and space when they were made.

2.12.3 Assume Space and Time

The notion of space and time were the object of extensive philosophical debate. Human life is in space and time and our bodily functions lead to space and time as two fundamental categories of our experience. Time is unidirectional and observed processes are, in general, not reversible in time. Space is characterized – here we can follow the famous definition of geometry by Hilbert – by invariant properties for groups of transformations [22].

2.12.4 Observations Are Necessarily Limited

Observations are necessarily with limited precision in the observed value as well as in the point in time and space they are related to. This is a physical fundamental limit (Heisenberg's uncertainty principle), but in all practical cases, the uncertainties are much larger than the physical limits and due to imprecise practical observation processes. Most of the data collected are just precise enough for the purpose, and therefore, in absolute terms, quite imprecise.

2.12.5 Processes Determine Objects

Humans do not usually consider the world as consisting of data values for individual point observations. For economy of cognitive processes, sets of similar observations are grouped to form objects. These objects have properties which remain invariant under common operations. The invariance of form of objects through translation and rotation (which form a group) is fundamental for our understanding of solid bodies, i.e., for most of the objects of our daily life. Objects seem to be "real", but one must always remember that they depend on the classification used for their formation and therefore alternative ways of "carving" the world in objects are possible. Some classifications are extremely closely related to fundamental operations of the human body and are therefore likely "universals" (i.e., the same for all human cultures); others are not. Objects endure in time and have an identity which links observations of an earlier time with observations at a later time. Even when we are not watching, our car parked in the road keeps its color (it actually fades slowly) and the height of the Himalayan mountains remains the same (it actually raises slowly). Object identity is usually represented with identifiers in the data collection.

2.12.6 Names of Objects

To keep track of the identity of objects, various methods for naming are used by the social system. People have names and cars have manufacturer numbers and license plates to make them unique and to make it easy to find the same object again without constant supervision. Unnamed objects, like the fork you are eating with, remain only "your fork" as long as you constantly keep track of its location to differentiate it from other forks.

2.12.7 Social, Especially Institutionally Constructed Reality

The social system constructs relations between objects, which are very important for human living and seem very real, but are not part of physical reality: marriage (as opposed to parenthood) or ownership (as opposed to physical possession) are social concepts and only related to physical reality. Much data used for modern administration concern such socially – especially institutionally – constructed reality [172].

2.12.8 Knowledge of an Agent Is Changing in Time

The knowledge an agent, a single human being, or an organization maintains, cannot be a true and perfect model of reality. Agents use their knowledge to make decisions about actions. The representation of reality is not only of limited precision (see above), but also varies in time. Facts in the world are later represented in the knowledge base. With each fact the time of occurrence and time of entry into the knowledge collection is associated. This is often called "valid time" and "database time" [187]. A specific decision was correct yesterday, but with the additional knowledge gained since, is wrong today. Social fairness dictates that decisions of agents are judged with respect to what they could have known, not the perfect knowledge available later.

2.13 Conclusions

For spatio-temporal databases a careful review of the ontological bases on which programs rest is necessary. The assumptions built-in must be documented carefully.

Integration of data from different sources is one of the dominant problems of today's practical use of spatial and spatio-temporal databases. Investigation into the ontological bases of such data collections is a research direction from which essential contributions to solve the real and immediate problems are expected. This is demonstrated by the commercial interest in ontologies. The division of ontology into tiers in this chapter identifies different levels of commitment and agreement between different data collections. Observation of physical reality is more likely to be similar for similar data sources; the constructions of social reality are necessarily context dependent, and will differ for collections originating in different sources. The tiers selected reflect these differences in expected agreement, respectively disagreement, between data sources. The description of the tiers point to methods to overcome the differences and to integrate different data sources within the limits of their ontological common base. The multi-tier ontology extends the difference between ontology and epistemology philosophers have argued for centuries.

Progress in the CHOROCHRONOS project in this respect evolved in various ways. One of them (see Chapter 4) is centered around the move from a first-order logic based investigation to a second-order architecture [112]. In this framework, time-varying "moving" values and points can be manipulated, and are properly typed. It appears that many of the ontological problems discussed currently [109] are related to the concept of type and the type systems of programming languages are not exactly the type systems necessary for databases. A type system for a programming language is generally defined as a set of type classes. Object instances have a specific type, which means that the operations defined in the type class are applicable to the object instance [34]. Type systems for programming language are developed with the goal to allow certain checks on the program code such that a large class of errors can be detected statically (i.e., during compilation) and static errors cannot occur during the execution of a program. Types in the sense of programming languages are static. Object instances cannot change their type during execution (unless they go through a specific "type conversion" operation) [33]. This concept of type is suitable for the short duration of a program run, but is not suitable for a temporal database, where long-term representation is the objective. It clashes with an ontological type concept: with graduation a student becomes an alumni and sometimes an employee – but his identity as a human being continues. A number of extensions of a type system with "mix-ins" etc., have been proposed to deal with such cases [7]. But none covers all the cases. A review of the five-tier ontology with a powerful type system, for example, based on Hindley-Millner type inference and classes [125], is likely to yield valuable insights into the connection between type system and ontology.

Linguists use the concept of "ontological categories" to describe types (in the sense of applicable operations) which cannot be changed: a solid body cannot become an event; a parcel (a piece of land) cannot become a material or a point. A boundary point, however, can cease to be a boundary point, but continues as a point: it can never become an event.

The ontology presented here is designed with the construction of a computational model in mind, following the suggestions in Section 2.4. The contribution possible from a strongly typed language and polymorphism based on a class structure will be explored then.

References

1. K. Al-Taha. *Temporal Reasoning in Cadastral Systems.* Ph.D., University of Maine, 1992.
2. J. Allen. Maintaining knowledge about temporal intervals. *Communications of the ACM*, 26(11):832–843, 1983.
3. ANSI/X3/SPARC. Study group on database management systems, interim report 75-02-08. *SIGMOD*, 7(2), 1975.
4. I. Asimov. *Earth is Room Enough.* Doubleday, New York, 1957.
5. A. Asperti and G. Longo. *Categories, Types and Structures - An Introduction to Category Theory for the Working Computer Scientist.* Foundations of Computing. The MIT Press, Cambridge, Mass., 1 edition, 1991.
6. D.G. Baird, R.H. Gertner, and R.C. Pickr. *Game Theory and the Law.* Harvard University Press, Cambridge, Mass., 1994.
7. F. Bancilhon, C. Delobel, and P. Kanellakis. *Building an Object-Oriented Database System - The Story of O2.* Morgan Kaufmann, San Mateo, CA, 1992.
8. A. Barr and E.A. Feigenbaum. *The Handbook of Artificial Intelligence.* W. Kaufman; HeirisTech Press, Los Altos, CA; Stanford, CA, 1981.
9. M. Barr and C. Wells. *Category Theory for Computing Science.* Prentice Hall, 1990.
10. Tim Berners-Lee and Mark Fischetti. *Weaving the Web: The original design and ultimate destiny of the World Wide Web, by its inventor.* Harper, 1999.
11. Tim Berners-Lee, James Hendler, and Ora Lassila. The Semantic Web. Scientific American, May 2001.
12. P.L. Berger and T. Luckmann. *The Social Construction of Reality.* Doubleday, New York, Anchor Books edition, 1996.
13. D. Bernasconi. *Rahmenkonzept zur Gestaltung eines Datenmanagementsystems Siedlungsentwässerung.* Doctoral thesis, ETH Zürich, 1999.
14. G. Birkhoff. Universal algebra. In *First Canadian Math. Congress*, pages 310–326. Toronto University Press, 1945.
15. G. Birkhoff and J.D. Lipson. Heterogeneous algebras. *Journal of Combinatorial Theory*, 8:115–133, 1970.
16. S. Bittner. *Die Modellierung eines Grundbuchsystems im Situationskalkül.* Diploma thesis, Universität Leipzig, 1998.

17. T. Bittner. *Rough Location.* PhD, Technical University, 1999.
18. T. Bittner. Approximate temporal reasoning. In *AAAI Workshop on Spatial and Temporal Granularity, Time2000*, 2000.
19. T. Bittner and J. Stell. A boundary-sensitive approach to qualitative location. *Annals of Mathematics and Artificial Intelligence*, 24:93–114, 1998.
20. T. Bittner and J. Stell. Rough sets in approximate spatial reasoning. In *Proceedings of RSCTC 2000*, Lecture Notes in Computer Science. Springer-Verlag, Berlin, 2000.
21. H.C. Black. *Black's Law Dictionary.* West Publishing, pocket edition, 1996.
22. L. Blumenthal and K. Menger. *Studies in Geometry.* W.H. Freeman, 1970.
23. G. Booch, J. Rumbaugh, and I. Jacobson. *Unified Modeling Language Semantics and Notation Guide 1.0.* Rational Software Corporation, San Jose, CA, 1997.
24. S. Borgo, N. Guarino, and C. Masolo. A pointless theory of space based on strong connection and congruence. In L. Carlucci and J. Doyle, editors, *Principles of Knowledge Representation and Reasoning (KR'96).* Morgan Kaufmann, 1996.
25. S. Borgo, N. Guarino, and C. Masolo. Stratified ontologies: The case of physical objects. In *ECAI'96, Workshop on Ontological Engineering*, Budapest (August 1996), 1996.
26. K. Buehler and L. McKee, editors. *OpenGIS Guide: An Introduction to Interoperable Geoprocessing, Part 1 of the Open Geodata Interoperability Specification (OGIS).* The Open GIS Consortium, Inc. (OGC), 35 Main Street, Suite 5, Wayland, MA 01778. Available from `http://ogis.org/`, 1996.
27. P.A. Burrough and A.U. Frank. Concepts and paradigms in spatial information: Are current geographic information systems truly generic? *International Journal of Geographical Information Systems*, 9(2):101–116, 1995.
28. P.A. Burrough and A.U. Frank, editors. *Geographic Objects with Indeterminate Boundaries*, volume 2 of *GISDATA Series.* Taylor & Francis, London, 1996.
29. T. Buyong and W. Kuhn. Local adjustment for measurement-based cadastral systems. *Journal of Surveying Engineering and Land Information Systems*, 52(1):25–33, 1992.
30. T. Buyong, W. Kuhn, and A.U. Frank. A conceptual model of measurement-based multipurpose cadastral systems. *Journal of the Urban and Regional Information Systems Association (URISA)*, 3(2):35–49, 1991.
31. I. Campari. Uncertain boundaries in urban space. In P.A. Burrough and A.U. Frank, editors, *Geographic Objects with Indeterminate Boundaries*, volume 2 of *GISDATA*, pages 57–69. Taylor & Francis, London, 1996.
32. I. Campari and A.U. Frank. Cultural aspects and cultural differences in geographic information systems. In T.L. Nyerges, D.M. Mark, R. Laurini, and M. Egenhofer, editors, *Cognitive Aspects of Human-Computer Interaction for Geographic Information Systems - Proceedings of the NATO Advanced Research Workshop, Palma de Mallorca*, volume 83 of *NATO ASI Series D*, pages 249–266. Kluwer Academic Publishers, Dordrecht, 1995.
33. L. Cardelli. Type systems. In A.B. Tucker, editor, *The Computer Science and Engineering Handook*, pages 2208–2236. CRC Press, 1997.
34. L. Cardelli and P. Wegner. On understanding types, data abstraction, and polymorphism. *ACM Computing Surveys*, 17(4):471–522, 1985.
35. R. Casati, B. Smith, and A.C. Varzi. Ontological tools for geographic representation. In N. Guarino, editor, *Formal Ontology in Information Systems*, pages 77–85. IOS Press, Amsterdam, 1998.
36. R. Casati and A.C. Varzi. *Holes and Other Superficialities.* MIT Press, Cambridge, Mass., 1994.

37. R. Casati and A.C. Varzi. *Parts and Places.* The MIT Press, Cambridge, Mass., 1999.
38. CEN. Geographic information - Data description: Metadata. Draft standard, CEN/TC 287/WG2, 1995.
39. S. Ceri, P. Fraternali, and S. Paraboschi. XML: Current development and future challenges for the database community. In C. Zaniolo, P.C. Lockemann, M.H. Scholl, and T. Grust, editors, *Advances in Database Technology - EDBT 2000 (7th Int. Conference on Extending Database Technology, Kontanz, Germany)*, volume 1777 of *Lecture Notes in Computer Science*, pages 3–17. Springer-Verlag, Berlin Heidelberg, 2000.
40. J. Chevallier. Land information systems - A global and system theoretic approach. In *FIG International Federation of Surveyors*, volume 3, page paper 301.2, Montreux, Switzerland, 1981.
41. N. Chrisman. *Exploring Geographic Information Systems.* John Wiley, New York, 1997.
42. J. Clifford and A. Croker. Objects in time. *Database Engineering*, 7(4):189–196, 1988.
43. CODASYL. Report of the data base task group. Technical report, April 1971 1971.
44. A. Cohn, Z. Cui, and D. Randell. Exploiting temporal continuity in qualitative spatial calculi. specialist meeting, NCGIA, May, 8–11 1993.
45. A. Cohn and N. Gotts. The 'egg-yolk' representation of regions with indeterminate boundaries. In P. Burrough and A.U. Frank, editors, *Geographic Objects with Indeterminate Boundaries*, volume GISDATA II. Taylor & Francis, London, 1996.
46. J. Corbett. Topological principles in cartography. Technical Paper 48, Bureau of the Census, US Department of Commerce, 1979.
47. R.S. Corp. UML notation guide, 9/1/97 1997.
48. H. Couclelis. People manipulate objects (but cultivate fields): Beyond the raster-vector debate in GIS. In A.U. Frank, I. Campari, and U. Formentini, editors, *Theories and Methods of Spatio-Temporal Reasoning in Geographic Space*, volume 639 of *Lecture Notes in Computer Science*, pages 65–77. Springer-Verlag, Berlin, 1992.
49. H. Couclelis and N. Gale. Space and spaces. *Geografiske Annaler*, 68B:1–12, 1986.
50. R. Cover. The XML cover pages, url: `http://www.oasis-open.org/cover/xmlandsemantics.html`, 1998.
51. N.A. Cressie. *Statistics for Spatial Data.* Wiley Series in Probability and Mathematical Statistics. John Wiley, New York, 1991.
52. H.B. Curry and R. Feys. *Combinatory Logic.* North Holland, Amsterdam, 1956.
53. CYC. The CYC corporation web page, url: `http://www.cyc.com/tech.html`, 2000.
54. P.F. Dale and J.D. McLaughlin. *Land Information Management - An introduction with special reference to cadastral problems in third World countries.* Oxford University Press, Oxford, 1988.
55. E. Davis. *Representation of Commonsense Knowledge.* Morgan Kaufmann Publishers, Inc., San Mateo, CA, 1990.
56. M.D. Davis. *Game Theory.* Dover Publications, Minneola, NY, revised edition, 1983.
57. T. Devogele, C. Parent, and S. Spaccapietra. On spatial database integration. *IJGIS*, 12(4 - Special Issue):335–352, 1998.
58. C. Eastman. *Database Facilities for Engineering Design*, volume 69. 1981.

59. M. Egenhofer. What's special about spatial - Database requirements for vehicle navigation in geographic space. *SIGMOD Record*, 22(2):398–402, 1993.
60. M. Egenhofer and K. Al-Taha. Reasoning about gradual changes of topological relationships. In A.U. Frank, I. Campari, and U. Formentini, editors, *Theories and Methods of Spatio-Temporal Reasoning in Geographic Space*, volume 639 of *Lecture Notes in Computer Science*, pages 196–219. Springer-Verlag, Heidelberg-Berlin, 1992.
61. M. Egenhofer and A.U. Frank. Connection between local and regional: Additional "intelligence" needed. In *FIG XVIII International Congress of Surveyors*, Toronto, Canada (June 1-11, 1986), 1986.
62. M.J. Egenhofer, A.U. Frank, and J. Jackson. A topological data model for spatial databases. In A. Buchmann, editor, *SSD'89 - Design and implementation of large spatial databases*, Lecture Notes in Computer Science 409, pages 271–186. Springer, 1989.
63. M.J. Egenhofer. A formal definition of binary topological relationships. In W. Litwin and H.-J. Schek, editors, *Third International Conference on Foundations of Data Organization and Algorithms (FODO)*, volume 367, pages 457–472, Paris, France, June 1989, 1989. Springer-Verlag, Berlin Heidelberg, Germany (FRG).
64. M.J. Egenhofer. *Spatial Query Languages.* Ph.D., University of Maine, 1989.
65. M.J. Egenhofer and R.D. Franzosa. Point-set topological spatial relations. *International Journal of Geographical Information Systems*, 5(2):161–174, 1991.
66. M.J. Egenhofer and D.M. Mark. Naive geography. In A.U. Frank and W. Kuhn, editors, *Spatial Information Theory - A Theoretical Basis for GIS*, volume 988 of *Lecture Notes in Computer Science*, pages 1–15. Springer-Verlag, Berlin, 1995.
67. H.-D. Ehrich. Key extensions of abstract data types, final algebras, and database semantics. In D. E. Pitt, editor, *Category Theory and Computer Programming; Tutorial and Workshop Proceedings*, volume 240 of *Lecture Notes in Computer Science.* Springer, 1985.
68. H.-D. Ehrich, M. Gogolla, and U. Lipeck. *Algebraische Spezifikation abstrakter Datentypen.* Leitfäden und Monographien der Informatik. B.G. Teubner, Stuttgart, 1989.
69. C. Eschenbach. *Zählangaben - Maßangaben.* Studien zur Kognitionswissenschaft. Deutscher Universitätsverlag, Wiesbaden, 1995.
70. C. Eschenbach. On representation and descriptions of spatial socio-economic units. In A.U. Frank, J. Raper, and J.-P. Cheylan, editors, *Life and Motion of Socio-Economic Units.* Taylor & Francis, ESF Series, London, 2000.
71. C. Eschenbach and L. Kulik. An axiomatic approach to the spatial relations underlying left-right and in front of-behind. In *KI'97*, 1997.
72. ESPRIT. The esprit project web page, url: `http://www.cordis.lu/esprit/src/intro.htm`, 2000.
73. G. Fauconnier. *Mappings in Thought and Language.* Cambridge University Press, Cambridge, UK, 1997.
74. C. Fellbaum, editor. *WordNet: An Electronic Lexical Database.* Language, Speech, and Communication. The MIT Press, Cambridge, Mass., 1998.
75. J. Ferber, editor. *Multi-Agent Systems - An Introduction to Distributed Artificial Intelligence.* Addison-Wesley, 1998.
76. F.T. Fonseca and M.J. Egenhofer. Ontology-driven geographic information systems. *Geographic Information Systems*, 8:385–399, 1994.
77. G. Franck. Time, actuality, novelty and history. In A.U. Frank, editor, *Life and Motion of Socio-Economic Units.* Taylor & Francis, London, to appear.

78. A.U. Frank. *Datenstrukturen für Landinformationssysteme - Semantische, Topologische und Räumliche Beziehungen in Daten der Geo-Wissenschaften.* Dissertation, ETH Zürich, Institut für Geodäsie und Photogrammetrie, 1983.
79. A.U. Frank. Qualitative temporal reasoning in gis-ordered time scales. In T.C. Waugh and R.G. Healey, editors, *Sixth International Symposium on Spatial Data Handling, SDH'94*, volume 1, pages 410–430, Edinburgh, Scotland, Sept. 5-9, 1994, 1994. IGU Commission on GIS.
80. A.U. Frank. An object-oriented, formal approach to the design of cadastral systems. In M.J. Kraak and M. Molenaar, editors, *7th Int. Symposium on Spatial Data Handling, SDH'96*, volume 1, pages 5A.19–5A.35, Delft, The Netherlands, 1996. IGU.
81. A.U. Frank. Spatial ontology: A geographical information point of view. In O. Stock, editor, *Spatial and Temporal Reasoning*, pages 135–153. Kluwer Academic Publishers, Dordrecht, 1997.
82. A.U. Frank. GIS for politics. In C.-R. Proceedings, editor, *GIS Planet'98*, Lisbon, Portugal (9 - 11 Sept. 1998), 1998. IMERSIV.
83. A.U. Frank. Metamodels for data quality description. In R. Jeansoulin and M. Goodchild, editors, *Data quality in Geographic Information - From Error to Uncertainty*, pages 15–29. Editions Hermes, Paris, 1998.
84. A.U. Frank. One step up the abstraction ladder: Combining algebras - from functional pieces to a whole. In C. Freksa and D.M. Mark, editors, *Spatial Information Theory - Cognitive and Computational Foundations of Geographic Information Science (Int. Conference COSIT'99, Stade, Germany)*, volume 1661 of *Lecture Notes in Computer Science*, pages 95–107. Springer-Verlag, Berlin, 1999.
85. A.U. Frank. Communication with maps: A formalized model. In C. Freksa, W. Brauer, C. Habel, and K.F. Wender, editors, *Spatial Cognition II (International Workshop on Maps and Diagrammatical Representations of the Environment, Hamburg, August 1999)*, volume 1849 of *Lecture Notes in Artificial Intelligence*, pages 80–99. Springer-Verlag, Berlin Heidelberg, 2000.
86. A.U. Frank. Socio-economic units: Their life and motion. In A.U. Frank, J. Raper and J.P. Cheylan, editors, *Life and Motion of Socio-Economic Units.* Taylor and Francis, ESF Series, London, 2000.
87. A.U. Frank and W. Kuhn. Cell graphs: A provable correct method for the storage of geometry. In D. Marble, editor, *Second International Symposium on Spatial Data Handling*, pages 411–436, Seattle, Wash., 1986.
88. A.U. Frank and W. Kuhn. A specification language for interoperable GIS. In M.F. Goodchild, M. Egenhofer, R. Fegeas, and C. Kottman, editors, *Interoperating Geographic Information Systems.* Kluwer, Norwell, MA, 1998.
89. A.U. Frank and D. Medak. Formal models of a spatiotemporal database. In K. Richta, editor, *DATASEM'99 - 19th Annual Conference on the Current Trends in Databases and Information Systems*, pages 117–130, Brno, Czech Republic (24-26 October, 1999), 1999. Dept. of Computer Science, Czech Technical University.
90. A.U. Frank, B. Palmer, and V. Robinson. Formal methods for accurate definition of some fundamental terms in physical geography. In D. Marble, editor, *Second International Symposium on Spatial Data Handling*, pages 583–599, Seattle, Wash., 1986.
91. A.U. Frank, J. Raper, and J.-P. Cheylan, editors. *Life and Motion of Socio-Economic Units.* ESF Series. Taylor & Francis, London, 2000.
92. A.U. Frank and M. Raubal. Formal specifications of image schemata - a step to interoperability in geographic information systems. *Spatial Cognition and Computation*, 1(1):67–101, 1999.

93. A.U. Frank, G.S. Volta, and M. McGranaghan. Formalization of families of categorical coverages. *IJGIS*, 11(3):215–231, 1997.
94. W. Franklin. Cartographic errors symptomatic of underlying algebra problems. In *First International Symposium on Spatial Data Handling*, pages 190–208, Zürich, Switzerland, 1984.
95. J.T. Fraser, editor. *The Voices of Time.* The University of Massachusetts Press, Amherst, second edition, 1981.
96. C. Freksa. Qualitative spatial reasoning. In D. M. Mark and A.U. Frank, editors, *Cognitive and Linguistic Aspects of Geographic Space*, NATO ASI Series D: Behavioural and Social Sciences, pages 361–372. Kluwer Academic Press, Dordrecht, The Netherlands, 1991.
97. A. Galton. Towards a qualitative theory of movement. In A.U. Frank and W. Kuhn, editors, *Spatial Information Theory (Proceedings of the European Conference on Spatial Information Theory COSIT'95)*, volume 988 of *Lecture Notes in Computer Science*, pages 377–396. Springer Verlag, Berlin, 1995.
98. A. Galton. Continuous change in spatial regions. In S. C. Hirtle and A.U. Frank, editors, *Spatial Information Theory - A Theoretical Basis for GIS (International Conference COSIT'97)*, volume 1329 of *Lecture Notes in Computer Science Vol.1329*, pages 1–14. Springer-Verlag, Berlin-Heidelberg, 1997.
99. R. Giblin. *Graphs, Surfaces and Homology.* Mathematics Series. Chapman and Hall, London, 1977.
100. J. Gibson. *The ecological approach to visual perception.* Erlbaum, Hillsdale, NJ, 1979.
101. J. Goguen. An introduction to algebraic semiotics, with applications to user interface design. In C. Nehaniv, editor, *Computation for Metaphor, Analogy and Agents*, volume 1562 of *Lecture Notes in Artificial Intelligence*, pages 242–291. Springer-Verlag, Berlin Heidelberg, 1999.
102. M. Goodchild. Geographical data modeling. *Computers and Geosciences*, 18(4):401– 408, 1992.
103. M.F. Goodchild, M. Egenhofer, R. Fegeas, and C. Kottmann, editors. *Interoperating Geographic Information Systems (Proceedings of Interop'97, Santa Barbara, CA).* Kluwer, Norwell, MA, 1998.
104. A. Gopnik and A.N. Meltzoff. *Words, Thoughts, and Theories.* Learning, Development, and Conceptual Change. The MIT Press, Cambridge, Mass., 1997.
105. D. Greene and F. Yao. Finite-resolution computational geometry. In *27th IEEE Symp. on Foundations of Computer Science*, pages 143–152. IEEE, 1986.
106. N. Guarino. Semantic matching: formal ontological distinctions for information organization, extraction, and integration. In M. T. Pazienza, editor, *Information Extraction: An Interdisciplinary Approach to an Emerging Information Technology*, pages 139–170. Springer-Verlag, Berlin Heidelberg.
107. N. Guarino. Formal ontology, conceptual analysis and knowledge representation. *International Journal of Human and Computer Studies. Special Issue on Formal Ontology, Conceptual Analysis and Knowledge Representation, edited by N. Guarino and R. Poli*, 43(5/6), 1995.
108. N. Guarino. Formal ontology and information systems. In N. Guarino, editor, *Formal Ontology in Information Systems (Proceedings of FOIS'98, Trento, Italy, 6-8 June, 1998)*, pages 3–15. IOS Press, Amsterdam, 1998.
109. N. Guarino and C. Welty. A formal ontology of properties. In R. Dieng and O. Corby, editors, *Proceedings of EKW-2000, 12th Int. Conference on Knowledge Engineering and Knowledge Management (2-6 October, 2000)*, Lecture Notes in Artificial Intelligence). Springer-Verlag, Berlin Heidelberg, 2000.

110. N. Guarino and C. Welty. Identity, unity and individuation: towards a formal toolkit for ontological analysis. In W. Horn, editor, *ECAI-2000, The European Conference on Artificial Intelligence (August 2000).* IOS, 2000.
111. O. Guenther, K.-P. Schulz, and J. Seggelke, editors. *Umweltanwendungen geographischer Informationssystem.* Wichmann, Karlsruhe (Germany), 1992.
112. R. Gueting. Second-order signature: a tool for specifying data models, query processing, and optimization. In *ACM Sigmod Conference*, pages 277–286, Washington, DC, 1993.
113. I. Hacking. *The Social Construction of What?* Harvard University Press, Cambridge, Mass., 1999.
114. T. Hägerstrand. The propagation of innovation waves. *Lund Studies in Geography, Series B*, (4), 1952.
115. P.J. Hayes. Naive physics I: Ontology for liquids. In J. R. Hobbs and R.C. Moore, editors, *Formal Theories of the Commonsense World*, Ablex Series in Artificial Intelligence, pages 71–107. Ablex Publishing, Norwood, NJ, 1985.
116. P.J. Hayes. The second naive physics manifesto. In J. R. Hobbs and R.C. Moore, editors, *Formal Theories of the Commonsense World*, pages 1–36. Ablex Publishing Corp., Norwood, N.J., 1985.
117. J. Herring, M.J. Egenhofer, and A.U. Frank. Using category theory to model GIS applications. In K. Brassel, editor, *4th International Symposium on Spatial Data Handling*, volume 2, pages 820–829, Zürich, Switzerland, 1990. IGU, Commission on Geographic Information Systems.
118. W. Hodges. *A Shorter Model Theory.* Cambridge University Press, Cambridge, UK, 1997.
119. W. Hölbling, W. Kuhn, and A.U. Frank. Finite-resolution simplicial complexes. *GeoInformatica*, 2(3):281–298, 1998.
120. K. Hornsby and M.J. Egenhofer. Qualitative representation of change. In S. C. Hirtle and A.U. Frank, editors, *Spatial Information Theory - A Theoretical Basis for GIS (International Conference COSIT'97)*, volume 1329 of *Lecture Notes in Computer Science Vol.1329*, pages 15–33. Springer-Verlag, Berlin-Heidelberg, 1997.
121. ICPR. Discussion at the international conference on the development and maintenance of property rights, vienna, 12-15 january, 2000, 2000.
122. ISO. The international standardization organization web page. Url: `http://www.iso.ch/infoe/text.htm`, 2000.
123. R. Jackendoff. *Semantics and Cognition.* MIT Press, Cambridge, Mass., 1983.
124. M. Johnson. *The Body in the Mind: The Bodily Basis of Meaning, Imagination, and Reason.* University of Chicago Press, Chicago, 1987.
125. M. Jones. *Qualified Types: Theory and Practice.* Cambridge University Press, 1994.
126. W. Kent. *Data and Reality - Basic Assumptions in Data Processing Reconsidered.* North-Holland, Amsterdam, 1978.
127. S. Kripke. Semantical analysis of modal logic. *Zeitschrift fuer Mathematische Logik und Grundlagen der Mathematik*, 9:67–96, 1963.
128. S. Kuan. *Geodetic Network Analysis and Optimal Design.* Ann Arbor Press, Chelsea, Mich., 1996.
129. B. Kuipers. A hierarchy of qualitative representations for space. In C. Freksa, C. Habel, and K. F. Wender, editors, *Spatial Cognition - An Interdisciplinary Approach to Representing and Processing Spatial Knowledge*, volume 1404 of *Lecture Notes in Artifical Intelligence*, pages 337–350. Springer-Verlag, Berlin Heidelberg, 1998.

130. G. Lakoff. *Women, Fire, and Dangerous Things: What Categories Reveal About the Mind.* University of Chicago Press, Chicago, IL, 1987.
131. G. Lakoff. Cognitive semantics. In U. Eco, M. Santambrogio, and P. Violi, editors, *Meaning and Mental Representations*, pages 119–154. Indiana University Press, Bloomington, 1988.
132. G. Lakoff and M. Johnson. *Metaphors We Live By.* University of Chicago Press, Chicago, 1980.
133. D.G. Lenat, R.V. Guha, K. Pittman, D. Pratt, and M. Shepherd. CYC: Toward programs with common sense. *Comm. ACM*, 33(8):30 – 49, 1990.
134. P. Lockemann and H. Mayr. *Rechnergestützte Informationssysteme.* Springer-Verlag, Berlin, 1978.
135. J. Loeckx, H.-D. Ehrich, and M. Wolf. *Specification of Abstract Data Types.* John Wiley and B.G. Teubner, Chichester, UK and Stuttgart, 1996.
136. S. MacLane and G. Birkhoff. *Algebra.* Macmillan, New York, 1967.
137. J. McCarthy and P.J. Hayes. Some philosophical problems from the standpoint of artificial intelligence. In B. Meltzer and D. Michie, editors, *Machine Intelligence 4*, pages 463–502. Edinburgh University Press, Edinburgh, 1969.
138. N. McCoy and T. Berger. *Algebra: Groups, Rings and other Topics.* Allyn and Bacon, London, 1977.
139. D. Medak. *Lifestyles - A Paradigm for the Description of Spatiotemporal Databases.* Ph.D. thesis, Technical University Vienna, 1999.
140. D. Medak. Lifestyles. In A.U. Frank, J. Raper, and J.-P. Cheylan, editors, *Life and Motion of Socio-Economic Units.* Taylor & Francis, London, 2000.
141. H. Meixner and A.U. Frank. GI policy - Study on policy issues relating to geographic information in europe. Final Report of a Project "GI Policy" in the IMPACT Program `ftp://ftp.echo.lu/pub/gi/gi_poli.zip`, European Commission DG XIII, July 1997.
142. M. Minsky. *The Society of Mind.* Simon & Schuster, New York, 1985.
143. D. Montello. How significant are cultural differences in spatial cognition? In A.U. Frank and W. Kuhn, editors, *Spatial Information Theory - A Theoretical Basis for GIS*, volume 988 of *Lecture Notes in Computer Science*, pages 485–500. Springer-Verlag, Berlin, 1995.
144. J. Morrison. The proposed standard for digital cartographic data. *Amer. Cartographer*, 15(1):9–140, 1988.
145. H. Mounsey and R.F. Tomlinson, editors. *Building Databases for Global Science - Proceedings of the IGU Global Database Planning Project, Tylney Hall, Hampshire, UK, 9-13 May 1988.* Taylor & Francis, London, 1988.
146. G. Navratil. *An object-oriented approach to a model of a cadaster.* M.Sc., Technical University of Vienna, 1998.
147. J. Neumann von and O. Morgenstern. *Theory of Games and Economic Behavior.* Princeton University Press, Princeton, NJ, 1944.
148. J. Nutter. Epistemology. In S. Shapiro, editor, *Encyclopedia of Artificial Intelligence.* John Wiley & Sons, New York, 1987.
149. OGC. The open GIS consortium web page. Url: `http://www.opengis.org`, 2000.
150. OMG. The object management group web page. Url: `http://www.omg.com`, 2000.
151. ONTEK. The ontek corporation web page url:
`http://www.ontek.com/mikey/current.html`, 2000.
152. J.A.d.C. Paiva and M. J. Egenhofer. Robust inference of the flow direction in river networks. *Algorithmica*, to appear.

153. J.A.d.C. Paiva, M.J. Egenhofer, and A.U. Frank. Spatial reasoning about flow directions: Towards an ontology for river networks. In L. Fritz and J. Lucas, editors, *International Society for Photogrammetry and Remote Sensing. XVII Congress*, volume 24/B3 Comission III of *International Archives of Photogrammetry and Remote Sensing*, pages 318–324, Washington, D.C., 1992.
154. J. Peterson, K. Hammond, L. Augustsson, B. Boutel, W. Burton, J. Fasel, A.D. Gordon, J. Hughes, P. Hudak, T. Johnsson, M. Jones, E. Meijer, S.P. Jones, A. Reid, and P. Wadler. The haskell 1.4 report. `http://haskell.org/report/index.html`, 1997.
155. D. Peuquet, B. Smith, and B. Brogaard. The ontology of fields: Report of the specialist meeting held under the auspices of the varenius project, bar harbour, maine, june 1998. Technical report, NCGIA, 1999.
156. S. Peyton Jones, J. Hughes, and L. Augustsson. Haskell 98: A non-strict, purely functional language, 1999.
157. B.C. Pierce. *Basic Category Theory for Computer Scientists.* Foundations of Computing. MIT Press, Cambridge, Massachussetts, 1991.
158. S. Pigot and B. Hazelton. The fundamentals of a topological model for a four-dimensional GIS. In P. Bresnahan, E. Corwin, and D. Cowen, editors, *Proceedings of the 5th International Symposium on Spatial Data Handling*, volume 2, pages 580–591, Charleston, 1992. IGU Commission of GIS.
159. Protege. The protege project web page. `http://www.smi.stanford.edu/projects/protege`, 2000.
160. D.A. Randell and A. Cohn. Modelling topological and metrical properties of physical processes. In R. Brachmann, H. Levesque, and R. Reiter, editors, *First International Conference on the Principles of Knowledge Representation and Reasoning*, pages 55–66. Los Altos, CA: Morgan-Kaufmann, 1989.
161. D.A. Randell, Z. Cui, and A. Cohn. A spatial logic based on regions and connection. In R. Brachmann, H. Levesque, and R. Reiter, editors, *Third International Conference on the Principles of Knowledge Representation and Reasoning*, pages 165–176. Los Altos, CA: Morgan-Kaufmann, 1992.
162. R. Reiter. Towards a logical reconstruction of relational database theory. In M. L. Brodie, M. Mylopoulos, and L. Schmidt, editors, *On Conceptual Modelling, Perspectives from Artificial Intelligence, Databases, and Programming Languages*, pages 191–233. Springer Verlag, New York, 1984.
163. R. Reiter. *Knowledge in Action: Logical Foundations for Describing and Implementing Dynamical Systems.* in preparation.
164. B.L. Rhoads. Beyond pragmatism: The value of philosophical discourse for physical geography. *The Annals of the Association of American Geographers*, 89(4):760–771, 1999.
165. E. Rosch. Natural categories. *Cognitive Psychology*, 4:328 – 350, 1973.
166. E. Rosch. Principles of categorization. In E. Rosch and B. B. Lloyd, editors, *Cognition and Categorization.* Erlbaum, Hillsdale, NJ, 1978.
167. H.J. Rossmeissl and R.D. Rugg. An approach to data exchange: The spatial data transfer standard. In A. I. Johnson, C. B. Petterson, and J. L. Fulton, editors, *Geographic Information Systems (GIS) and Mapping – Practices and Standards*, pages 38–44. ASTM, Philadelphia, 1992.
168. F. Saussure de. *Cours linguistique generale.* Payot & Rivages, Paris, 1995.
169. M. Schneider. *Spatial Data Types for Database Systems*, volume 1288 of *Lecture Notes in Computer Science.* Springer-Verlag, Berlin-Heidelberg, 1997.

170. F.A. Schreiber. Is time a real time? An overview of time ontology in informatics. In W. A. Halang and A. D. Stoyenko, editors, *Real Time Computing*, NATO ASI Series. Heidelberg, 1994.
171. J.R. Searle. *Speech Acts.* Cambridge University Press, 1969.
172. J.R. Searle. *The Construction of Social Reality.* The Free Press, New York, 1995.
173. P. Simon. *Parts - A Study in Ontology.* Clarendon Press, Oxford, 1987.
174. D. Sinton. The inherent structure of information as a constraint to analysis: Mapped thematic data as a case study. In G. Dutton, editor, *Harvard Papers on GIS*, volume Vol.7. Addison-Wesley, Reading, MA., 1978.
175. B. Smith. An essay in formal ontology. *Grazer Philosophische Studien*, 6:39–62, 1978.
176. B. Smith. On drawing lines on a map. In A.U. Frank and W. Kuhn, editors, *Spatial Information Theory - A Theoretical Basis for GIS (Int. Conference COSIT'95)*, volume 988, pages 475–484. Springer-Verlag, Berlin Heidelberg, 1995.
177. B. Smith. Mereotopology: A theory of parts and boundaries. *Data and Knowledge Engineering*, 20 (1996):287–303, 1996.
178. B. Smith. Basic concepts of formal ontolgy. In N. Guarino, editor, *Formal Ontology in Information Systems*, pages 19–28. IOS Press, Amsterdam Oxford Tokyo, 1998.
179. B. Smith. Objects and their environments: from aristotle to ecological ontology. In A.U. Frank, J. Raper, and J.-P. Cheylan, editors, *Life and Motion of Socio-Economic Units.* Taylor & Francis, ESF Series, London, 2000.
180. B. Smith. Ontology: Philosophical and computational. In L. Floridi, editor, *The Blackwell Guide to the Philosophy of Computing and Information.* Blackwell, Oxford, January 2003.
181. B. Smith and B. Brogaard. Quantum mereotopology. In *Spatial and Temporal Granularity - Papers from the AAAI Workshop*, volume AAAI Technical Report WS-00-08, pages 25–31. AAAI Press, 2000.
182. B. Smith and B. Brogaard. Sixteen days, 2000.
183. B. Smith and K. Mulligan. Framework for formal ontology. *Topoi*, (2):73–85, 1983.
184. B. Smith and J. Searle. The construction of social reality: An exchange. *American Journal of Economics and Sociology*, 60, 2001.
185. H. Smith and K. Poulter. The role of shared ontology in XML-based trading architectures. *Communications of the ACM, Special Issue on Agent Software*, 1999.
186. R. Snodgrass. Temporal databases: Status and research directions. *SIGMOD RECORD*, 19(4, December):83–89, 1990.
187. R. Snodgrass. Temporal databases. In A.U. Frank, I. Campari, and U. Formentini, editors, *Theories and Methods of Spatio-Temporal Reasoning in Geographic Space (Int. Conference GIS - From Space to Territory, Pisa, Italy)*, volume 639 of *Lecture Notes in Computer Science*, pages 22–64. Springer-Verlag, Berlin, 1992.
188. J.G. Stell. A lattice theoretic account of spatial regions, 1997.
189. S.S. Stevens. On the theory of scales of measurement. *Science*, 103(2684):677 – 680, 1946.
190. A. Tansel, J. Clifford, S. Gadia, S. Jajodia, A. Segev, and R. Snodgrass. *Temporal Databases.* Benjamin Cummings, Redwood City, CA, 1993.
191. A. Tarski. *Introduction to Logic and to the Methodology of Deductive Sciences.* Dover Publications, Mineola, N.Y., 1995.
192. S. Thompson. *Haskell - The Craft of Functional Programming.* International Computer Science Series. Addison-Wesley, Harlow, UK, 1996.

193. S. Timpf. *Hierarchical Structures in Map Series.* PhD, Technical University Vienna, 1998.
194. S. Timpf, G. Volta, D. Pollock, and M.J. Egenhofer. A conceptual model of wayfinding using multiple levels of abstractions. In A.U. Frank, I. Campari, and U. Formentini, editors, *Theories and Methods of Spatio-Temporal Reasoning in Geographic Space*, volume 639 of *Lecture Notes in Computer Science*, pages 348–367. Springer Verlag, Heidelberg-Berlin, 1992.
195. C.D. Tomlin. *Digital Cartographic Modeling Techniques in Environmental Planning.* Doctoral dissertation, Yale Graduate School, Division of Forestry and Environmental Studies, 1983.
196. C.D. Tomlin. *Geographic Information Systems and Cartographic Modeling.* Prentice Hall, New York, 1990.
197. M. Turner. *The Literary Mind.* Oxford University Press, New York, 1996.
198. B. Tversky. Spatial perspectives in descriptions. In P. Bloom, A. Peterson, L. Nadel, and M. Garrett, editors, *Language and Space - Language, Speech, and Communciation.* MIT Press, Cambridge, MA, 1996.
199. A. Vckovski, K.E. Brassel, and H.-J. Scheck, editors. *Interoperating Geographic Information Systems (Interop'99, Zürich)*, volume 1580 of *Lecture Notes in Computer Science.* Springer-Verlag, Berlin, 1999.
200. A. Vckovski and F. Bucher. Virtual data sets - Smart data for environmental applications, Dec. 1995 1998.
201. L. Vieu. A logical framework for reasoning about space. In A.U. Frank and I. Campari, editors, *Spatial Information Theory: Theoretical Basis for GIS*, volume 716 of *Lecture Notes in Computer Science*, pages 25–35. Springer Verlag, Heidelberg-Berlin, 1993.
202. P. Wadler. Monads for functional programming. In J. Jeuring and E. Meijer, editors, *Advanced Functional Programming*, volume 925 of *Lecture Notes in Computer Science.* Springer-Verlag, Berlin Heidelberg, 1995.
203. G. Weiss. *Multi-Agent Systems: A Modern Approach to Distributed Artificial Intelligence.* The MIT Press, Cambridge, Mass., 1999.
204. A. Wierzbicka. *Semantics - Primes and Universals.* Oxford University Press, Oxford, 1996.
205. M. Worboys. A model for spatio-temporal information. In P. Bresnahan, E. Corwin, and D. Cowen, editors, *Proceedings of the 5th International Symposium on Spatial Data Handling*, volume 2, pages 602–611, Charleston, 1992. IGU Commission of GIS.
206. C. Zaniolo, P.C. Lockemann, M.H. Scholl, and T. Grust, editors. *Advances in Database Technology - Proceedings of EDBT 2000 (Konstanz, Germany, March 2000)*, volume 1777 of *Lecture Notes in Computer Science.* Springer-Verlag, Berlin Heidelberg, 2000.

3 Conceptual Models for Spatio-temporal Applications

Nectaria Tryfona[1], Rosanne Price[2], and Christian S. Jensen[1]

[1] Aalborg University, Denmark
[2] Monash University, Caulfield East, Australia

3.1 Motivation

Improved support for modeling information systems involving time-varying, geo-referenced information, termed spatio-temporal information, has been a long-term user requirement in a variety of areas, such as cadastral systems that capture the histories of landparcels, routing systems computing possible routes of vehicles, and weather forecasting systems. This chapter concerns the conceptual database design phase for such spatio-temporal information systems and presents two models, namely the spatio-temporal Entity Relationship (ER) Model and the Extended spatio-temporal Unified Modeling Language (UML) as proposed in [33,34] and [26], respectively.

The conceptual design phase focuses on expressing application requirements without the use of computer metaphors. The design should be understandable to the user and complete, so that it can be translated into the logical phase that follows without any further user input. Popular conceptual models include the ER model [6], IFO [2], OMT [30], and UML [17].

For conventional administrative systems, exemplified by the "supplier-supplies-parts" paradigm, the available modeling notations and techniques that support the conceptual and logical modeling phases are mature and adequate. However, this is not the case in non-standard systems managing spatio-temporal, multimedia, VLSI, image, and voice data. Rather, these lead to new and unmet requirements for modeling techniques.

The basic rationale behind the work presented here is to introduce new modeling techniques, based on minimal extensions of existing models, developed to accommodate the peculiarities of the combined spatial and temporal information. The ER Model and the UML have been extended as prototypical examples for this purpose.

First, we present the fundamental aspects of the spatio-temporal domain, covering concepts such as objects, properties, and relationships. Based on these, in the ER approach, we present a small set of constructs aimed at improving the ability to conveniently model spatio-temporal information at the conceptual level. These constructs may be included in a wide range of existing conceptual data models, improving their modeling capabilities without fundamentally changing the models. We incorporate the proposed modeling constructs into the ER model ([5]), resulting in the semantically richer Spatio-Temporal ER (STER) model [33].

T. Sellis et al. (Eds.): Spatio-temporal Databases, LNCS 2520, pp. 79–116, 2003.
© Springer-Verlag Berlin Heidelberg 2003

In the UML approach, an extension of UML, Extended spatio-temporal UML [26], is presented that addresses spatio-temporal modeling requirements. Extending the Object Management Group standard for object-oriented (OO) modeling was selected as the best approach, given UML's high level of acceptance, tool support, understandability, and extensibility. In order to satisfy the requirement for a clear, simple, and consistent notation, the extension introduces a small base set of modeling constructs for spatio-temporal data. These can be combined and applied to attributes, objects and associations in a consistent manner, guided by the same simple rules.

3.2 Spatio-temporal Foundations

In this section we discuss fundamental spatial and temporal aspects that should be considered for the conceptual modeling of spatio-temporal applications. Later on, we show the enhancement of the ER model and UML with constructs to support these concepts.

Objects, Properties, and Relationships. We perceive reality as a collection of objects, characterized by a set of properties. Objects are interrelated via relationships.

Spatial Aspects. Most real-world objects have a *position* or a *spatial extent.* In a spatio-temporal application, the positions of some objects *matter* and should be recorded, while the positions of other objects are not to be recorded. The former objects, we term *spatial*, or *geographic objects* (GO).

The function p (position) takes spatial objects as arguments and returns the positions of the objects. Positions are parts of space and may be points, lines, regions, or combinations thereof, and are called *geometric figures.* So, function p is defined as follows.

$$p : GO \rightarrow G$$

where G is the domain of geometric figures.

The embedding *space* must also be modeled in order to locate the objects in it. Capturing space is a rather complicated issue. Many researchers see space as what human beings experience [32] or see it as the description of reality based on geometric factors [21].

Philosophical discussions aside, we model space as a set, and we term the elements of space (*points*). Many different sets will do for space, but for practical reasons, space is modeled as a subset of R^2 or R^3 in current spatial applications and we use R^2 as our space; this choice does not affect the generality of the proposed approach.

It is also an inherent property of spatial objects that their positions may be viewed at different granularities and that the granularity affects the concrete data type of the position. For example, a "landparcel" may be seen as a point, a region, or both, depending on the granularity requirements of the application at hand. Such different *object views* have to be integrated into one conceptual description.

Spatial objects have *descriptive properties*, such as the "owner's name" or the "cadastral-id" of a landparcel, and *spatial properties*, such as the "soil type" of a landparcel. Spatial properties are properties of the embedding space that indirectly become properties of the spatial objects via their position in space, i.e., the spatial objects inherit them from space. For example, although one application may view the "soil type" of a landparcel as a property of the landparcel, it is clear that: (a) the "soil type" is defined whether or not the landparcel exists at that position in space, and (b) when the landparcel moves (or changes shape), the landparcel's "soil type" will not remain unchanged; rather the "soil type" attribute inherits (or, obtains) new values from the new position.

The spatial properties of objects may be captured independently of the objects using so-called fields (the term *layer* is also used). Formally speaking, a field can be seen as a function from geometric figures to a domain of descriptive attribute values [8].

$$f_1 : G \rightarrow D_1 \times D_2 \times \ldots \times D_k$$

where G is the set of geometric figures and the D_i are (not necessarily distinct) domains. In other words, a field is a set of geometric figures with associated values.

There are two basic types of fields:

(a) those that are continuous functions, e.g., "temperature," or "erosion," and
(b) those that are discrete functions, e.g., "county divisions" represented as regions.

In case (a), we visualize a field as a homogeneous (or continuous) area consisting of points, while in case (b), a field represents a set of areas with different values of the same attribute or positions of objects in space.

Finally, geographic objects may be related to each other in *space* via spatial (or *geographic*) *relationships*. For example, "the fjord Limfjorden *traverses* the city of Aalborg." Spatial relationships among geographic objects are actually relations on the objects' positions.

The set of spatial relationships is subdivided into three subsets: *topological* (e.g., "inside," "outside," etc.), *directional* (e.g., "North of," "North-East of," etc.), and *metric* (e.g., "5 km away from") relationships [10,14,15]. Spatial relationships are further translated into spatial integrity constraints on the database.

Temporal Aspects. Information about objects' properties and relationships among objects can be considered as *statements*, or *facts*, about the objects.

For example, an application involving countries may include a "capital" property for the countries. The "Copenhagen" value of the property "capital" associated with "country" Denmark denotes the fact that "Copenhagen is the capital of Denmark." Precisely everything that can be assigned a truth value is a fact. For example, "Denmark is south of Greece" is a fact; it can be assigned the truth value "false." The sentence "Denmark and Greece are south" is not a fact.

Three temporal aspects have been the focus of attention in the research literature; they are universal, and applications frequently require that these be captured in the database:

1. The *valid time* aspect applies to facts: the valid time of a fact is the time when the fact is true in the modeled reality. For example, the valid time of "Copenhagen is the capital of Denmark" is the time from year 1445 until the present day.
2. The *transaction time* aspect applies not only to facts, but to any "element" that may be stored in a database: the transaction time of a database element is the time when the element is part of the current state of the database. Put differently, the transaction time of element e is the valid time of "e is current in the database." Transaction time is important in applications that demand traceability and accountability.
3. The *existence time* aspect applies to objects: the existence time of an object is the time when the object exists. Again, this aspect can be formulated in terms of valid time. The existence time of object o is thus the valid time of "o exists."

We will assume that it only makes sense for an object to have properties and participate in relationships *when the object exists*. This implies that the valid times of facts associated with objects must be contained in the existence times for those objects.

Time values are drawn from a domain of time values, with the individual values being termed *chronons*. All three temporal aspects have duration, and they may be captured using time intervals, where a time interval $[t_s, t_e]$ is defined to be a set of consecutive chronons. We call t_s and t_e the start and the end chronon of the interval, respectively.

The following section extends the ER model in accordance with the foundations outlined here.

3.3 Spatio-temporal Entity-Relationship Model

In this section we present an extension of the ER model, namely the Spatio-temporal ER model (STER), to accommodate spatio-temporal peculiarities [33,34]. STER combines spatial and temporal aspects in a meaningful way. The following section serves to explicitly state which aspects are independent and, by implication, which are not. It thus provides a guide for applying the general design criteria.

3.3.1 Extending the ER with Spatio-temporal Constructs

The Entity Relationship (ER) model [6] is arguably the first conceptual model that appeared in the literature. This easy-to-use model, consisting of very few modeling constructs, has gained an unparalleled, widespread popularity in industry. The model's basic constructs include the following: (a) entity sets that represent objects are depicted by rectangles; (b) relationship sets that represent associations among entity sets are illustrated as diamonds; and (c) attributes of entity sets and relationship sets that capture properties of the objects and associations and are represented graphically as ovals. Relationships can be 1:1 (one to one), 1:M (one to many), and N:M (many to many). Furthermore, in some extensions of ER (reference [12] is both recent and accessible) a very useful, special kind of relationship, the ISA relationship, is proposed to model an entity set as a subset of another.

The Spatio-Temporal Entity Relationship Model (STER) [33] includes constructs with built-in spatial, temporal, and spatio-temporal functionality. A construct that captures a temporal aspect is called *temporal*; if it has built-in support only for a spatial aspect, it is termed *spatial*; and if it has both, it is *spatio-temporal*. The upper-right corner of each extended construct indicates its temporal support. The bottom-right corner indicates the spatial support. For each STER construct, we give its corresponding representation in the ER model.

While all basic constructs of the ER model can have spatial and temporal extents, not all temporal aspects are semantically meaningful for each construct. The aspect termed existence time when "something" is considered to exist. So existence time is applicable precisely to the entities in entity sets, which are the only elements in the ER model with independent existence.

An entity set may be given attributes that describe the properties of the set's entities. Earlier, we stated that valid time is meaningful only for *facts*. When assigning valid time to an attribute of an entity set, we indicate that the valid times of the facts—that specific entities in the set are associated with specific values for this attribute—are to be captured in the database. The same applies to attributes of relationship sets in place of entity sets. Finally, valid time may be assigned to a relationship set, indicating that the time when each relation in the set is true in the miniworld that is to be captured in the database.

Transaction time applies to any "element" stored in the database, regardless of whether or not it may be assigned a *truth* value. So unlike valid time, transaction time applies to entities in entity sets. Table 3.1 shows the meaningful combinations of temporal aspects and modeling constructs.

Table 3.1. Assigning temporal aspects to ER constructs

	entity set	attributes	relationship
existence time	Yes	No	No
valid time	No	Yes	Yes
transaction time	Yes	Yes	Yes

As the next step, we illustrate in more detail how to assign existence and valid time to the ER constructs. Transaction time is covered separately, in the next section. The abbreviations "et," "vt," "tt," and "bt" are used for "existence time," "valid time," "transaction time," and "bitemporal time," respectively. Abbreviation "bt" is a shorthand for the combination of "vt" and "tt" that occurs often in a spatio-temporal database.

Entity Sets. Entity sets represent objects.

(i) Temporal Entity Sets. Entities in an entity set can be assigned existence and transaction time. We term the former support for existence time, and this is indicated by placing an "et" in a circle in the upper-right corner of the entity set's rectangle as indicated in Figure 3.1. Figure 3.1(b) shows that for "car" entities, we keep track of their existence time. This notation is in effect shorthand for a

Fig. 3.1. (a) Capturing existence time, and (b) representing car entities and their existence times

larger ER diagram. This shorthand is convenient because it concisely states that the existence times of the entities in the entity set should be captured in the database. The more verbose ER diagram corresponding to the STER diagram in Figure 3.1(a) is given in Figure 3.2. Attributes connected to each other denote composite attributes, i.e., attributes resulting of the combination of other attributes [12]. Thus "existence time/id" values consist of pairs of "existence time" and "id" (i.e., identification number) values.

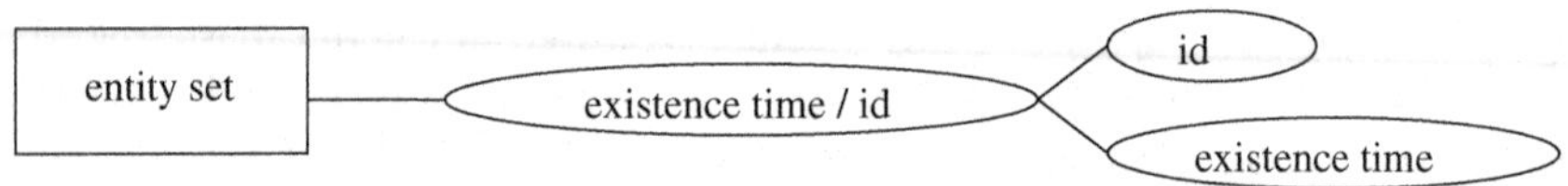

Fig. 3.2. ER diagram corresponding to Figure 3.1(a)

(ii) Spatial Entity Sets. Spatial objects have a position in space, and it is frequently necessary to capture this position in the database. The first step to support this is to provide means for representing the space in which the objects are embedded. The next step is to provide means for indicating that the objects' positions in this space are to be captured. For these purposes, we introduce the following special entity and relationship sets.

1. The special entity sets SPACE, GEOMETRY, POINT (or "P"), LINE (or "L"), REGION (or "R"). Entity set GEOMETRY is used for capturing

the shapes of entities and can be specialized as (i.e., is-a) POINT, LINE, REGION, or any other geometric type (or geometry). For simplicity we use only POINT, LINE, REGION, and their combinations.

2. The special relationship set "is_located_at" that associates a spatial entity set with its geometry. The cardinality of this set is 1:M, meaning that a spatial entity may have more than one geometry when multiple granularities are employed. Assuming that in each application we deal only with one space, then the relationship set "belongs_to" between GEOMETRY and SPACE with cardinality M:1 is also included. When GEOMETRY is connected to SPACE then it captures also the *locations* of objects. In this case, GEOMETRY describes objects positions, i.e., shapes and locations [9].

The letters "s," "P," "L," or "R" in a circle in the lower-right corner of an entity set rectangle specify the spatial support. Letter "s" stands for SPATIAL and is used to indicate a spatial entity set whose exact geometric type is unknown. Letters "P," "L," "R," and their combinations specify geometric types as indicated above. These annotations may occur simultaneously and represent then different views of the same object. A spatial entity set is depicted as shown in Figure 3.3(a), and its meaning in terms of the ER model is given in Figure 3.4. Figure 3.3(b) illustrates the spatial entity set "landparcel" with simultaneous geometries point and region; in this case, the representation in the ER model will have only REGION and POINT as geometries.

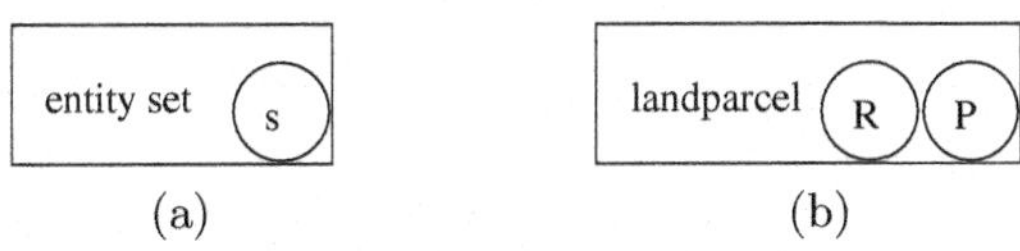

Fig. 3.3. (a) Spatial entity sets, (b) a landparcel as POINT or REGION

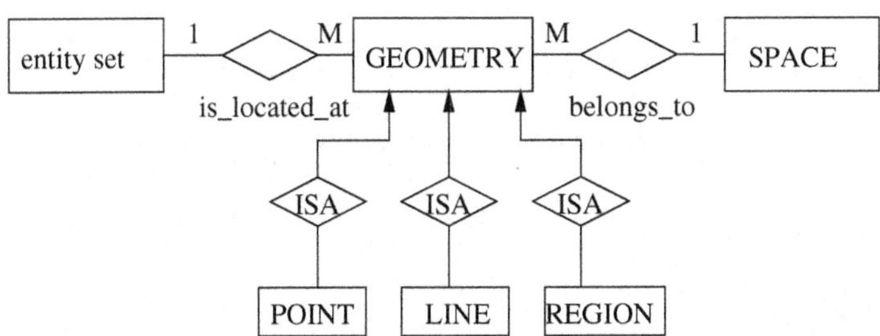

Fig. 3.4. ER diagram corresponding to Figure 3.3(a)

(iii) Spatio-temporal Entity Sets. For a geo-referenced object, GEOMETRY captures both its shape and location. When an object changes its position over time,

i.e., the object moves (change of location) or the object changes shape, it is the GEOMETRY aspect of the object that changes rather than the object itself. For example, a car moving on a road changes its location, but this is not considered to change the car's identity. To capture a temporal aspect of the positions of the objects in an entity set, an "svt," an "stt," or an "sbt" is placed in a circle in the lower-right corner of the entity set's rectangle. The first annotation indicates valid-time support: the objects' current positions as well as their past and future positions are to be captured. This is illustrated in Figure 3.5(a). The second annotation (i.e., "stt") indicates transaction-time support: the current positions as well as all positions previously recorded as current in the database are to be captured. The third annotation (i.e., "sbt") indicates support for both valid and transaction time. Figure 3.5(b) shows that, when a "car" changes position we record both the car's position in time (i.e., "Pvt") and the time this is recorded in the database (i.e., "Ptt"); we indicate this by "Pbt". If the geometric type of the entity set is known, the "s"-part is replaced by "P," "L," "R," or a combination of these. The meaning of the spatio-temporal entity in Figure 3.5(a) is

Fig. 3.5. (a) A spatio-temporal entity set with valid-time support, (b) recording car position with valid-time and transaction-time support

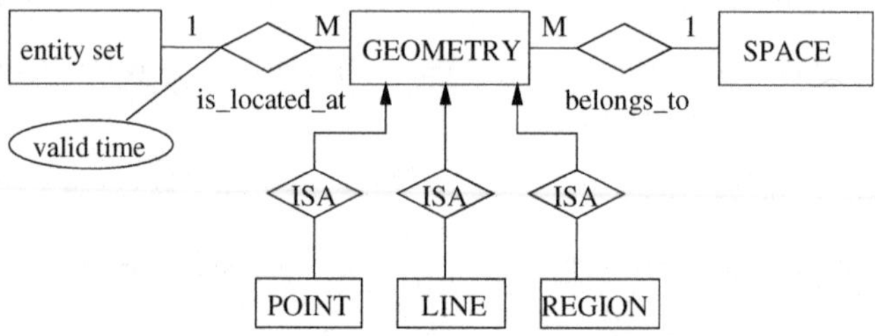

Fig. 3.6. ER diagram corresponding to Figure 3.5(a)

given in Figure 3.6 in terms of the ER model.

It is important to point out the difference between keeping track of (a) a spatial entity set in time, and (b) the position of a spatial entity set in time. In case (a) the temporal support refers to an entity's existence and recording in time. Figure 3.6 models case (a). Figures 3.7(a) and (b) show the entity set "landparcel" as a spatial entity set (indicated by an "s"-part). Figure 3.7(a) illustrates that a cadastral database captures the existence time (i.e., when it first existed-indicated by "et") as well as the transaction time (i.e., when it was recorded in the database-indicated by "tt"). In contrast, Figure 3.7(b) shows that for "landparcel," the database captures transaction time as before (indicated by "tt"), as well as its geometry over time (indicated by "svt"). This means that if a landparcel changes shape in time, the (current) shape and the time this shape is true is recorded (valid time).

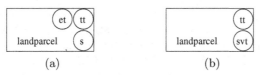

Fig. 3.7. (a) A landparcel in space and time and (b) a landparcel in time, with position in time

Attributes of Entity and Relationship Sets. In a spatial environment, entity sets have two types of attributes: (a) *descriptive* attributes, such as the "cadastral-id" of a landparcel, and (b) *spatial* attributes, such as the "soil type" of a landparcel. The values of descriptive attributes for an entity (or a relationship) often change over time, and it is often necessary to capture this in the database. Spatial attributes for which a temporal aspect is captured are termed *spatio-temporal attributes.*

(i) Temporal Descriptive Attributes. Values of attributes of entities denote facts about the entities and thus have both valid- and transaction-time aspects. A circle with a "vt" or a "tt" in the upper-right corner of an oval denoting an attribute indicates that valid or transaction time, respectively, is to be captured. A circle with "bt" (bitemporal) indicates that both temporal aspects are to be captured. The sample STER diagram in Figure 3.8(a) contains an attribute with valid-time support and Figure 3.9 gives the equivalent ER diagram. Figure 3.8(b) shows an example keeping track of cars' colors and their valid time periods.

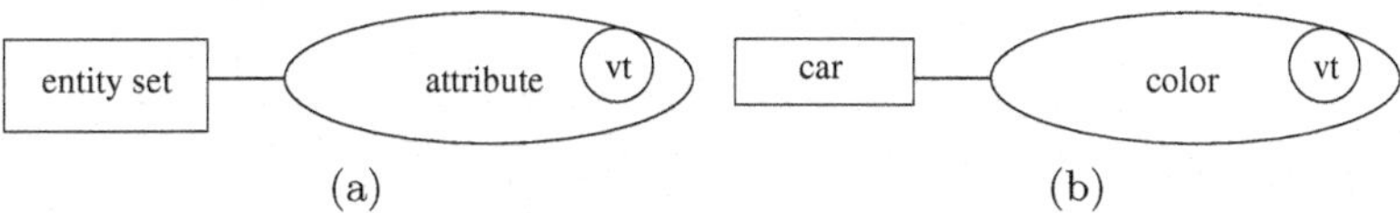

Fig. 3.8. (a) An attribute with valid-time support and (b) car "color" with valid-time support

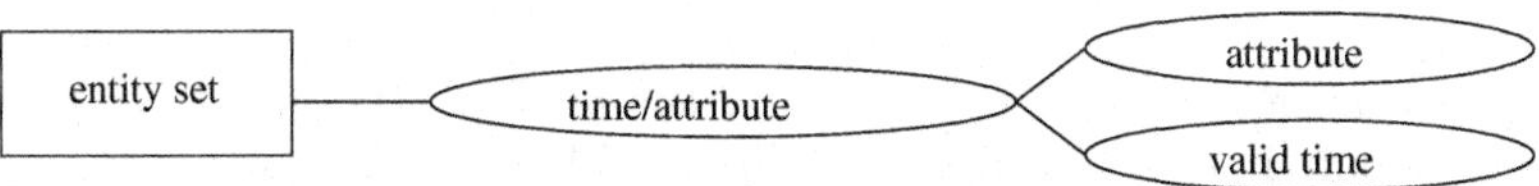

Fig. 3.9. ER diagram corresponding to Figure 3.8(a)

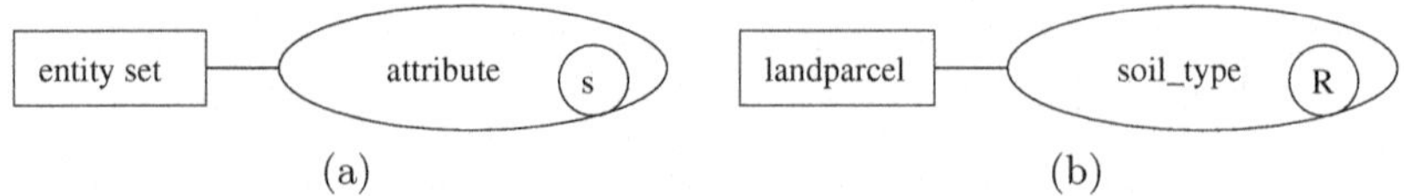

Fig. 3.10. (a) A spatial attribute in STER and (b) "soil type" as a spatial attribute

(ii) Spatial Attributes. Facts captured by attributes may also have associated locations in space, which are described as sets of geometric figures. To capture this spatial aspect of an attribute, a circle with an "s" is used, as shown in Figure 3.10. Figure 3.10(a) depicts the general representation of a spatial attribute, while Figure 3.10(b) shows that the "soil type" value of a landparcel is associated with a set of spatial regions ("R"). In terms of the ER model, a spatial attribute (Figure 3.11) is modeled as an entity set with a composite attribute "attribute/spatial_unit." This consists of the "attribute value" and the "spatial-unit," where the unit represents the geometry in which the "attribute value" is constant. So, the spatial unit can be "POINT" (or "P"), "LINE" (or "L") and "REGION" (or "R") geometric type. The spatial attribute is further connected to SPACE via the relationship set "has_spatial_attribute." In this way, each part of space is assigned a specific value of the attribute. By connecting a spatial entity set to GEOMETRY (via the special relationship "is_located_at," see previous figures) and GEOMETRY to SPACE (via "belongs_to"), an object inherits spatial attributes. So, spatial attributes of entities are *derived properties* ([11] uses *propagated* instead of *derived*) from space (indicated as shaded).

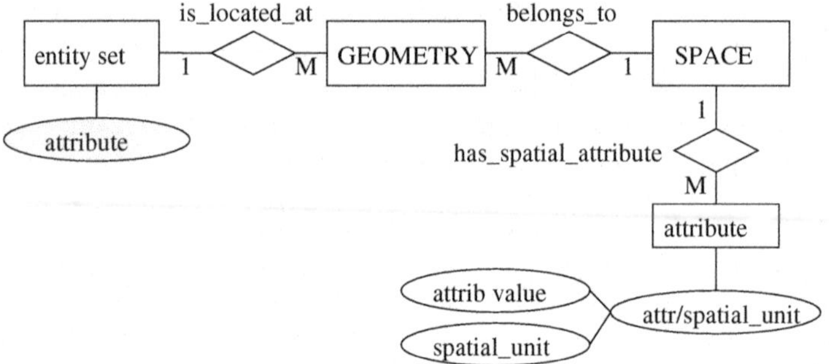

Fig. 3.11. ER diagram corresponding to Figure 3.10(a)

(iii) Spatio-temporal Attributes. Two cases are distinguished here. The first case concerns spatial attributes with temporal support that refers to the attributes' valid- and transaction-time periods (i.e., the spatial attribute is treated as a normal attribute in time). This is illustrated in Figure 3.12(a) and in Figure 3.13, which gives the equivalent ER diagram. Figure 3.12(b) gives an example. In the second case, the temporal aspects (valid and transaction time) of spatial attributes are recorded by placing "svt," "stt," or "sbt" (and replacing the "s"

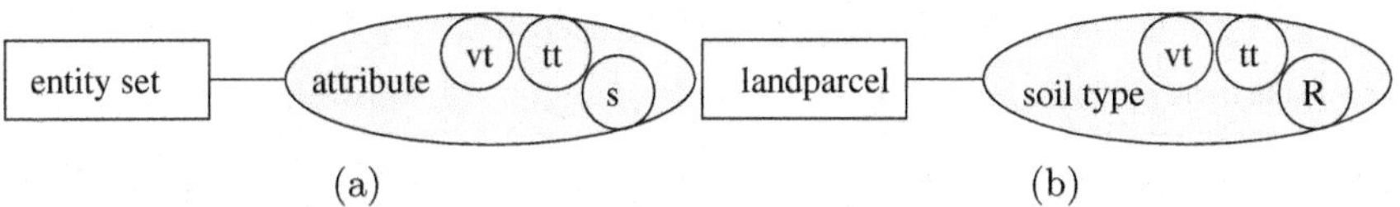

Fig. 3.12. (a) A spatial attribute with temporal support and (b) "soil type" with temporal support

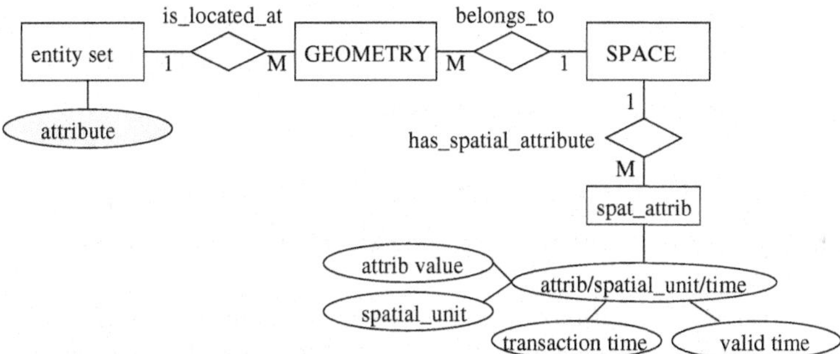

Fig. 3.13. ER diagram corresponding to Figure 3.12(a)

with "P," "L," or "R," or a combination of these if the geometric types of the geometric figures of the attributes are known) in same way as for entity sets. This is illustrated in Figure 3.14(a), and Figure 3.15 gives the equivalent ER diagram. Figure 3.14(b) gives an example. The difference between the two cases may be seen by comparing Figure 3.13 and Figure 3.15.

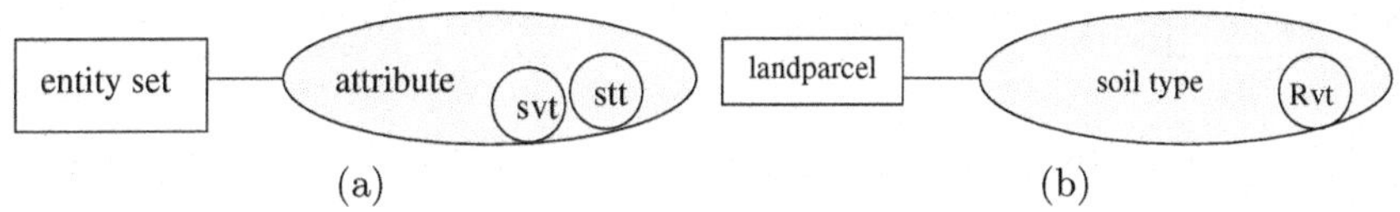

Fig. 3.14. (a) A spatial attribute with temporal support for its spatial part and (b) "soil type" as a spatio-temporal attribute

Relationship Sets

(i) Temporal Relationship Sets. By annotating a relationship set with a temporal aspect (valid time, transaction time, or both), we capture the changes of the set's relationships with respect to that aspect.

(ii) Spatial Relationship Sets. Spatial relationship sets are special kinds of relationship sets. In particular, they are associations among the geometries of the spatial entities they relate. For reasons of simplicity and ease of understanding, spatial relationship sets are given as relationships among the spatial entity sets

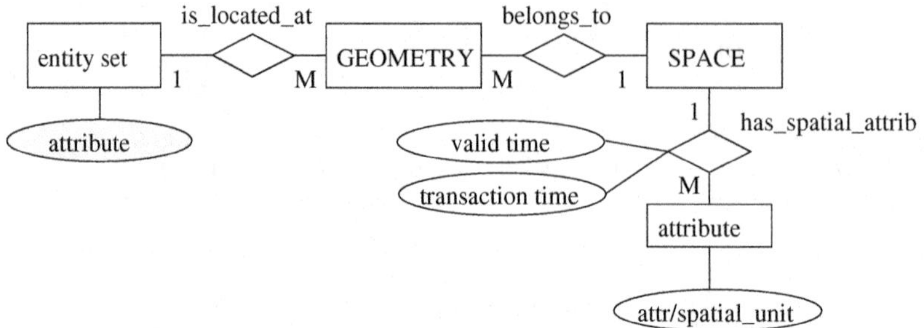

Fig. 3.15. ER diagram corresponding to Figure 3.14(a)

themselves. For example, the relationship "traverses" between cities and rivers relates the geometries of entities of these two spatial entity sets.

(iii) Spatio-temporal Relationship Sets. A spatio-temporal relationship set is a spatial relationship set with time support. In particular, by annotating a spatial relationship set with a temporal aspect, we capture the changes of the spatial relationship with respect to that aspect. Figure 3.16(a) shows the general representation of a spatio-temporal relationship set, while Figure 3.16(b) depicts changes of the relationship "traverses" between cities and rivers are recorded in time. Figure 3.17 gives the equivalent of Figure 3.16(a) in terms of an ER diagram. Finally, the previous discussion about temporal, spatial, and spatio-

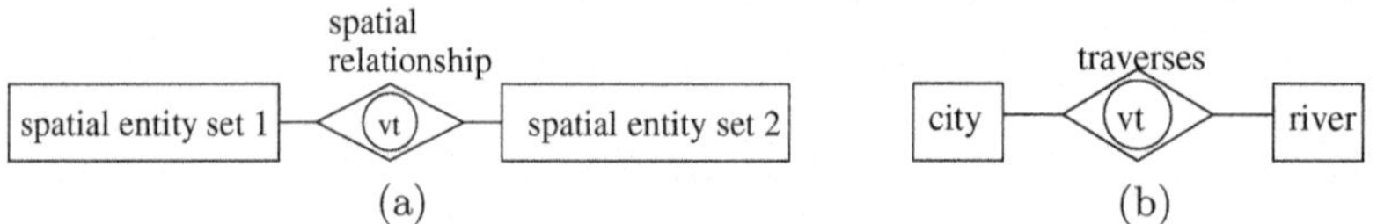

Fig. 3.16. (a) A spatio-temporal relationship set in STER, and "traverses" as a spatio-temporal relationship set

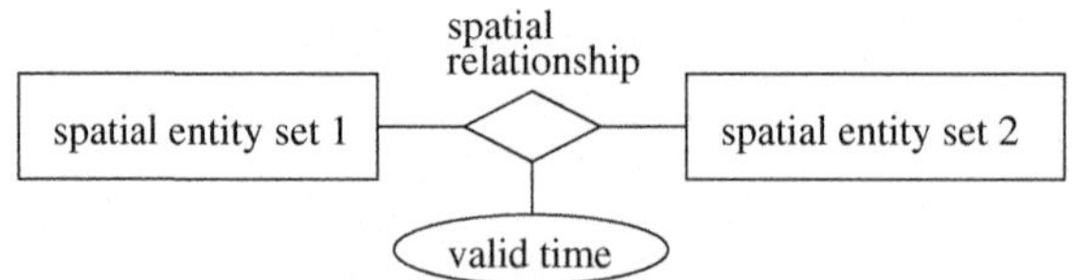

Fig. 3.17. ER diagram corresponding to Figure 3.16(a)

temporal attributes applies also to regular attributes of relationship sets.

In the description of STER so far, we have primarily focused on adding existence and valid time to the ER model constructs. The temporal aspect of transaction time is not strictly part of capturing the modeled reality. However, capturing transaction time, as reflected in the systems requirements, is important when designing a real-world information system, and an ER model should also provide built-in support for the capture of this aspect. Transaction time is applicable in STER exactly where valid time or existence time is applicable. For example, to capture the transaction time of a relationship set, in Figure 3.16(a), all that is needed is to add "tt" to the relationship set construct.

3.3.2 A Textual Notation for STER

In order to fully support the STER model, its graphical notation as well as its textual notation has to be provided. In this way all the constructs of the diagrammatic STER can be presented and/or transformed into the textual notation.

The definitions use the following meta syntax. A `typewriter-like` font indicates elements of the data definition language; syntactic definitions are given with the normal font; *italics* are used for semantic explanations. Upper case words are reserved words, and lower case words are variables that represent arbitrary names. For example, `attr_name_i` stands for any alphanumeric string that is not a reserved word. Capitalized words in lower case denote variables with restricted range. For example, "`Domain`" is one of INTEGER, STRING, DATE, etc. Optional elements are enclosed in "<>"'s, and "(, ...)" denotes repeatable elements; the notation "{...|...}" denotes selection (one of), and "( )" simply indicates grouping of arguments. A longhand notation (e.g., DEFINE instead of DEF) is used to facilitate a first reading; in practice one uses the shorthand.

Definition of Entity Sets

```
DEFINE ENTITY SET entity_set_name
<TYPE Entity_construct (entity_set_name_i, ...)>
<ATTRIBUTES
   ((attribute_name_j
      <VALID_TIME> <TRANSACTION_TIME>
      <GEOMETRY Geometric_type
         <VALID_TIME> <TRANSACTION_TIME>>), ...)>
<GEOMETRY Geometric_type
         <VALID_TIME> <TRANSACTION_TIME>>
<EXISTENCE_TIME> <TRANSACTION_TIME>
<AS ISA OF (entity_set_name_k, ...)>>
```

- `entity_set_name` is an identifying alphanumeric string, i.e., different from any other used in the same syntactic position. `entity_set_name` *is used as the name of the entity set being defined.*
- `Entity_construct` is one of PART_OF, GROUP_OF, SPATIAL_PART_OF, and SPATIAL_MEMBER_OF. `Entity_construct` *is optional and is used to define an entity's complex type.*
- `entity_set_name_i` is an identifying alphanumeric string. `i` is an integer. `entity_set_name_i` *is used to define the constructs of the complex entity set.*

- `attribute_name_j` is an identifying alphanumeric string. `j` is an integer. `attribute_name_j` *is used to define the name of an attribute of an entity set.*
- `Geometric_type` is one of P, L, or R or combination thereof. *This optional clause is used to define the geometric type of the entity set or the attribute (spatial) defined.*
- `entity_set_name_k` is an identifying alphanumeric string and k is an integer. `entity_set_name_k` *defines pre-existing classes used to construct the superset.*

Definition of Composite Attributes

```
DEFINE ATTRIBUTE attribute_name_m
   <AGGREGATION_OF ((attribute_name_n <VALID_TIME Valid_time>
      <TRANSACTION_TIME Transaction_time>
      <GEOMETRY Geometric_type
      <VALID_TIME> <TRANSACTION_TIME>>), ...)>
```

- `attribute_name_m` is an identifying alphanumeric string. `m` is an integer. *It is used to define the name of the composite attribute.*
- `attribute_name_n` is an identifying alphanumeric string. `n` is an integer. *It is used to define the names of the attributes that compose the complex one.*
- `Geometric_type` is one of P, L, or R or combination thereof. *This optional clause is used to define the geometric type of the entity set or the attribute (spatial) being defined.*

Definition of Relationship Sets

```
DEFINE <SPATIAL> RELATIONSHIP SET
   relationship_set_name (entity_set_name_i, ...)
   TYPE Relationship_type
   <ATTRIBUTES ((attribute_name_k
      <VALID_TIME> <TRANSACTION_TIME>), ...)>
   <VALID_TIME> <TRANSACTION_TIME>
```

- `relationship_set_name` is an identifying alphanumeric string. *It defines the functional relationship between entity sets.*
- `entity_set_name_i` is an identifying alphanumeric string and `i` is an integer. *It defines the entity sets which are related through the relationship.*
- `Relationship_type` is one of `ONE_TO_ONE`, `ONE_TO_MANY`, `MANY_TO_ONE`, `MANY_TO_MANY`. *It is used to define the relationship type.*
- `attribute_name_k` is an identifying alphanumeric string and `k` is an integer. *It is used to indicate attributes of the defined relationship set.*

3.3.3 Example of Usage of STER

The following example presents an excerpt of the conceptual schema for a cadastral application based on user requirements [22]. This excerpt is a substantial simplification of a real-world situation. We used STER in an intermediate phase to translate user requirements (which were expressed in natural language) to a formal logical schema consisting of relations and maps.

Figure 3.18 states that for "landparcels," existence time is recorded; the positions of landparcels can be either points or regions. Moreover, "landparcels" have two spatio-temporal attributes: (a) "soil type," which is of type REGIONS ("R"), and for which both valid and transaction time are captured, and (b) "elevation," which is recorded in terms of points. Additionally, "landparcels" may be traversed by "rivers." For the spatial relationship "traverses," we capture transaction time. For "rivers," transaction and existence time is captured. Finally, "rivers" are represented as lines ("L") and are distinguished by their "cadastral number." "landparcels" may have different "land use": (a) "agricultural" use is recorded in regions, based on its "vegetation" (which is also captured in regions), and (b) "industrial" use which is defined in regions and is characterized by different "types," such as heavy industry. For the type of industry we keep track of the time this is valid. For the relationship "land use," we record valid and transaction time (this could also be indicated by the symbol "bt"). Finally, "buildings" may be inside "landparcels," at different time periods, which are recorded in the database. A "building" occupies a region and belongs to "owners." Valid time is captured for "ownership"; for owners, Social Security Number ("SSN") and "name" are known. In the following, we informally explain an excerpt (the first five constructs) of the description line by line, showing the one-to-one translation of the constructs of the diagrammatic representation.

We express that the entity set "landparcel" has "soil type" and "elevation" as attributes. For "soil type" we record the valid and the transaction time of its geometry, which is REGION. For "elevation" we record only its geometry, which is POINT. "landparcel" is of type REGION or POINT and we keep track of its existence time. Additionally, we have the entity set "river" keeping track of its existence and transaction time in the database. A "river" is of type LINE and has its identification ("cadastral number") as attribute. Landparcels and rivers are related via the spatial relationship "traverses," for which we record the transaction time ("tt") of the changes.

```
DEFINE ENTITY SET landparcel
ATTRIBUTES
 ((soil type (GEOMETRY REGION VALID_TIME TRANSACTION_TIME))
 (elevation GEOMETRY POINT))
GEOMETRY POINT REGION
EXISTENCE_TIME

DEFINE ENTITY SET river
ATTRIBUTES (cadastral number)
```

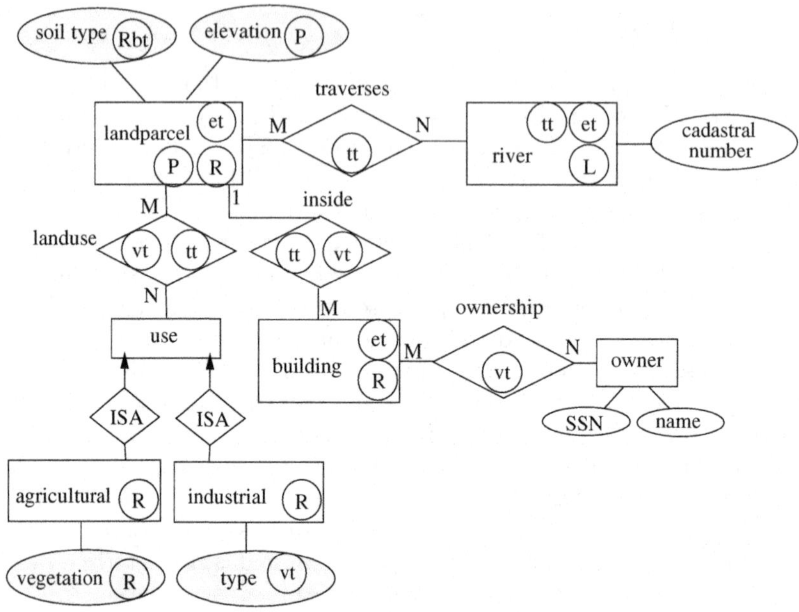

Fig. 3.18. Using STER to model an excerpt of a cadastral application

```
GEOMETRY LINE
EXISTENCE_TIME TRANSACTION_TIME

DEFINE RELATIONSHIP SET traverses (landparcel, river)
TYPE MANY_TO_MANY
TRANSACTION_TIME

DEFINE ENTITY SET agricultural
ATTRIBUTES(vegetation GEOMETRY REGION)
GEOMETRY REGION

DEFINE ENTITY SET industrial
ATTRIBUTES(type  VALID_TIME)
GEOMETRY REGION

DEFINE ENTITY SET use
AS ISA OF agricultural, industrial

DEFINE RELATIONSHIP SET landuse (landparcel, use)
TYPE MANY_TO_MANY
VALID_TIME TRANSACTION_TIME

DEFINE ENTITY SET building
```

```
GEOMETRY POINT REGION
EXISTENCE_TIME

DEFINE RELATIONSHIP SET inside (landparcel, building)
TYPE ONE_TO_MANY
VALID_TIME TRANSACTION_TIME

DEFINE ENTITY SET owner
ATTRIBUTES(SSN, name)

DEFINE RELATIONSHIP SET ownership (building, owner)
TYPE ONE_TO_MANY
VALID_TIME
```

3.4 Spatio-temporal Unified Modeling Language

In this section, an extension of UML, Extended spatio-temporal UML [26], is presented to meet spatio-temporal requirements by using a clear, simple and consistent notation to capture alternative semantics for time, space and change processes (i.e., discrete vs. continuous). The focus is on providing support for different categories of spatio-temporal data. These include temporal changes in spatial extents, changes in the value of thematic (i.e., alphanumeric) data across time or space, and composite data whose components vary depending on time or location.

3.4.1 Using UML Core Constructs for Spatio-temporal Data

In order to motivate the need for a spatio-temporal extension to UML, we first evaluate the core constructs of UML [4,17,31] in terms of their suitability for modeling spatio-temporal data. UML consists of a set of nine types of diagrams and notational conventions based on the OO paradigm. Diagrams can be categorized as specifying either structure or behavior of a system (business or software system) or its data elements (classes or object instances). The Class diagram, exemplified in Figure 3.19, is the most relevant in the current context since it describes the structure of object classes.

The fundamental element of the Class diagram is a class description, consisting of the class name, attribute descriptions, and operation signatures graphically represented in separate compartments of the class box. Attribute descriptions include the name and type (regarded as equivalent to domain for the purposes of this chapter) for each attribute. The classes can be connected by different types of standard OO links including generalization (sub-classes based on a super-class) and association (structural relationships between different object classes). These are represented graphically by a line between classes, terminated with a triangle next to the super-class in the case of generalization. Associations that have their own properties are represented in UML by promoting the association to a class, i.e., an association class, with attributes. UML also defines the specialized associations aggregation (with whole-part semantics) and composition (aggregation with the additional constraint that parts cannot be shared); however, their semantics are not standardized [16,18]. These relationships are represented by an association line terminating in a diamond next to the "whole" class, which is shaded in the case of composition. Any element of a class diagram can also be annotated by ad hoc notes (in a rectangle with a "folded" corner) or constraints (in curly braces) giving further details of that element's semantics.

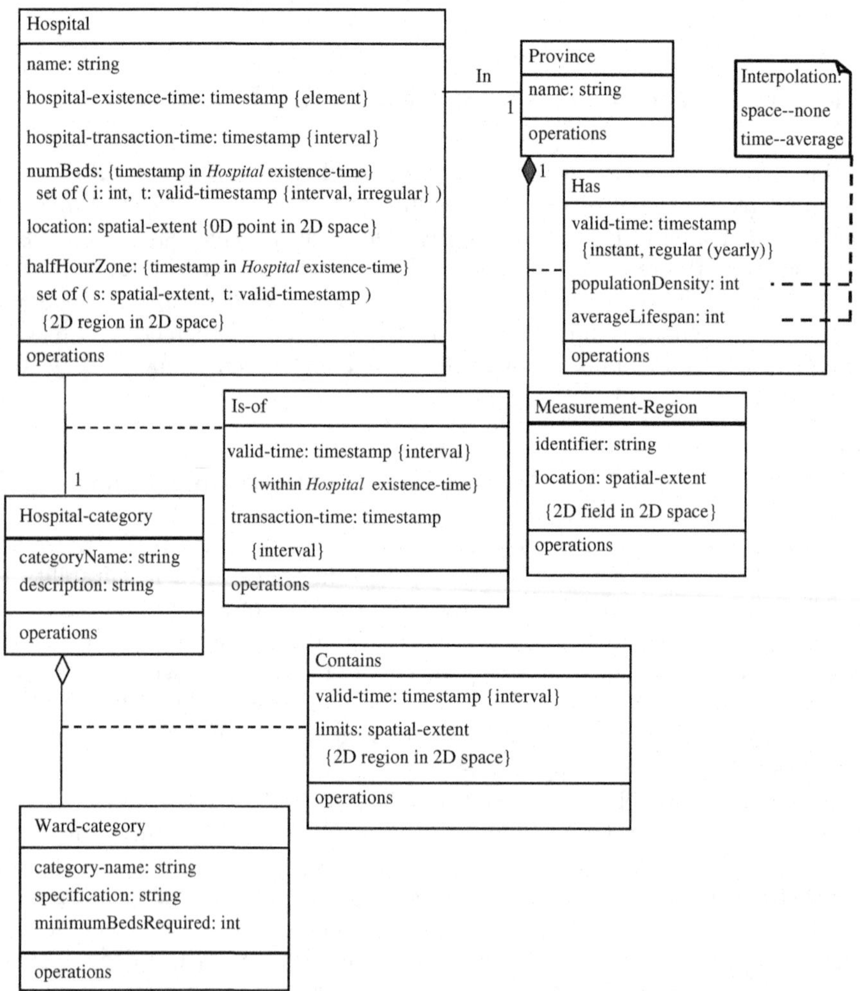

Fig. 3.19. Regional health application in UML

A regional health application will be used to illustrate the use of the standard UML class diagram for spatio-temporal semantics, as follows. Assume an application measuring health statistics of different provinces, in terms of average lifespan, as related to the location (i.e., a point in 2D space), number of beds, accessibility (i.e., a half-hour travel zone around the hospital), and surrounding population densities of a province's hospitals. A hospital is classified by category, where a given category is required to have a minimum number of beds in specific kinds of wards. However, category definitions may differ between regions due to local regulations.

For properties dependent on time, location, or both, we want to record information about when (using time intervals unless otherwise specified) or where a given value is valid in the miniworld (i.e., valid time) or is current in the database (i.e., transaction time). For example, a province's population densities and average lifespans can vary and are recorded yearly at the same time instants (values are averaged between yearly measurements) and for the same regions. The number of beds, the half-hour travel zone, a hospital's category, and the regional definition of hospital categories may change over time as well. We want to record existence and transaction time for hospitals, valid time and transaction time for a hospital's category, and valid time for all of the other time dependent properties. The time unit for the half-hour travel zone is not yet specified, demonstrating incremental design specification. Time elements (the union of a set of time intervals) are used to model hospital existence time since hospitals may sometimes be closed and later re-opened based on changes in local population density. Note that the number of beds, half-hour travel zone, and hospital category are only defined when the hospital is open.

This example demonstrates three categories of spatio-temporal data. For example, changes to the hospital's half-hour travel zone illustrate temporal changes in spatial extents, changes in population density demonstrate changes in the value of thematic data across time and space, and changing definitions of hospital categories show composite data whose components vary depending on time or location.

Representation of spatio-temporal concepts using the core constructs of UML is not straightforward. This is illustrated using the regional health example in Figure 3.19, which assumes that spatial extents can be a combination of points, lines, regions, and volumes, and that timestamps can be instants, intervals, and elements.

Attributes with spatial, temporal, or spatio-temporal properties (e.g., the half-hour travel zone) can be modeled (e.g., with the halfHourZone attribute) using composite attribute domains consisting of a set of tuples, where each tuple consists of a thematic value, possibly a spatial extent, and possibly one or more timestamps. Alternatively, an attribute with spatial and temporal properties (e.g., population density or average lifespan) could be promoted to a separate but associated class with the same information added to the new class. Although not required by the semantics of the example application, we must also create an artificial identifier attribute for this class because its instances must be uniquely identified [31]. Of more concern, this approach will lead to redundancy whenever the same attribute value is repeated for different object instances, times, or spatial extents. This is especially significant for spatial data because of their size. The extra classes also complicate the schema.

A more correct approach, in general, would be to promote the association to an association class (e.g., Has) with spatial data in the associated class (e.g., Measurement-Region) and thematic data, timestamp data (e.g., populationDensity, averageLifespan, and valid-time), or both in the association class. Classes and associations with temporal, spatial, or spatio-temporal properties (e.g., Hospital and hospital Is-of category respectively) can be treated similarly, by adding the timestamp and spatial attributes after promoting the association to an association class in the latter case. However, this still does not solve the problem of the artificial identifier or the extra complexity introduced.

Constraints are used to indicate the time units for timestamps, the time model, the dimensions of spatial extents, and the existence-dependencies described for the application example. Notes are used to show interpolation semantics. Association, rather than generalization, is used to represent the hospital category, since its definition varies regionally and does not affect the properties defined for the hospital class.

Figure 3.19 indicates that it is necessary to create a new association class for each association with spatial or temporal properties. As it can be seen, this leads to the creation of a host of artificial constructs that significantly complicate the schema diagram. Furthermore, there is no single, easily visible notation to represent spatio-temporal properties. This violates the requirement that the notation be simple, clear, and consistent. A better approach is to extend the fundamental characteristics of the existing UML elements to directly support spatio-temporal requirements ([26] compares different approaches to extending a data model). This would involve changes to the structure of instantiated UML elements (i.e., object, association, and attribute instances) to provide for associated time periods or spatial extents. Stereotypes (indicating a variation in usage or meaning), tagged values (adding properties), and constraints (adding semantics) are extension mechanisms defined within UML to extend any core element in the UML metamodel. However, none of the mechanisms support the structural changes in element instances necessary to model spatio-temporal semantics. Other limitations, inconsistencies, and ambiguities in the definitions of these extension mechanisms ([26,18] have a detailed discussion of the problems) further reinforce the argument that spatio-temporal extension constructs must necessarily go beyond the extension mechanisms defined for UML. Although these mechanisms may be used as a guide, a strict adherence to their UML definitions is not desirable. We then proceed to describe the proposed extension, based on a small set of orthogonal constructs and consistent rules for combining them.

3.4.2 Overview of Extended Spatio-temporal UML

The proposed extension to UML is based on the addition of five new symbols, illustrated in Figure 3.20, and a specification box describing the detailed semantics of the spatio-temporal data represented using the five symbols. The basic approach is to extend UML by adding a minimal set of constructs for spatial, temporal, and thematic data, represented respectively by spatial, temporal, and thematic symbols. These constructs can then be applied at different levels of the UML class diagram and in different combinations to add spatio-temporal semantics to a UML model element. In addition, the group symbol is used to group attributes with common spatio-temporal properties or inter-attribute constraints and the existence-dependent symbol is used to describe attributes and associations dependent on object existence.

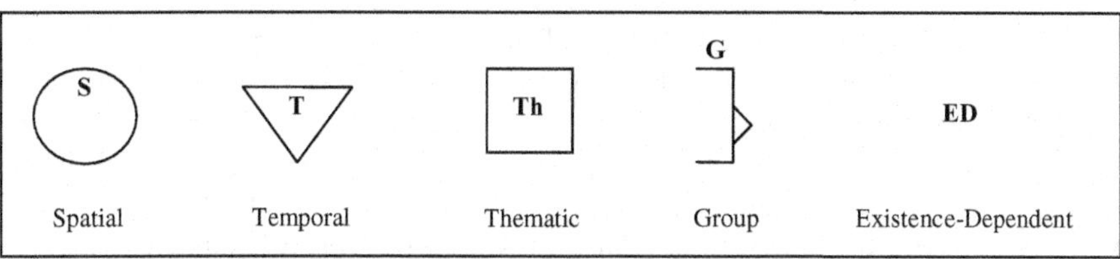

Fig. 3.20. Extended spatio-temporal UML symbols

As discussed previously, although these new symbols can be roughly described as stereotypes they do not adhere strictly to the UML definition. For improved readability, we use the alternative graphical notation for stereotypes described in [31]. These symbols can be annotated with a unique label used to reference the associated specification box. The first four symbols can optionally be used without the abbreviations shown in the figure (i.e., S, T, Th, and G respectively). The specific alphanumeric domain can be optionally indicated, e.g., Th: int, which denotes a thematic attribute of type integer. We first discuss the basic spatial, temporal, and thematic constructs.

3.4.3 Basic Constructs: Spatial, Temporal, Thematic

These constructs can be used to model spatial extents, object existence or transaction time, and the three different types of spatio-temporal data previously discussed (i.e., temporal changes in spatial extents; changes in the values of thematic data across time or space; and composite data whose components vary depending on time or location). To understand the use and semantics of the spatial, temporal, and thematic constructs, we first describe the notation used in this section, then discuss the interpretation of each individual symbol separately, and finally give rules for combining symbols.

The primitives used in this section to denote various time, space, and model elements are listed below. The subset of these that have graphic equivalents are illustrated in Figure 3.21.

- $\langle \mathrm{T} \rangle$ ::= domain of time instants
- $\langle 2^{\mathrm{T}} \rangle$::= arbitrary set of time instants, i.e., a timestamp or set of timestamps
- $\langle \mathrm{S} \rangle$ ::= domain of points in space
- $\langle 2^{\mathrm{S}} \rangle$ ::= arbitrary set of points in space, i.e., a spatial extent or set of spatial extents
- $\langle \mathrm{oid} \rangle$ ::= domain of object-identifiers
- $\langle \mathrm{aid} \rangle$ ::= domain of association-instance identifiers, essentially $\{\langle \mathrm{oid} \rangle\}^n$
- $\langle \mathrm{id} \rangle$ ::= domain of object and association identifiers, essentially $\{\langle \mathrm{oid} \rangle | \langle \mathrm{aid} \rangle\}$
- $\langle \mathrm{D} \rangle$ ::= thematic, i.e., alphanumeric, domain (e.g., integer, string)
- $\langle \mathrm{t} \rangle$ ::= temporal symbol, graphically illustrated by a triangle, see Figure 3.21
- $\langle \mathrm{s} \rangle$ ::= spatial symbol, graphically illustrated by a circle, see Figure 3.21
- $\langle \mathrm{s}\langle \mathrm{t} \rangle\rangle$ ::= temporal symbol inside a spatial symbol, see Figure 3.21
- $\langle \mathrm{d}\langle \mathrm{s} \rangle\rangle$ ::= spatial symbol inside a thematic symbol (illustrated by a rectangle), see Figure 3.21
- $\langle \mathrm{d}\langle \mathrm{t} \rangle\rangle$ ::= temporal symbol inside a thematic symbol, see Figure 3.21
- $\langle \mathrm{d}\langle \mathrm{s}\langle \mathrm{t} \rangle\rangle\rangle$ ::= temporal symbol inside a spatial symbol; both inside a thematic symbol, see Figure 3.21
- $\langle \mathrm{s\&t} \rangle$ ::= any nested combination of a spatial and a temporal symbol
- $\langle \mathrm{s\&d} \rangle$ ::= any nested combination of a spatial and a thematic symbol
- $\langle \mathrm{t\&d} \rangle$ ::= any nested combination of a temporal and a thematic symbol
- $\langle \mathrm{s\&t\&d} \rangle$ ::= any nested combination of a spatial, a temporal, and a thematic symbol
- $\langle \mathrm{ED} \rangle$ ::= existence-dependent symbol

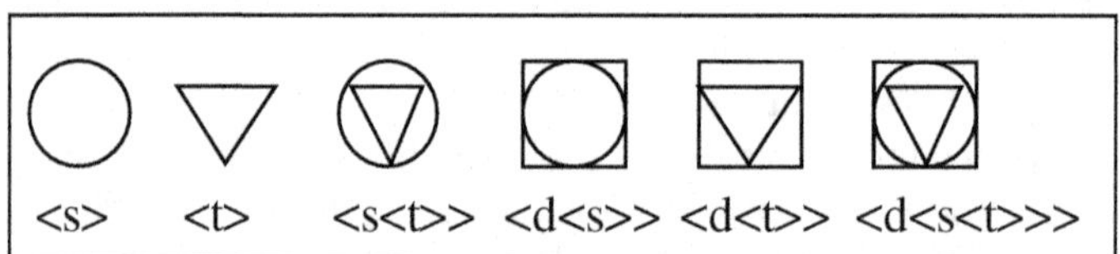

Fig. 3.21. Primitives with graphical equivalents

The spatial symbol represents a spatial extent, which consists of an arbitrary set of points, lines, regions, or volumes. The spatial extent may be associated with thematic or composite data, or may be used to define an attribute domain. The temporal symbol represents a temporal extent, or timestamp, which may be associated with thematic, spatial, or composite data. Timestamps may represent existence time for objects, valid time for associations and attributes, and transaction time for objects, associations, and attributes. The thematic symbol represents thematic data.

The thematic symbol can only be used at the attribute level and only in conjunction with one of the other two symbols to describe an attribute with temporal or spatial properties. A thematic attribute domain with no spatial or temporal properties uses standard UML notation, i.e., $\langle$attribute-name$\rangle$: $\langle$domain$\rangle$. When there are such properties, either this notation can be used for the thematic domain or the specific thematic domain can be indicated inside the thematic symbol. Figure 3.22 illustrates the four possible cases for a thematic attribute: attributes with a thematic domain and (a) no spatial or temporal properties, (b) temporal properties, (c) spatial properties, or (d) spatio-temporal properties. Adjectives are used to describe the attribute domain (e.g., thematic attribute) and adverbs with the word dependent to describe additional attribute properties for composite attribute domains (e.g., temporally-dependent thematic attribute). Therefore, the four possible cases for thematic attributes are called (a) thematic, (b) temporally-dependent thematic, (c) spatially-dependent thematic, or (d) spatio-temporally-dependent thematic attributes respectively.

The semantics of Extended spatio-temporal UML-depend on three factors: (a) the symbol used, (b) the model element described by the symbol (i.e., object, association, or attribute), and (c) whether the symbol is combined with other symbols. The general rules for combining symbols can be summarized as follows:

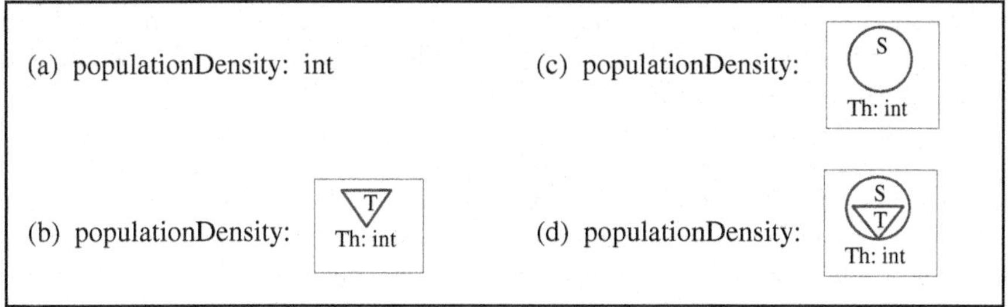

Fig. 3.22. Thematic attribute examples

- Nesting one symbol inside another represents mathematically a function from the domain represented by the inner symbol to the domain represented by the outer symbol. Therefore, different orders of nesting symbols correspond to different functional expressions and represent different perspectives of the data.
  For example, Figure 3.22(b) represents a function from the time to the integer domain for a given object or association instance as follows:

  $$f : \langle \text{id} \rangle \rightarrow (\langle \text{T} \rangle \rightarrow \langle \text{D} \rangle)$$

  If we reverse symbol nesting, this would represent the inverse function, from the integer domain to the powerset of the time domain (this follows directly from the standard definition of a function):

  $$f : \langle \text{id} \rangle \rightarrow (\langle \text{D} \rangle \rightarrow \langle 2^{\text{T}} \rangle)$$

  However, from the conceptual design perspective, both represent the same semantic modeling category and would result in the same conceptual and logical schema, i.e., a temporally-dependent, thematic attribute. Similarly, although both represent a spatio-temporally-dependent thematic attribute, Figure 3.22(d) and its inverse (rectangle inside a circle, both inside a triangle) correspond to different mathematical functions, respectively:

$$f : \langle \text{id} \rangle \to (\langle \text{T} \rangle \to (\langle \text{S} \rangle \to \langle \text{D} \rangle))$$
$$f : \langle \text{id} \rangle \to (\langle \text{D} \rangle \to (\langle \text{S} \rangle \to \langle 2^{\text{T}} \rangle)).$$

Allowing different orders of nesting allows users to select the one that best matches their perspective of the application data and could potentially be exploited for a graphical query language or to indicate preferred clustering patterns to the database management system when generating the physical schema.

Note also that in Figure 3.22(b), only one integer value is associated with each timestamp; however, several different timestamps may be associated with the same integer value. In Figure 3.22(d), several integer values will be associated with each timestamp, one for each spatial location.

- Placing one symbol next to another symbol represents mathematically two separate functions, one for each symbol. The order in which the two symbols are written is not significant.

We now give the rule for which symbolic combinations are legal at each model level, the semantic modeling constructs defined at each level, and a mapping between the two. For a given semantic modeling construct, a symbolic representation, mathematical definition in terms of function(s), and textual definition is given for the nesting used in this chapter's figures. However, as discussed previously, any other nesting order for the same semantic modeling construct is allowed and is described in detail in [26]. Note that any reference to a timestamp, timestamps, a time point, or time validity in the definitions for a given symbol nesting could be for any time dimension, i.e., one or both of transaction and valid time (for attributes and associations) or existence time (for objects).

The Attribute (and Attribute Group) Level. At the attribute level, we can model temporal changes in spatial extents, where the spatial extent represents a property of an object (i.e., spatial attribute), and changes in the value of thematic data across time, space, or both (i.e., temporally, spatially, or spatio-temporally-dependent thematic attributes).

Legal combinations of symbols at the attribute level are include any nested combination of spatial temporal and thematic symbols, except that neither the thematic nor the temporal symbol may be used alone. A thematic attribute uses standard UML notation as discussed earlier. The only exception is that the temporal symbol cannot be used alone. An attribute with a temporal domain is treated as thematic data since temporal data types are pre-defined for popular standard query languages such as SQL. The attribute domain can optionally be followed by an existence-dependent symbol. The rule for notation at this level can be defined using BNF notation and the primitives defined previously:

attribName: [$\langle$D$\rangle$ | $\langle$s&d$\rangle$ | $\langle$t&d$\rangle$ | $\langle$s&t&d$\rangle$ | $\langle$s$\rangle$ | $\langle$s&t$\rangle$] [$\langle$ED$\rangle$]

Six different attribute domains are possible, corresponding to the semantic categories of attributes (i.e., modeling constructs). Reading the domain symbols left to right, we have: thematic attributes; spatially, temporally, and spatio-temporally-dependent thematic attributes; spatial attributes; and temporally-dependent spatial attributes. Except for thematic attributes, these domains represent extensions for spatio-temporal data modeling. A general textual description and symbolic representation (with its corresponding mathematical and textual definition) is given below for each semantic attribute category (or for each semantic object or association category respectively in the following two sections). Note that each one of the definitions below applies to the identified object or association instance: therefore, we do not state this explicitly in the definitions.

- Thematic Attribute: This is an attribute with thematic values.

  $\langle \mathrm{D} \rangle \qquad f : \langle \mathrm{id} \rangle \rightarrow \langle \mathrm{D} \rangle$

  Returns the thematic attribute value.
- Spatially-dependent Thematic Attribute: This is a set of thematic attribute values, each associated with a spatial extent representing the location where that attribute value is valid. This implies that the attribute values may change over space and their changed values may be retained.

  $\langle \mathrm{d}\langle \mathrm{s} \rangle \rangle \qquad f : \langle \mathrm{id} \rangle \rightarrow (\langle \mathrm{S} \rangle \rightarrow \langle \mathrm{D} \rangle)$

  Returns a set of spatial points, each with its associated thematic attribute value (valid for that spatial point).
- Temporally-dependent Thematic Attribute: This is a set of thematic attribute values, each associated with one or more timestamps, representing the valid time, the transaction time, or both times of the attribute value. This implies that the attribute values may change over time and their changed values may be retained.

  $\langle \mathrm{d}\langle \mathrm{t} \rangle \rangle \qquad f : \langle \mathrm{id} \rangle \rightarrow (\langle \mathrm{T} \rangle \rightarrow \langle \mathrm{D} \rangle)$

  Returns a set of time points, each with its associated thematic attribute value (i.e., valid for that time point).
- Spatio-temporally-dependent Thematic Attribute: This is a combination of spatially and temporally-dependent thematic attributes as defined above, i.e., a set of thematic attribute values, each associated with a spatial extent and one or more timestamps.

  $\langle \mathrm{d}\langle \mathrm{s}\langle \mathrm{t} \rangle \rangle \rangle \qquad f : \langle \mathrm{id} \rangle \rightarrow (\langle \mathrm{T} \rangle \rightarrow (\langle \mathrm{S} \rangle \rightarrow \langle \mathrm{D} \rangle))$

  Returns a set of time points, each with its associated set of spatial points, and, for each spatial point, its associated thematic attribute value (i.e., valid for that time and spatial point).

- Spatial Attribute: This is an attribute with a spatial domain, i.e., the attribute value is a spatial extent.

  $\langle \mathrm{s} \rangle \qquad\qquad f : \langle \mathrm{id} \rangle \rightarrow \langle 2^{\mathrm{S}} \rangle$

  Returns the spatial attribute value.
- Temporally-dependent Spatial Attribute: A spatial attribute is associated with one or more timestamps, representing the valid time, transaction time, or both times of the spatial extent.

  $\langle \mathrm{s} \langle \mathrm{t} \rangle \rangle \qquad\qquad f : \langle \mathrm{id} \rangle \rightarrow (\langle \mathrm{T} \rangle \rightarrow \langle 2^{\mathrm{S}} \rangle)$

  Returns a set of time points, each with its associated spatial attribute value (i.e., spatial extent).

The use of these symbols at the attribute level is illustrated in Figure 3.23. The difference between (a) thematic attributes, (b) temporally-dependent thematic attributes, (c) spatio-temporally-dependent thematic attributes, (d) spatial attributes, and (e) temporally-dependent spatial attributes is illustrated by (a) name (for Hospital and Province), (b) numBeds, (c) populationDensity, (d) location, and (e) halfHourZone respectively.

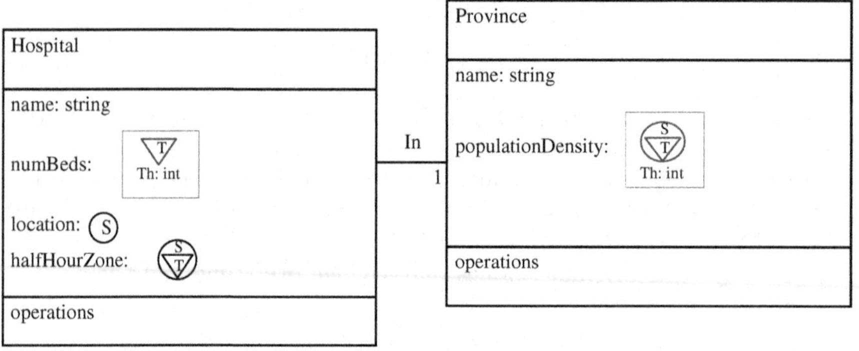

Fig. 3.23. Using extended spatio-temporal UML at the attribute level

A thematic attribute domain is indicated as a string after the attribute or-if that attribute also has temporal or spatial properties-by the use of a thematic symbol. If no domain is explicitly specified for an attribute, then the use of the spatial symbol indicates that the attribute has a spatial domain. Thus, the Hospital location and halfHourZone attributes represent spatial data. The nested temporal symbol used for halfHourZone indicates that the spatial extent associated with this attribute may change over time and thus should be timestamped. Therefore, an attribute marked by a spatio-temporal symbol (and no thematic domain) represents a spatial extent that changes over time. In this case, as transport networks change, the geometry of the half-hour travel zone must be updated.

In contrast, an attribute that has a thematic domain and a spatial and/or temporal symbol represents a spatially- and/or temporally-dependent thematic attribute. This is indicated graphically by using the thematic symbol; thus this symbol is used to differentiate two different types of spatio-temporal data: temporal changes in spatial extents and changes in the value of thematic data across time and space. Therefore, the fact that numBeds has an integer domain associated with a temporal symbol indicates that the integer value of numBeds may change over time and should be timestamped. Analogously, the integer value of populationDensity may change over time or space and thus each value is associated with a timestamp and spatial extent.

The Object Class Level. At the object class level, we can model temporal changes in spatial extents, where the spatial extent is associated with an object instance. We can also model the time an object exists in the real world (i.e., existence time) or is part of the current database state (i.e., transaction time).

An object class can be marked by a temporal symbol, a spatial symbol, or any nested combination of these. In addition, this is the only level where the symbols can be paired; i.e., a temporal symbol can be paired with either a spatial symbol or a nested combination of the two symbols. The separate temporal symbol represents the existence or transaction time of the object. The spatial symbol represents the spatial extent associated with that object. If the spatial symbol is combined with a nested temporal symbol, then the spatial extent is timestamped to show the valid or transaction time of the spatial extent. Since the object can exist or be current even when not actually associated with a spatial extent, separate timestamps are required for the object instance and for the object instance's spatial extent. The rule for object level notation can be given in BNF as follows:

className [$\langle$s$\rangle$ | $\langle$s&t$\rangle$] [$\langle$t$\rangle$]

Corresponding to the five possible instantiations of this rule, $\langle$s$\rangle$; $\langle$s&t$\rangle$; $\langle$t$\rangle$; $\langle$s$\rangle\langle$t$\rangle$; and $\langle$s&t$\rangle\langle$t$\rangle$, there are five different categories of object classes as defined below.

- Spatial Object (Class): An object is associated with a spatial extent. This is equivalent to an object having a single spatial attribute except that there is no separate identifier for the spatial extent.

  $\langle$s$\rangle$ $\qquad f : \langle \text{oid} \rangle \rightarrow \langle 2^{\text{S}} \rangle$

  Returns the spatial extent of the identified object.
- Temporally-dependent Spatial Object (Class): The spatial extent associated with a spatial object is also associated with one or more timestamps, representing the spatial extent's valid time or transaction time (or both).

$\langle s\langle t\rangle\rangle$ $\quad f : \langle oid\rangle \rightarrow (\langle T\rangle \rightarrow \langle 2^S\rangle)$

Returns a set of timepoints, each associated with the spatial extent of the identified object at that timepoint.

- Temporal Object (Class): An object is associated with one or more timestamps, representing the object's existence time or transaction time (or both).

$\langle t\rangle$ $\quad f : \langle oid\rangle \rightarrow \langle 2^T\rangle$

Returns the timestamp of the identified object.

- Spatio-temporal Object (Class): This is a combination of a spatial and temporal object as defined above, i.e., each object instance is associated with a spatial extent and one or more timestamps representing the object's existence time or transaction time (or both).

$\langle t\rangle \langle s\rangle$ $\quad f : \langle oid\rangle \rightarrow \langle 2^T\rangle$ and $f : \langle oid\rangle \rightarrow \langle 2^S\rangle$

Returns the timestamp and the spatial extent of the identified object.

- Temporally-dependent Spatio-temporal Object (Class): This is a combination of a temporally-dependent spatial object and a temporal object as defined above, i.e., an object is associated with a spatial extent, one or more timestamps representing the spatial extent's valid time or transaction time (or both), and one or more timestamps representing the object's existence time or transaction time (or both).

$\langle t\rangle \langle s\langle t\rangle\rangle$ $\quad f : \langle oid\rangle \rightarrow \langle 2^T\rangle$ and $f : \langle oid\rangle \rightarrow (\langle T\rangle \rightarrow \langle 2^S\rangle)$

Returns the timestamp of the identified object and a set of timepoints, each with its associated spatial extent (i.e., valid at that timepoint), for the identified object.

The use of symbols at the object class level is illustrated in Figure 3.24. In Figure 3.24(a), the temporal symbol at the Hospital object level represents a temporal object class with existence and transaction time (discussed in Section 3.4.4). In Figure 3.24(b), we give an example of a temporally-dependent spatial object. This example assumes that there is no need to represent hospital location separately from the half-hour travel zone. Instead, a hospital object is treated as a spatial object with a single associated spatial extent, showing the half-hour travel zone around that hospital. The temporal symbol indicates that the spatial extent should be timestamped, since the half-hour travel zone can change over time. Finally, Figure 3.24(c) combines (a) and (b), illustrating a temporally-dependent spatio-temporal object. The object is spatio-temporal because it has a timestamp and a spatial extent; and temporally-dependent because the spatial extent also has a timestamp.

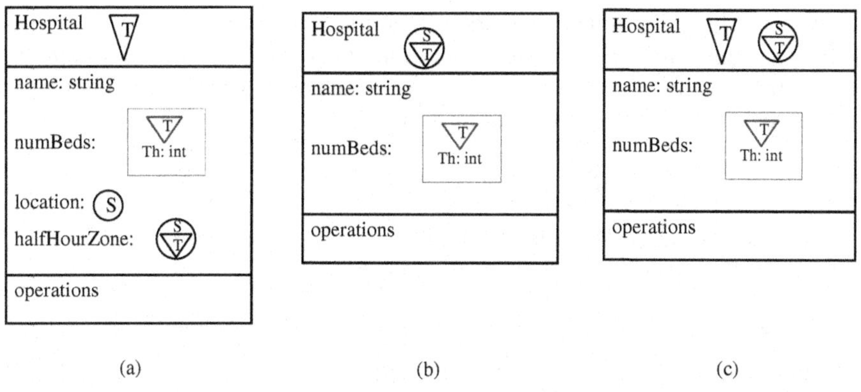

Fig. 3.24. Using extended spatio-temporal UML at the object class level

The Association Level. At the association level, we can model temporal changes in spatial extents, where the spatial extent is associated with a relationship between object instances, and composite data whose components vary depending on time or location. The following discussion applies to any type of association, including aggregation and composition.

At the association level, any nested combination of a spatial and a temporal symbol represents a legal combination describing spatio-temporal properties of the association. Except for the omission of the thematic symbol, the association level is similar to the attribute level. The association spatio-temporal properties can optionally be followed by an existence-dependent symbol (discussed in Section 3.4.4). The rule for the association level notation can be given in BNF as follows:

assoc-line [⟨s⟩ | ⟨t⟩ | ⟨s&t⟩] [⟨ED⟩]

Reading the BNF rule from left to right, three different categories of associations are possible, as defined below.

- Spatially-dependent association: An association instance is associated with a spatial extent representing the location where the association instance is valid. This implies that the association instances may change over space and their changed instances may be retained.

  $\langle \mathrm{s} \rangle \qquad f : \langle \mathrm{aid} \rangle \rightarrow \langle 2^{\mathrm{S}} \rangle$

  Returns the spatial extent of the identified association.
- Temporally-dependent association: An association instance is associated with one or more timestamps, representing the association's valid time or transaction time (or both). This implies that association instances may change over time and the changed instances may be retained.

  $\langle \mathrm{t} \rangle \qquad f : \langle \mathrm{aid} \rangle \rightarrow \langle 2^{\mathrm{T}} \rangle$

Returns the timestamp of the identified association.

- Spatio-temporally-dependent association: This is a combination of spatially and temporally-dependent associations as defined above, i.e., an association is associated with a spatial extent and one or more timestamps.

$$\langle \mathrm{s}\langle \mathrm{t}\rangle\rangle \qquad f : \langle \mathrm{aid}\rangle \rightarrow (\langle \mathrm{T}\rangle \rightarrow \langle 2^{\mathrm{S}}\rangle)$$

Returns a set of time points, each with the associated spatial extent for the identified association at that time point.

The use of these symbols at the association level is shown in Figure 3.25. Marking the Is-of association with a temporal symbol signifies that the category of a hospital may change over time, as local health needs change and wards are opened or closed. Therefore, association instances should be timestamped.

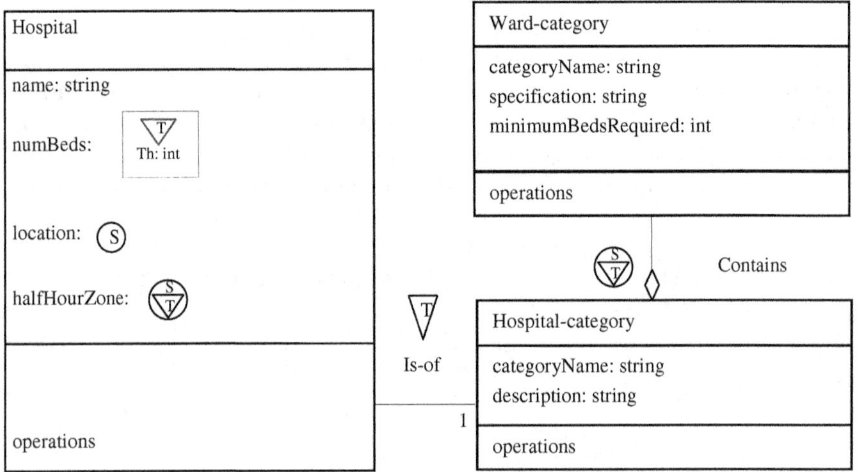

Fig. 3.25. Using extended spatio-temporal UML at the association level

A spatially-dependent association is one where an association instance is associated with a spatial extent to show where that instance is valid. For example, the same category of hospital may require different categories of wards in different areas depending on local regulations. Therefore, the Contains aggregation association must be spatially-dependent. In fact, since the local categories may also change over time, the Contains aggregation association is actually spatio-temporally-dependent. In this case, both of the associated object classes are purely conceptual. An association between two physical object classes can also be spatio-temporally-dependent; e.g., a consultation of a ward doctor with a specialist is scheduled for a specific location and period of time in the hospital.

Since constraints are often application-dependent, no constraints are assumed between the timestamps or spatial extents of participating objects with those of a temporally, spatially or spatio-temporally-dependent association. They can be specified explicitly for individual applications either (1) on an ad hoc basis as required using UML constraints or (2) by defining explicit modeling constructs for commonly used constraint patterns. The latter approach is illustrated by the introduction of the existence-dependent symbol, described in the next section, to support the semantics of temporal-dependency in composite model elements. Explicit modeling constructs for common spatial constraints have been defined in [27].

3.4.4 Additional Constructs: Specification Box, Existence Time, and Groups

The previous section described the different types of timestamps that can be associated with an attribute, association, or object class: but where do we specify which types are required for a given application? Detailed spatio-temporal semantics are specified in a specification box, which can be associated with any of the icons or combinations using a unique naming convention (used here) or label. The specification box includes information on the time units and the time and space dimensions, models, and interpolation. Users can specify regular (recorded at regular intervals) or irregular time models and object- or field-based space models. Interpolation functions can be specified to derive values in-between those explicitly recorded for spatially, temporally, and spatio-temporally-dependent thematic attributes. The time dimensions and units (i.e., instant, interval, element) used are defined in [19]. Specification boxes can be inherited from parent classes as with any other class property. The specification box syntax is illustrated in Figure 3.26.

```
SPECIFICATION BOX <Identifier>:

TimeDimen. := [existence |valid] [transaction]

TimeInterpolation := discrete |step |min |max |avg |linear |spline |<user-defined>

TimeModel := irregular | (regular{<frequency> [,<beginning>,<end>]} )

TimeUnit [(<TimeDimen.>)] := instant |interval |element

SpaceInterpolation := <same as TimeInterpolation>

SpaceModel := '('<max object/field dim>,<max search space dim> ')':object |field

Group := independent | (dependent (formula)* )]
```

Fig. 3.26. Specification box syntax in extended spatio-temporal UML

Time dimensions include existence time (for objects), valid time (for attributes and associations), and transaction time (for objects, attributes, or associations), as defined in [19]. However, object existence time is more precisely defined as the time during which existence-dependent attributes and associations can be defined (i.e., have legal values) for that object. In other words, existence-dependent attributes (e.g., person's home phone number) and associations are those that are defined only when the related object(s) (e.g., the person) exist. This implies that attributes and associations that are not existence-dependent (e.g., a person's social-security number) may be defined even when the related object(s) no longer exist. Object identifiers can never be existence-dependent, as they can be used to refer to historical objects. The superscript ED is added to those attributes and associations defined only when the corresponding object (or objects, in the case of an association) exists, which means that existence time must be recorded for the object (or objects). This further implies that any valid timestamps recorded for the existence-dependent attributes and associations must be within the existence time of the corresponding object(s).

The time model applies only to valid time and time interpolation only to the valid time of temporally-dependent thematic attributes. Space dimensions include the dimensions of the spatial extent(s) being specified, followed by the dimensions of the underlying search space. The object-based spatial model is used for a spatial attribute, where an attribute instance consists of a single spatial extent. The field-based spatial model is used for a spatially-dependent thematic attribute; where an attribute instance is a set of thematic values, each associated with a different spatial extent. Space interpolation applies only to spatially-dependent thematic attributes.

The specification box can also be used to specify spatio-temporal constraints, including those within an attribute group. The group symbol is used to group attributes sharing the same timestamps or spatial extents, thus requiring only one specification for the group. Thus, this symbol graphically illustrates associated sets of attributes and ensures that spatial extents and timestamps are not specified redundantly. Note that a group's attributes never share thematic values, even if the thematic symbol is used in the group specification. If the group's attributes have different thematic domains, then these can be indicated next to each attribute using standard UML text notation.

Following UML convention, another compartment is added to the object class to accommodate its specification boxes, i.e., the specification compartment. The specification compartment can be used to specify spatio-temporal semantics for the object, its attributes, and any associations in which the object class participates. Alternatively, a specification compartment can be added to an association class to specify spatio-temporal semantics for that association and its attributes (see [26,23] for further discussion of the specification box).

3.4.5 Example of Usage

Figure 3.27 shows the same full regional health application described previously as it would be represented using the proposed extension and illustrates the use of the specification box, group symbol, and existence-dependent symbol.

For example, Hospital location is specified as a single point in the 2D space. Hospital halfHourZone and Contains are specified as a region in 2D space. In contrast, the Province populationDensity and averageLifespan group is associated with a 2D field in 2D space. This means that, for a single object instance, the two attributes in the group are associated with a set of regions and have a separate attribute value for each region for a given point in time. Since these two attributes share common timestamps and spatial extents, they are grouped. Since both attributes are integers, we can specify the thematic domain in the group symbol. If the attributes had different thematic domains, then we would specify them for each attribute rather than for the group.

The group is then associated with a single symbol and specification box. Here we specify that any attribute in the group uses average interpolation in time and no interpolation in space, has a valid time dimension using instant as the time unit, and is measured yearly (i.e., a new set of values is recorded for the attribute each year). This means that the population density and average lifespan between recorded time instants is assumed to be the average of the values at the two nearest time instants and undefined outside of recorded spatial regions. No inter-attribute constraints are defined for the group, as shown by the keyword independent.

The temporal symbol at the Hospital object level is used to indicate existence and transaction time. Existence time is used to model the periods when the hospital is open, i.e., when the existence-dependent numBeds, halfHourZone and Is-of are defined. Since these model elements are temporally-dependent, the valid timestamps of all their instances must be included within the Hospital existence time. Attribute numBeds is specified as irregular because this attribute is not recorded periodically: whenever it changes the new value is recorded.

The specification box for an association (e.g., Is-of) can be placed in the specification compartment of either of its participating object classes (e.g., Hospital or Hospital-category). Note that since Hospital-category is not temporal and therefore does not have existence time defined, the only constraint on the valid-time timestamps of the Is-of association comes from the Hospital class existence time.

3.5 Related Work

A range of aspects of spatial and temporal databases have been studied in isolation for more than a decade. Only more recently has the combination of spatial and temporal data, i.e., spatio-temporal data, been subject to scrutiny.

Story et al. [28] propose a design support environment for spatio-temporal databases, focusing on the integration of time with application data. Temporal classes, events, and states are emphasized as components of an ideal environment that supports the developer's better understanding of spatio-temporal data and applications. Claramunt and Theriault [7] integrate time in GIS's by presenting a systematic typology of spatio-temporal processes, leading to an event-oriented model. Allen et al. in [1] present a generic model consisting of objects, states, events, and conditions for explicitly representing links within a spatio-temporal GIS. This work focuses on the interaction among the com-

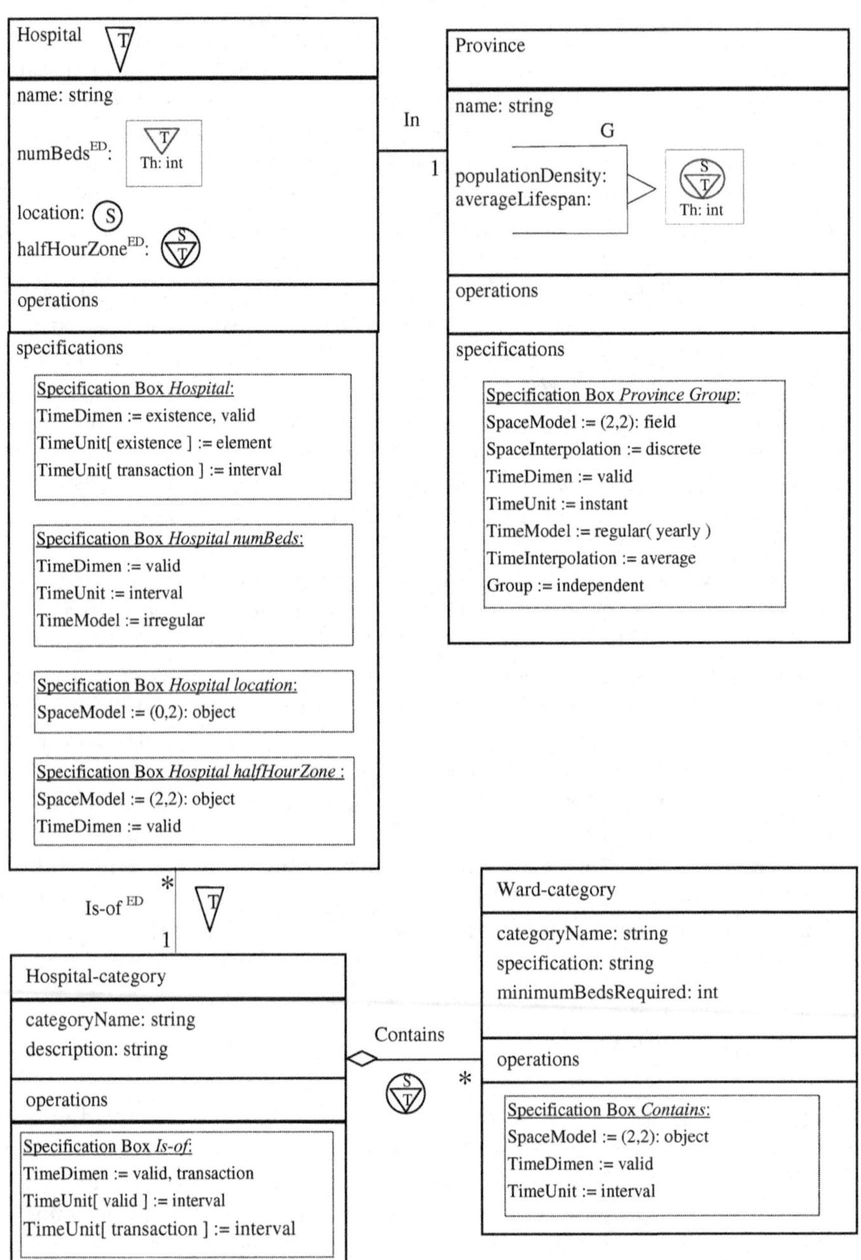

Fig. 3.27. Regional health application in extended spatio-temporal UML

ponents over time rather than on presenting them as the appropriate mechanisms for representing spatio-temporal data. Worboys in [37] proposes a unified model for spatio-temporal data with two spatial and two temporal dimensions (database and event times). This model is not aimed at conceptual design, but is a mathematically-oriented model. Claramunt et al. [25] present a set of design patterns for spatio-temporal processes expressed in an object-relationship data model. The focus is on the analysis of spatio-temporal processes and on the properties of object-oriented and entity-relationship data models. This work focuses on the design of processes rather than on the design of modeling constructs. Moreover, spatial issues such as the representation of time-varying fields (or spatial attributes) are not considered. Langran in [20] provides a broad survey of research on spatio-temporality, the philosophy of time, temporal databases, and spatial data structuring. Finally, mathematical models have been developed based on a systems perspective [29] with the purpose of understanding and, sometimes, predicting a wide variety of natural phenomena. Such models almost always incorporate notions of change over time.

Becker et al. and Faria et al. [5,13] propose OO models based on extensions of ObjectStore and O2 respectively. [5] considers both object- and field-based spatial models, defining a hierarchy of elementary spatial classes with both geometric and parameterized thematic attributes. Temporal properties are incorporated by adding instant and interval timestamp keywords to the query language. In [13], spatial and temporal properties are added to an object class definition by associating it with pre-defined temporal and spatial object classes. This solution is not suitable for representing temporal or spatial variation at the attribute level, as the timestamp and spatial locations are defined only at the object component level. In addition, both offer text-based query languages; the non-graphical query languages of these models reduce their suitability as conceptual modeling languages.

Perceptory [3] is a spatio-temporal model aligned with international geographic standards. It is a CASE tool that stores the conceptual content of object-oriented application schemas and dictionaries. The basic approach of the underlying model is to add spatial and temporal stereotypes to objects in UML. The emphasis is on supporting geographic objects rather than fields or associations between objects.

The MADS model [24] extends an object-based model with pre-defined hierarchies of spatial and temporal abstract data types and special complex data types to describe all of an attribute's properties, i.e., name, cardinality, domain, and temporal or spatial dimensions. The use of a non-standard, hybrid ER/OO model and the definition of new composite data structures for spatio-temporal properties, rather than exploiting existing features of the ER or OO models, increase the complexity of the model syntax. There is no provision for attributes having a spatial domain, therefore any data element associated directly with several different spatial extents must be modeled as an association of spatial objects.

3.6 Conclusions

In this chapter, we view the design of spatio-temporal applications from the perspective of information systems development. STER and Extended spatio-temporal UML focus on the special modeling needs of spatio-temporal applications at the conceptual phase. They both provide a small set of new constructs for spatio-temporal data and thus reduce the complexity of the resulting schemas. This allows the application developer to concentrate on the characteristics of the specific application domain of interest.

Both models facilitate the integration of two disjoint views [36] of spatial information, namely the (a) field-based view, suitable for the representation of properties like "temperature" or "vegetation" and (b) the object-based view, which captures spatial information over time in terms of identified objects, like "landparcel."

STER was applied in real, large-scale applications ([35] for the development of a utility network management system; [22] for the design of the Greek cadastral system), as well as in small prototypical examples with quite encouraging results: diagrams were more easily understood by the users, and the data modeling process proceeded more rapidly.

In STER, topological relationships between spatial objects can be explicitly represented in the schema diagram using spatial relationships. STER also addresses the problem of modeling multiple granularities.

In contrast, Extended STUML supports associations having their own spatial extent separate from those of participating objects, e.g., the location of an accident involving two cars, using spatially-dependent associations. It further provides formal mathematical definitions of the modeling constructs and offers a more detailed representation of the spatio-temporal semantics using the specification box, which provides a standard notation for their description, including interpolation and alternative time models (e.g., periodic versus aperiodic recording of data values). This facilitates effective communication and consistent design documentation. Analogously, the use of an explicit existence dependent construct, rather then UML constraints or notes, to represent application defined temporal dependencies between associations, attributes, and objects further contributes to standardization and consistency.

References

1. E. Allen, G. Edwards, and Y. Bedard. Qualitative casual modeling in temporal gis. In *International Conference on Spatial Information Theory (COSIT)*, volume 988 of *Lecture Notes in Computer Science*. Springer-Verlag, 1995.
2. A. Abiteboul and R. Hull. Ifo: A formal semantic database model. *ACM TODS*, 12(4):525–565, 1987.
3. J. Brodeur, Y. Bedard, and M.J. Proulx. Modeling geospatial application databases using uml-based repositories aligned with international standards in geomatics. In *Eighth International Symposium of ACMGIS*, pages 39–46, Washington DC, November 2000.

4. G. Booch, J. Rumbaugh, and I. Jacobson. *The Unified Modeling Language User Guide*. Addison-Wesley, 1999.
5. L. Becker, A. Voigtmann, and K. Hinrichs. Temporal support for geo-data in object-oriented databases. In *International Conference on Database and Expert Systems (DEXA)*, volume 1134 of *Lecture Notes in Computer Science*. Springer-Verlag, 1996.
6. P.S. Chen. The entity-relationship model: Toward a unified view of data. *ACM TODS*, 1(1):9–36, 1976.
7. C. Claramunt and M. Theriault. Managing time in gis: An event-oriented approach. In *International Workshop on temporal Databases*, Zurich, September 1995.
8. V. Delis, T. Hadzilacos, and N. Tryfona. An introduction to layer algebra. In *6th International Conference on Spatial Data Handling (SDH)*. Taylor and Francis, 1995.
9. B. David, M. van den Herrewegen, and F. Salge. *Geographic Objects with Indeterminate Boundaries*, chapter 13. Taylor and Francis, 1996.
10. M.J. Egenhofer and R. Franzosa. Point-Set Topological Spatial Relationships. *International Journal of GIS*, 5(2):161–174, 1991.
11. M.J. Egenhofer and A.U. Frank. Connection between Local and Regional: Additional Intelligence Needed. In *18th International Congress of FIG*, Toronto, 1986.
12. R. Elmasri and S. Navathe. *Fundamentals of Database Systems*. Benjamin/Cummings, 1994.
13. G. Faria, C.B. Medeiros, and M.A. Nascimento. An extensible framework for spatio-temporal database applications. Technical report, TimeCenter, 1998.
14. A.U. Frank. Qualitative Spatial Reasoning: Cardinal Directions as an Example. *International Journal of GIS*, 10(3):269–290, 1996.
15. A.U. Frank, J. Raper, and J.P. Cheylan. Life and Motion of Socio-Economic Units. ESF Series, Taylor and Francis, 2000.
16. M. Fowler and K. Scott. *UML Distilled*. Addison-Wesley, 1997.
17. Object Management Group. Omg unified modeling language specifications, version 1.3. In `http://www.ftp.omg.org/pub/docs/ad/99-06-08.pdf`, pages 1–808, World Wide Web, Retrieved 19 May, 2000 1999.
18. B. Henderson-Sellers and F. Barbier. Black and white diamonds. In *Conference on the Unified Modeling Language*, pages 550–565, 1999.
19. C.S. Jensen and C.E. Dyreson (with multiple contributors). The consensus glossary of temporal database concepts. In *Temporal Databases: Research and Practice*, pages 367–405, 1998.
20. G. Langran. *Time in Geographic Information Systems*. Taylor and Francis, 1993.
21. D.R. Montello. The geometry of environmental knowledge. In *International Conference on Spatial Information Theory (COSIT'92)*, volume 639 of *Lecture Notes in Computer Science*, 1992.
22. National Technical University of Athens NTUA. Definition of a standard for the exchange of digital cadastral data. Technical report, National Technical University of Athens, 1997.
23. R. Price, B. Srinivasan, and K. Ramamohanarao. Extending the unified modeling language to support spatiotemporal applications. In *Conference on Technology of Object Oriented Languages and Systems, Asia*, pages 163–174, Melbourne, Australia, November 1999.
24. C. Parent, S. Spaccapietra, and E. Zimanyi. Spatio-temporal conceptual models: Data structure+space+time. In *Seventh International Symposium of ACMGIS*, Dallas, Texas, November 1999.
25. C. Claramunt C. Parent and M. Theriault. Design patterns for spatiotemporal processes. In *IFIP*. Chapman and Hall, 1997.

26. R. Price, N. Tryfona, and C.S. Jensen. Extended spatiotemporal UML: Motivations, requirements, and constructs. *Journal of Database Management*, 11(4):14–27, October/December 2000.
27. R. Price, N. Tryfona, and C.S. Jensen. Modeling part-whole relationships for spatial data. In *Eighth International Symposium of ACMGIS*, pages 1–8, Washington DC, November 2000.
28. P.A. Story and M. Worboys. A design support environment for spatio-temporal database applications. In *International Conference on Spatial Information Theory (COSIT)*, volume 988 of *Lecture Notes in Computer Science*. Springer-Verlag, 1995.
29. N. Roberts, D. Andersen, M. Garet, and W. Shaffer. *Introduction to Computer Simulation: A System Dynamics Modeling Approach.* Addison-Wesley, 1983.
30. J. Rumbaugh, M. Blaha, W. Premerlani, F. Eddy, and W. Lorensen. *Object-Oriented Modeling and Design.* Prentice-Hall, 1991.
31. J. Rumbaugh, I. Jacobson, and G. Booch. *The Unified Modeling Language Reference Manual.* Addison-Wesley, 1999.
32. M. Raubal and M. Worboys. A formal model of the process of wayfinding in built environments. In *International Conference on Spatial Information Theory (COSIT'99)*, volume 1626 of *Lecture Notes in Computer Science*, 1999.
33. N. Tryfona and C.S. Jensen. Conceptual data modeling for spatiotemporal applications. *Geoinformatica*, 3(3), 1999.
34. N. Tryfona and C.S. Jensen. Using abstractions for spatio-temporal conceptual modeling. In *14th Annual Symposium on Applied Computing (SAC 2000)*, 2000.
35. UTILNETS. A network unitlity management system. Technical report, European Commission, Brite-EURAM project 7120, 1994.
36. M. Worboys. Object-oriented approaches to geo-referenced information. *International Journal of GIS*, 8(4), 1994.
37. M. Worboys. A unified model for spatial and temporal information. *The Computer Journal*, 37(1), 1994.

4 Spatio-temporal Models and Languages: An Approach Based on Data Types

Ralf Hartmut Güting[1], Michael H. Böhlen[2], Martin Erwig[1], Christian S. Jensen[2], Nikos Lorentzos[3], Enrico Nardelli[4], Markus Schneider[1], and Jose R.R. Viqueira[3]

[1] Fern Universität, Hagen, Germany
[2] Aalborg University, Denmark
[3] Agricultural University of Athens, Greece
[4] Universita Degli Studi di L'Aquila, Italy

4.1 Introduction

In this chapter we develop DBMS data models and query languages to deal with geometries changing over time. In contrast to most of the earlier work on this subject, these models and languages are capable of handling continuously changing geometries, or *moving objects*. We focus on two basic abstractions called *moving point* and *moving region*. A *moving point* can represent an entity for which only the position in space is relevant. A *moving region* captures moving as well as growing or shrinking regions. Examples for moving points are people, polar bears, cars, trains, or air planes; examples for moving regions are hurricanes, forest fires, or oil spills in the sea.

The main line of research presented in this chapter takes a *data type oriented* approach. The idea is to view moving points and moving regions as three-dimensional (2D-space + time) or higher-dimensional entities whose structure and behavior is captured by modeling them as abstract data types. These data types can then be integrated as attribute types into relational, object-oriented, or other DBMS data models; they can be implemented as extension packages ("data blades") for suitable extensible DBMSs. Section 4.2 explains this idea in more detail and discusses some of the basic questions related to it.

Once the basic idea is established, the next task is to design precisely a collection of types and operations that adequately reflects the objects of the real world to be modeled and is capable of expressing all (or at least, many) of the questions one would like to ask about these objects. It turns out that besides the main types of interest, moving point and moving region, a relatively large number of auxiliary data types is needed. For example, one needs a line type to represent the projection of a moving point into the plane, or a "moving real" to represent the time-dependent distance of two moving points. It then becomes crucial to achieve (i) orthogonality in the design of the type system, i.e., type constructors can be applied uniformly, (ii) genericity and consistency of operations, i.e., operations range over as many types as possible and behave consistently, and (iii) closure and consistency between structure and operations of related non-temporal and temporal types. Examples of the last aspect are

T. Sellis et al. (Eds.): Spatio-temporal Databases, LNCS 2520, pp. 117–176, 2003.
© Springer-Verlag Berlin Heidelberg 2003

that the value of a moving region, evaluated at a certain instant of time, should be consistent with the definition of a static (non-temporal) region type, or that the time-dependent distance function between two moving points, evaluated at instant t_0, yields the same distance value as determining for each of the two moving points their positions p_1 and p_2 at instant t_0, and then taking the distance between p_1 and p_2. Section 4.3 presents such a design of types and operations in some detail; it also illustrates the expressivity of the resulting query language by example applications and queries.

Of course, when we design data types and operations, we have to specify their semantics in some way. For each type, one has to define a suitable domain (the set of values allowed for the type), and for operations one needs to define functions mapping the argument domains into the result domain. One of the fundamental questions coming up is at what *level of abstraction* one should define semantics. For example, a moving point can be defined either as a function from time into the 2D plane, or as a polyline in the three-dimensional (2D + time) space. A (static) region can be defined either as a connected subset of the plane with non-empty interior, or as a polygon with polygonal holes. The essential difference is that in the first case we define the domains of the data types just in terms of infinite sets whereas in the second case we describe certain finite representations for the types.

We will discuss this issue in a bit more depth in Section 4.2 and introduce the terms *abstract model* for the first and *discrete model* for the second level abstraction. Both levels have their respective advantages. An abstract model is relatively clean and simple; it allows one to focus on the essential concepts and not get bogged down by implementation details. However, it has no straightforward implementation. A discrete model fixes representations and is generally far more complex. It makes particular choices and thereby restricts the range of values of the abstract model that can be represented. For example, a moving point could be represented not only by a 3D polyline but also by higher order polynomial splines. Both cases (and many more) are included in the abstract model. On the other hand, once such a finite representation has been selected, it can be translated directly to data structures.

Our conclusion is that both levels of modeling are needed and that one should first design an abstract model of spatio-temporal data types and then continue by defining a corresponding discrete model. Section 4.3 describes in fact an abstract model in this sense. The definitions of semantics are given generally in terms of infinite sets.

Section 4.4 then proceeds to develop a corresponding discrete model. Finite representations for all the data types of the abstract model are introduced. Spatial objects and moving spatial objects are described by linear approximations such as polygons or polyhedra. For all the "moving" types, a *sliced representation* is introduced which represents a temporal development as a set of *units* where a unit describes the development as a certain "simple" function of time during a given time interval. In Chapter 6 of this book it is shown how the representations of the discrete model can be mapped into data structures that can be realistically used in a DBMS environment and how example algorithms can work on these data structures efficiently.

Section 4.4 concludes the main line of research presented in this chapter. Section 4.5 entitled "Outlook" presents four other pieces of work carried out within project CHOROCHRONOS. For lack of space, these developments are presented in the form of relatively brief summaries. The first two can be viewed as extensions of the approach described above, dealing with "spatio-temporal developments" and time varying partitions of the plane. The latter two have a different focus of interest and do not deal with moving objects; they develop a spatio-temporal model over a rasterized space, and address the problem of treating legacy databases and applications when a given database is changed to include the time dimension.

4.2 The Data Type Approach

In this section we describe the basic idea of representing moving objects by spatio-temporal data types. After some motivation, the approach to modeling is explained, and some example queries are shown. In the last subsection, we discuss two basic issues related to the approach. This section (4.2) is based on [7].

4.2.1 Motivation

We are interested in *geometries changing over time*, and in particular in geometries that can change continuously, and hence in *moving objects*. In spatial databases, three fundamental abstractions of spatial objects have been identified: A *point* describes an object whose location, but not extent, is relevant, e.g. a city on a large scale map. A *line* (meaning a curve in space, usually represented as a polyline) describes facilities for moving through space or connections in space (roads, rivers, power lines, etc.). A *region* is the abstraction for an object whose extent is relevant (e.g. a forest or a lake). These terms refer to two-dimensional space, but the same abstractions are valid in three or higher-dimensional spaces.

Since lines (curves) are themselves abstractions or projections of movements, it appears that they are not the primary entities whose movements should be considered. From a practical point of view, although line values can change over time, not too many examples for moving lines come into mind. Hence it seems justified to focus first[1] on *moving points* and *moving regions*. Table 4.1 shows a list of entities that can move, and questions one might ask about their movements.

Although we focus on the general case of geometries that may change in a continuous manner (i.e. move), one should note that there is a class of applications where geometries change only in discrete steps. Examples are boundaries of states, or cadastral applications, where e.g. changes of ownership of a piece of land can only happen through specific legal actions. Our proposed way of

[1] Nevertheless, in the systematic design of Section 4.2, time-dependent line values will come into play for reasons of closure.

Table 4.1. Moving objects and related queries

Moving Points	Moving Regions
People • Movements of a terrorist / spy / criminal Animals • Determine trajectories of birds, whales, ... • Which distance do they traverse, at which speed? How often do they stop? • Where are the whales now? • Did their habitats move in the last 20 years? Satellites, spacecraft, planets • Which satellites will get close to the route of this spacecraft within the next 4 hours? Cars • Taxis: Which one is closest to a passenger request position? • Trucks: Which routes are used regularly? • Did the trucks with dangerous goods come close to a high risk facility? Planes • Were any two planes close to a collision? • Are two planes heading towards each other (going to crash)? • Did planes cross the air territory of state X? • At what speed does this plane move? What is its top speed? • Did Iraqi planes cross the 39th degree? Ships • Are any ships heading towards shallow areas? • Find "strange" movements of ships indicating illegal dumping of waste. Rockets, missiles, tanks, submarines • All kinds of military analyses	Countries • What was the largest extent ever of the Roman empire? • On which occasions did any two states merge? (Reunification, etc). • Which states split into two or more parts? • How did the Serb-occupied areas in former Yugoslavia develop over time? When was the maximal extent reached? Was Ghorazde ever part of their territory? Forests, Lakes • How fast is the Amazone rain forest shrinking? • Is the dead sea shrinking? What is the minimal and maximal extent of river X during the year? Glaciers • Does the polar ice cap grow? Does it move? • Where must glacier X have been at time Y (backward projection)? Storms • Where is the tornado heading? When will it reach Florida? High/low pressure areas • Where do they go? Where will they be tomorrow? Scalar functions over space, e.g. temperature • Where has the 0-degree boundary been last midnight? People • Movements of the celts etc. Troops, armies • Hannibal going over the alps. Show his trajectory. When did he pass village X? Cancer • Can we find in a series of X-ray images a growing cancer? How fast does it grow? How big was it on June 1, 1995? Why was it not discovered then? Continents • History of continental shift.

modeling is general and includes these cases, but for them also more traditional strategies could be used.

Also, if we consider transaction time (or bitemporal) databases, it is clear that changes to geometries happen only in discrete steps through updates to the database. Hence it is clear that the description of moving objects refers first of all to valid time. So we assume that complete descriptions of moving objects are put into the database by the applications, which means we are in the framework of historical databases reflecting the current knowledge about the past[2] of the real world. Transaction time databases about moving objects may be feasible, but will not be considered initially.

There is also an interesting class of applications that can be characterized as artifacts involving space and time, such as interactive multimedia documents, virtual reality scenarios, animations, etc. The techniques developed here might be useful to keep such documents in databases and ask queries related to the space and time occurring in these documents.

4.2.2 Modeling

Let us assume that a database consists of a set of *object classes* (of different *types* or *schemas*). Each object class has an associated set of *objects*; each object has a number of *attributes* with values drawn from certain *domains* or *atomic data types*. Of course, there may be additional features, such as object (or oid-) valued attributes, methods, object class hierarchies, etc. But the essential features are the ones mentioned above; these are common to all data models and already given in the relational model.

We now consider extensions to the basic model to capture time and space. As far as objects are concerned, an object may be created at some time and destroyed at some later time. So we can associate a validity interval with it. As a simplification, and to be able to work with standard data models, we can even omit this validity interval, and just rely on time-dependent attribute values described next.

Besides objects, attributes describing *geometries changing over time* are of particular interest. Hence we would like to define collections of *abstract data types*, or in fact *many-sorted algebras* containing several related types and their operations, for spatial values changing over time. Two basic types are *mpoint* and *mregion*, representing a moving point and a moving region, respectively. Let us assume that purely spatial data types called *point* and *region* are given that describe a point and a region in the 2D-plane[3] (a region may consist of several disjoint areas which may have holes) as well as a type *time* that describes the valid time dimension. Then we can view the types *mpoint* and *mregion* as

[2] For certain kinds of moving objects with predetermined schedules or trajectories (e.g. spacecraft, air planes, trains) the expected future can also be recorded in the database.

[3] We restrict attention to movements in 2D space, but the approach can, of course, be used as well to describe time-dependent 3D space.

mappings from time into space, that is

$$\underline{mpoint} = \underline{time} \rightarrow \underline{point}$$
$$\underline{mregion} = \underline{time} \rightarrow \underline{region}$$

More generally, we can introduce a type constructor τ which transforms any given atomic data type α into a type $\tau(\alpha)$ with semantics

$$\tau(\alpha) = \underline{time} \rightarrow \alpha$$

and we can denote the types *mpoint* and *mregion* also as $\tau(\underline{point})$ and $\tau(\underline{region})$, respectively.

A value of type *mpoint* describing a position as a function of time can be represented as a curve in the three-dimensional space (x, y, t) shown in Figure 4.1. We assume that space as well as time dimensions are continuous, i.e., isomorphic to the real numbers. (It should be possible to insert a point in time between any two given times and ask for e.g. a position at that time.)

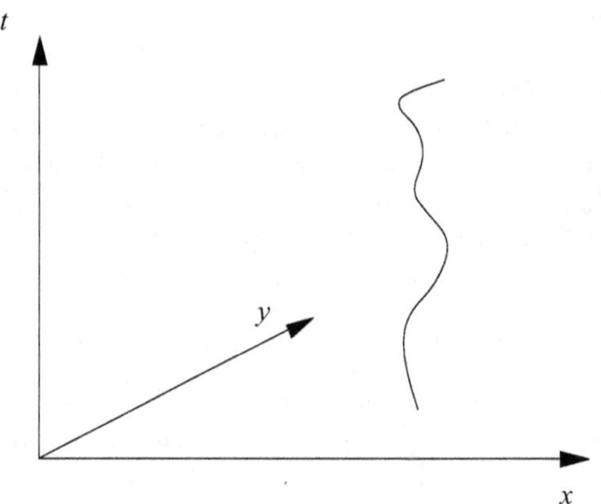

Fig. 4.1. A moving point

A value of type *mregion* is a set of volumes in the 3D space (x, y, t). Any intersection of that set of volumes with a plane $t = t_0$ yields a region value, describing the moving region at time t_0. Of course, it is possible that this intersection is empty, and an empty region is also a proper region value.

We now describe a few example operations for these data types. For the moment, these are purely for illustrative purposes; this is in no way intended to be a closed or complete design. Such a complete design is developed in Section 4.3.

Generic operations for moving objects are, for example:

$\tau(\alpha) \times \underline{time}$	$\rightarrow \alpha$	**at**
$\tau(\alpha)$	$\rightarrow \alpha$	**minvalue, maxvalue**
$\tau(\alpha)$	$\rightarrow \underline{time}$	**start, stop**
$\tau(\alpha)$	$\rightarrow \underline{real}$	**duration**
α	$\rightarrow \tau(\alpha)$	**const**

Operation **at** gives the value of a moving object at a particular point in time. **Minvalue** and **maxvalue** give the minimum and maximum values of a moving object. Both functions are only defined for types α on which a total order exists. **Start** and **stop** return the minimum and maximum of a moving value's (time) domain, and **duration** gives the total length of time intervals a moving object is defined. We shall also use the functions *startvalue(x)* and *stopvalue(x)* as an abbreviation for **at**(x, **start**(x)) and **at**(x, **stop**(x)), respectively. Whereas all these operations assume the existence of moving objects, **const** offers a canonical way to build spatio-temporal objects: **const**(x) is the "moving" object that yields x at any time.

In particular, for moving spatial objects we may have operations such as

$$\begin{array}{lll} \underline{mpoint} \times \underline{mpoint} & \rightarrow \underline{mreal} & \textbf{mdistance} \\ \underline{mpoint} \times \underline{mregion} & \rightarrow \underline{mpoint} & \textbf{visits} \end{array}$$

Mdistance computes the distance between the two moving points at all times and hence returns a time changing real number, a type that we call $\underline{mreal}$ ("moving real"; $\underline{mreal} = \tau(\underline{real})$), and **visits** returns the positions of the moving point given as a first argument at the times when it was inside the moving region provided as a second argument. Here it becomes clear that a value of type $\underline{mpoint}$ may also be a partial function, in the extreme case a function where the point is undefined at all times.

Operations may also involve pure spatial or pure temporal types and other auxiliary types. For the following examples, let $\underline{line}$ be a data type describing a curve in 2D space which may consist of several disjoint pieces; it may also be self-intersecting. Let $\underline{region}$ be a type for regions in the plane which may consist of several disjoint faces with holes. Let us also have operations

$$\begin{array}{lll} \underline{mpoint} & \rightarrow \underline{line} & \textbf{trajectory} \\ \underline{mregion} & \rightarrow \underline{region} & \textbf{traversed} \\ \underline{point} \times \underline{region} & \rightarrow \underline{bool} & \textbf{inside} \\ \underline{line} & \rightarrow \underline{real} & \textbf{length} \end{array}$$

Here **trajectory** is the projection of a moving point onto the plane. The corresponding projection for moving regions is the operation **traversed** that gives the total area the moving region ever has covered. **Inside** checks whether a point lies inside a region, and **length** returns the total length of a line value.

4.2.3 Some Example Queries

The presented data types can now be embedded into any DBMS data model as attribute data types, and the operations be used in queries. For example, we can integrate them into the relational model and have a relation

```
flights (id:string, from:string, to:string, route:mpoint)
```

We can then ask a query "Give me all flights from Düsseldorf that are longer than 5000 km":

```
SELECT id
FROM flights
WHERE from = "DUS" AND length(trajectory(route)) > 5000
```

This query uses projection into space. Dually, we can also formulate queries projecting into time. For example, "Which destinations can be reached from San Francisco within 2 hours?":

```
SELECT to
FROM flights
WHERE from = "SFO" AND duration(route) <= 2.0
```

Beyond projections into space and time, there are also genuine spatio-temporal questions that cannot be solved on projections. For example, "Find all pairs of planes that during their flight came closer to each other than 500 meters!":

```
SELECT A.id, B.id
FROM flights A, flights B
WHERE A.id <> B.id AND
  minvalue(mdistance(A.route, B.route)) < 0.5
```

This is in fact an instance of a spatio-temporal join.

The information contained in spatio-temporal data types is very rich. In particular, relations that would be used in traditional or spatial databases can be readily derived. For instance, we can easily define views for flight schedules and airports:

```
CREATE VIEW schedule AS
SELECT id, from, to, start(route) AS departure,
  stop(route) AS arrival
FROM flights

CREATE VIEW airport AS
SELECT DISTINCT from AS code, startvalue(route) AS location
FROM flights
```

The above examples use only one spatio-temporal relation. Even more interesting examples arise if we consider relationships between two or more different kinds of moving objects. To demonstrate this we use a further relation consisting of weather information, such as high pressure areas, storms, or temperature maps.

```
weather (kind:string, area:mregion)
```

The attribute "kind" gives the type of weather event, such as, "snow-cloud" or "tornado", and the "area" attribute provides the evolving extents of the weather features.

We can now ask, for instance, "Which flights went through a snow storm?"

```
SELECT id
FROM flights, weather
WHERE kind = "snow storm" AND duration(visits(route, area)) > 0
```

Here the expression **visits**(*route, area*) computes for each flight/storm combination a moving point that gives the movement of the plane inside this particular storm. If a flight passed a storm, this moving point is not empty, that is, it exists for a certain amount of time, which is checked by comparing the duration with 0. Similarly, we can find out which airports were affected by snow storms:

```
SELECT DISTINCT from
FROM airport, weather
WHERE kind = "snow storm" AND inside(location, traversed(area))
```

Finally, we can extend the previous query to find out which airports are most affected by snow storms. We can intersect the locations of airports with all snow storms by means of **visits** and determine the total durations:

```
SELECT code, SUM(duration(visits(const(location), area)))
  AS storm_hours
FROM airport, weather
WHERE kind = "snow storm"
GROUP BY code
HAVING storm_hours > 0
ORDER BY storm_hours
```

4.2.4 Some Basic Issues

Given this approach to spatio-temporal modeling and querying, several basic questions arise:

- We have seen spatio-temporal data types that are mappings from time into spatial data types. Is this realistic? How can we store them? Don't we need finite, discrete representations?
- If we use discrete representations, what do they mean? Are they observations of the moving objects?
- If we use discrete representations, how do we get the infinite entities from them that we really want to model? What kind of interpolation should be used?

In the following subsections we discuss these questions.

Abstract vs. Discrete Modeling. What does it mean to develop a data model with spatio-temporal data types? Actually, this is a design of a *many-sorted algebra.* There are two steps:

1. Invent a number of types and operations between them that appear to be suitable for querying. So far these are just names, which means one gives a *signature.* Formally, the signature consists of *sorts* (names for the types) and *operators* (names for the operations).
2. Define semantics for this signature, that is, associate an algebra, by defining *carrier sets* for the sorts and *functions* for the operators. So the carrier set for a type α contains the possible values for α, and the functions are mappings between the carrier sets.

For a formal definition of many-sorted signature and algebra see [24] or [18]. Now one can make such designs at two different levels of abstraction, namely as *abstract* or as *discrete models.*

Abstract models allow us to make definitions in terms of infinite sets, without worrying whether finite representations of these sets exist. This allows us to view a moving point as a continuous curve in the 3D space, as an *arbitrary* mapping from an infinite time domain into an also infinite space domain. All the types that we get by application of the type constructor τ are functions over an infinite domain, hence each value is an infinite set.

This abstract view is the conceptual model that we are interested in. The curve described by a plane flying over space is continuous; for any point in time there exists a value, regardless of whether we are able to give a finite description for this mapping (or relation). In Section 4.2.2 we have in fact described the types mentioned under this view. In an abstract model, we have no problem in using types like "moving real", *mreal*, and operations like

$$\underline{mpoint} \times \underline{mpoint} \to \underline{mreal} \ \textbf{mdistance}$$

since it is quite clear that at any time some distance between the moving points exists (when both are defined).

The only trouble with abstract models is that we cannot store and manipulate them in computers. Only finite and in fact reasonably small sets can be stored; data structures and algorithms have to work with *discrete (finite) representations* of the infinite point sets. From this point of view, abstract models are entirely unrealistic; only discrete models are usable.

This means we somehow need discrete models for moving points and moving regions as well as for all other involved types (*mreal*, *region*, ...). We can view discrete models as *approximations*, finite descriptions of the infinite shapes we are interested in. In spatial databases there is the same problem of giving discrete representations for in principle continuous shapes; there almost always *linear approximations* have been used. Hence, a region is described in terms of polygons and a curve in space (e.g. a river) by a polyline. Linear approximations are attractive because they are easy to handle mathematically; most algorithms in computational geometry work on linear shapes such as rectangles, polyhedra, etc. A linear approximation for a moving point is a polyline in 3D space; a linear approximation for a moving region is a set of polyhedra (see Figure 4.2). Remember that a moving point can be a partial function, hence it may disappear at times, the same is true for the moving region.

Suppose now we wish to define the type *mreal* and the operation **mdistance**. What is a discrete representation of the type *mreal*? Since we like linear approximations for the reasons mentioned above, the obvious answer would be to use a sequence of pairs (*value, time*) and use linear interpolation between the given values, similarly as for the moving point. If we now try to define the **mdistance** operator

$$\underline{mpoint} \times \underline{mpoint} \to \underline{mreal} \ \textbf{mdistance}$$

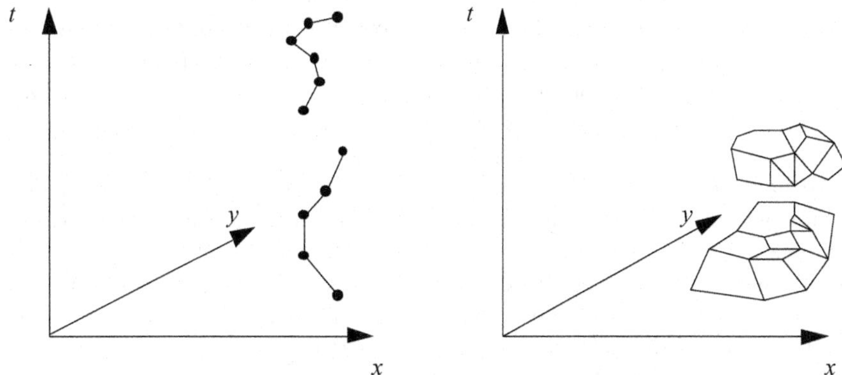

Fig. 4.2. Discrete representations for moving points and moving regions

we have to determine the time-dependent distance between two moving points represented as polylines. To see what that means, imagine that through each vertex of each of the two polylines we put a plane $t = t_i$ parallel to the xy-plane. Within each plane $t = t_i$ we can easily compute the distance; this will result in one of the vertices for the resulting *mreal* value. Between two adjacent planes we have to consider the distance between two line segments in 3D space (see Figure 4.3). However, this is not a linear but a quadratic function.

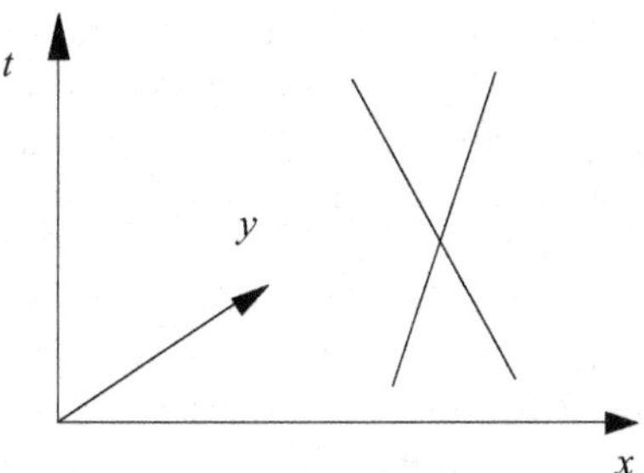

Fig. 4.3. Distance between segments of two moving points represented by polylines

So it seems that linear functions are not enough to represent moving reals. Maybe quadratic polynomials need to be introduced to represent the development between two vertices. But this immediately raises other questions. Why just quadratic functions motivated by the **mdistance** operation, perhaps other operations need other functions? And all kinds of operations that we need on moving reals must then be able to deal with these functions.

This example illustrates that choosing finite representations leads into difficult tradeoffs. Other choices for a moving point could be polynomial splines which are capable of describing changes in speed or acceleration much better (with polylines, speed is stepwise constant, and acceleration is either 0 or infinite, which seems quite unnatural). For moving regions, an alternative to the polyhedral representation could be sequences of affine mappings (where each transition from one state of a region to the next can be described by translation, rotation, and scaling). This model can describe rotations much better, but does not support arbitrary changes of shape.

We have concluded from such considerations that both levels of modeling are indispensable. For the discrete model this is clear anyway, as only discrete models can be implemented. However, if we restrict attention directly to discrete models, there is a danger that a conceptually simple, elegant design of query operations is missed. This is because the representational problems might lead us to prematurely discard some options for modeling.

For example, from the discussion above one might conclude that moving reals are a problem and no such type should be introduced. But then, instead of operations **minvalue**, **maxvalue**, etc. on moving reals one has to introduce corresponding operations for each time-dependent numeric property of a moving object. Suppose we are interested in **distance** between two moving points, **speed** of a moving point, and **size** and **perimeter** of a moving region. Then we need operators **mindistance**, **maxdistance**, **minspeed**, **maxspeed**, and so forth. Clearly, this leads to a proliferation of operators and to a bad design of a query language. So the better strategy is to start with a design at the abstract level, and then to aim for that target when designing discrete models.

Observations vs. Description of Shape. Looking at the sequence of 3D points describing a moving point in a discrete model, one may believe that these are observations of the moving object at a certain position at a specific time. This may or may not be the case. Our view is that it is, first of all, an adequate description of the shape of a continuous curve (i.e., an approximation of that curve). We assume that the application has complete knowledge about the curve, and puts into the database a discrete description of that curve.

What is the difference to observations? Observations could mean that there are far too many points in the representation, for example, because a linear movement over an hour happens to have been observed every second. Observations could also be too few so that arbitrarily complex movements have happened between two recorded points; in that case our (linear or other) interpolation between these points could be arbitrarily wrong. Hence we assume that the application, even if it does make observations, employs some preprocessing of observations and also makes sure that enough observations are taken. Note that it is quite possible that the application adds points other than observations to a curve description, as it may know some physical laws governing the movements of this particular class of objects.

The difference in view becomes even more striking if we consider moving regions. We require the application to have complete knowledge about the 3D shape of a moving region so that it can enter into the database the polyhedron (or set of polyhedra) as a good approximation. In contrast, observations could only be a sequence of region values. But whereas for moving points it is always possible to make a straight line interpolation between two adjacent positions, there is no way that a database system could, in general, deduce the shape of a region between two arbitrary successive observations. Hence, it is the job of the application to make enough observations and otherwise have some knowledge how regions of this kind can behave and then apply some preprocessing in order to produce a reasonable polyhedral description. How to get polyhedra from a sequence of observations, and what rules must hold to guarantee that the sequence of observations is "good enough" may be a research issue in its own right. We assume this is solved when data are put into the database.

The next two sections of this chapter will present first a careful and formal design of an abstract model, and then a discrete model offering finite representations for the types of the abstract model.

4.3 An Abstract Model: A Foundation for Representing and Querying Moving Objects

This section aims to offer a precise and conceptually clean foundation for implementing a spatio-temporal DBMS. It presents a simple and expressive system of abstract data types, comprising data types and encapsulating operations, that may be integrated into a query language, to yield a powerful language for querying spatio-temporal data such as moving objects. In addition to presenting the data types and operations, insight into the considerations that went into the design is offered, and the use of the abstract data types is exemplified using SQL. This section is based on [21] where complete formal definitions of all the concepts presented here can be found.

The next section defines the foundation's data types. As a precursor to defining the operations on these, Section 4.3.2 briefly presents the SQL-like language, the abstract data types are embedded into. An overview of the operations is provided in Section 4.3.3, and Sections 4.3.4 and 4.3.5 present the specific operations. Section 4.3.6 demonstrates the use of the abstract data types in a forest management application. Finally, Section 4.3.7 summarizes the section.

4.3.1 Spatio-temporal Data Types

This section presents a type system, constructed by introducing basic types and type constructors. Following an overview, the specific types are presented.

Overview. The signature (see, e.g., [18,24]) given in Table 4.2 is used in defining the type system. In this signature, *kinds* are capitalized and denote sets of types, and type constructors are in italics. This signature generates a set of terms, which are the types in the system. Terms include *int*, *region*, *moving*(*point*), *range*(*int*), etc. Type constructor *range* is applicable to all types in kinds *BASE* and *TIME*, and hence the types that can be constructed by it are *range*(*int*), *range*(*real*), *range*(*string*), *range*(*bool*), and *range*(*instant*). Type constructors with no arguments, for example *region*, are types already and are called *constant*.

Table 4.2. Signature describing the type system

	→ BASE	<u>*int*</u>, <u>*real*</u>, <u>*string*</u>, <u>*bool*</u>
	→ SPATIAL	<u>*point*</u>, <u>*points*</u>, <u>*line*</u>, <u>*region*</u>
	→ TIME	<u>*instant*</u>
BASE ∪ TIME	→ RANGE	<u>*range*</u>
BASE ∪ SPATIAL	→ TEMPORAL	<u>*intime*</u>, <u>*moving*</u>

Although the focus is on spatio-temporal types, especially <u>*moving*</u>(<u>*point*</u>) and <u>*moving*</u>(<u>*region*</u>), to obtain a closed system, it is necessary to include the other types given in the table. We proceed to describe these types in more detail, covering also their semantics.

Base Types. The base types are <u>*int*</u>, <u>*real*</u>, <u>*string*</u>, and <u>*bool*</u>. These have the usual semantics, and they all include an undefined value. The semantics of a type α is given by its carrier set, A_α. For example, $A_{string} \triangleq V^* \cup \{\perp\}$, where V is a finite alphabet. As a shorthand, we define $\bar{A}_\alpha$ to mean $A_\alpha \setminus \{\perp\}$, i.e., the carrier set without the undefined value.

Spatial Types. The four spatial types in the system, <u>*point*</u>, <u>*points*</u>, <u>*line*</u>, and <u>*region*</u> (cf. [19]), are illustrated in Figure 4.4. Informally, these types have the following meaning. A value of type <u>*point*</u> represents a point in the Euclidean plane or is undefined. A <u>*points*</u> value is a finite set of points. A <u>*line*</u> value is a finite set of continuous curves in the plane. A <u>*region*</u> is a finite set of disjoint parts, termed *faces*, each of which may have holes. A face may lie within a hole of another face. Each of the three set types may be empty.

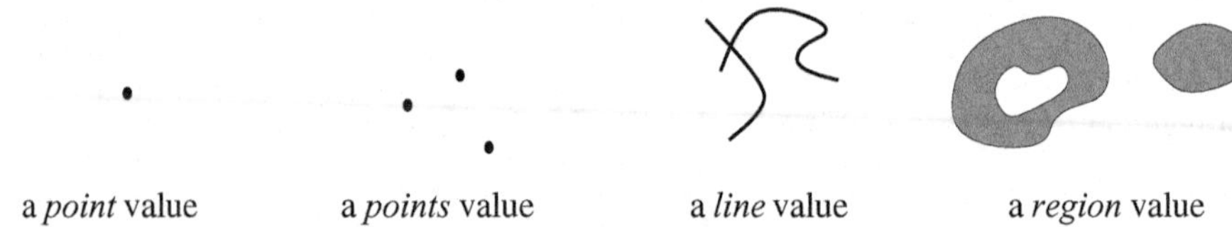

Fig. 4.4. The Spatial Data Types

The formal definitions, given elsewhere [21], are based on the point set paradigm and on point set topology, which provides concepts of continuity and closeness, as well as allows us to identify special topological structures of a point set, such as its interior, closure, boundary, and exterior. We assume that the reader has some familiarity with basic concepts of topology [17].

The point and point set types are quite simple to define formally. Specifically, $A_{point} \triangleq \mathbb{R}^2 \cup \{\perp\}$ and $A_{points} \triangleq \{P \subseteq \mathbb{R}^2 \mid P \text{ is finite}\}$.

The definition of <u>*line*</u> is based on curves (continuous mappings from $[0, 1]$ to $\mathbb{R}^2$) that are simple in the sense that the intersection of two curves yields only a finite number of proper intersection points (disregarding common parts that are curves themselves). The *line* data type is to represent any finite union of such simple curves. When the abstract design of data types given here is implemented by some discrete design, some class of curves will be selected for representation, for example polygonal lines, curves described by cubic functions, etc. We just require that the class of curves selected has this simplicity property. This is needed, for example, to ensure that the **intersection** operation between two *line* values yields a finite set of points, representable by the *points* data type.

A finite union of curves basically yields a planar graph structure (whose nodes are intersections of curves and whose edges are intersection-free pieces of curves). Given a set of points of such a graph, there are many different sets of curves resulting in this point set. For example, a path over the graph could be interpreted as a single curve or as being composed of several curves. A design is chosen where (i) a *line* value is a point set in the plane that can be described as a finite union of curves, and (ii) there is a unique collection of curves that can serve as a "canonical" representation of this *line* value. For a *line* Q, we let $sc(Q)$ (the *simple curves* of Q) denote this representation.

When defining certain operations, we need a notion of components of a *line* value. Let $meet^*$ denote the transitive closure of the *meet* relationship on curves. This relation partitions the components of a *line* value Q into connected components, denoted as $components(sc(Q))$. The decomposition into corresponding point sets is defined as $blocks(Q) = \{points(C) \mid C \in components(sc(Q))\}$, where $points(C)$ returns the points in a partition.

A *region* value is defined as a point set in the plane with a certain structure. *region* values do not have geometric anomalies such as isolated or dangling line or point features, and missing lines and points in the form of cuts and punctures. Specifically, a *region* can be viewed as a finite set of so-called *faces*, where any two faces are disjoint except for finitely many "touching points" at their boundaries. Moreover, boundaries of faces are simple (as for lines). For example, the intersection of two regions will also produce only finitely many isolated intersection points. The boundary of a face may have outer as well as inner parts, i.e., a face may have holes.

We require that the same class of curves is used in defining the *line* and the *region* type. We can denote a given *region* value Q by $faces(Q)$. We also extend the shorthand $\bar{A}$ to the spatial data types and all other types α whose carrier set contains sets of values. For these, we define $\bar{A}_\alpha \triangleq A_\alpha \setminus \{\emptyset\}$.

Time Type. Type *instant* represents time points. Time is considered to be linear and continuous, i.e., isomorphic to the real numbers. Specifically, the carrier set for *instant* is $A_{instant} \triangleq \mathbb{R} \cup \{\bot\}$.

Temporal Types. From the base and spatial types, we derive corresponding temporal types. Type constructor *moving* yields, for any given type α, a mapping from time to α.

Definition 1. Let α, with carrier set A_α, be a data type to which the <u>*moving*</u> type constructor is applicable. Then the carrier set for <u>*moving*</u>(α) is defined as follows:

$$A_{moving(\alpha)} \triangleq \{f | f : \bar{A}_{instant} \to \bar{A}_\alpha \text{ is a partial function} \wedge \Gamma(f) \text{ is finite}\}$$

Hence, each value f is a function describing the development over time of a value from the carrier set of α. The condition "$\Gamma(f)$ is finite" ensures that f consists of only a finite number of continuous components (the notion of continuity used here is defined elsewhere [21]). As a result, projections of moving objects (e.g., into the 2D plane) have only a finite number of components. This is needed in the **decompose** operation (for lack of space not defined in this section, but in [21]), and it serves to making the design implementable.

For all "moving" types we introduce extra names by prefixing the argument type with an "*m*", that is, <u>*mpoint*</u>, <u>*mpoints*</u>, <u>*mline*</u>, <u>*mregion*</u>, <u>*mint*</u>, <u>*mreal*</u>, <u>*mstring*</u>, and <u>*mbool*</u>. This is just to shorten some signatures.

The temporal types obtained through the <u>*moving*</u> type constructor are functions, or infinite sets of pairs (instant, value). It is practical to have a type for representing any single element of such a function, i.e., a single (instant, value)-pair, for example, to represent the result of a time-slice operation. The <u>*intime*</u> type constructor converts a given type α into a type that associates <u>*instant*</u> values with values of α.

Definition 2. Let α be a data type to which the <u>*intime*</u> type constructor is applicable, with carrier set A_α. The carrier set for <u>*intime*</u>(α), is defined as follows:

$$A_{intime(\alpha)} \triangleq A_{instant} \times A_\alpha$$

Range Types (Sets of Intervals). For all temporal types, we desire operations that project into their domains and ranges. For the moving counterparts of the base types, e.g., <u>*moving*</u>(<u>*real*</u>) (whose values come from a one-dimensional domain), the projections are, or can be compactly represented as, sets of intervals over the one-dimensional domain. Hence, we are interested in types to represent sets of intervals over the real numbers, over the integers, etc. The <u>*range*</u> type constructor provides these types.

Definition 3. Let α be a data type to which the <u>*range*</u> type constructor is applicable (and hence on which a total order $<$ exists). An *α-interval* is a set $X \subseteq \bar{A}_\alpha$ such that $\forall x, y \in X \; \forall z \in \bar{A}_\alpha \; (x < z < y \Rightarrow z \in X)$.

Two α-intervals are *adjacent* if they are disjoint and their union is an α-interval. An *α-range* is a finite set of disjoint, non-adjacent intervals. For an α-range R, *points*(R) denotes the union of all its intervals.

Definition 4. Let α be any data type to which the <u>*range*</u> type constructor is applicable. Then the carrier set for <u>*range*</u>(α) is:

$$A_{range(\alpha)} \triangleq \{X \subseteq \bar{A}_\alpha \mid \exists \text{ an } \alpha\text{-range } R \; (X = points(R))\}$$

A <u>*range*</u> value X has a unique associated α-range denoted by *intvls*(X). We use <u>*periods*</u> as a shorthand for ranges over the time domain, <u>*range*</u>(<u>*instant*</u>).

Design Rationale. We have attempted to ensure consistency and closure between non-temporal and temporal types. The former is ensured by introducing temporal types for all base types and all spatial types through the *moving* constructor. Closure under projection is ensured: For all temporal types, data types are available to represent the results of projections into (time) domain and range.

The type system offers uniform support for the point vs. point set view. All data types belong to either a one- or two-dimensional space. This principle requires that each space includes data types that represent a single value (a "point") and a set of values (a "point set"). This is the basis for the definitions of generic operations in the next sections and is explained in more detail there.

Additional discussions of design considerations may be found in [20].

4.3.2 Language Embedding of Abstract Data Types

In order to illustrate the use in queries of the operations to be defined in the next section, these must be embedded in a query language. We use an SQL-like language, with which most readers should be familiar. It is convenient to employ a few constructs, which are expressible in one form or another in most object-oriented or object-relational query languages. We briefly explain these constructs.

Assignment. The construct `LET <name> = <query>` assigns the result of `query` to `name`.

Multistep Queries. A query may encompass a list of initial assignments and one or more subsequent query expressions.

Conversions between Sets of Objects and Atomic Values. A single-attribute, single-tuple relation may be converted into a typed, atomic value and vice-versa, using the notations `ELEMENT(<query>)` and `SET(<attrname>, <value>)`. For example, expression `SET(name, "John Smith")` returns a relation with an attribute `name` and a single tuple with value `John Smith`.

Defining Derived Attributes. We allow arbitrary abstract data type operations in the `WHERE` clause, where they form predicates, and in the `SELECT` clause, where they produce new attributes. The notation `<new attrname> AS <expression>` is used.

Defining Operations. New operations may be derived from existing ones, using `LET <name> = <functional expression>`. A functional expression has the form `FUN (<parameter list>) <expression>` and corresponds to the lambda abstraction in functional languages. For example, a new operation **square** can be defined and used as follows: LET **square** = FUN (m:*integer*) m * m; **square**(5)

Defining Aggregate Functions. Any binary, associative, and commutative operation defined on a data type can be used as an aggregate function over a column of that data type, using the notation `AGGR(<attrname>, <operator>, <neutral element>)`. In case the argument relation is empty, the neutral element is returned. In case it has a single tuple, then that single attribute value is returned; otherwise, the existing values are combined by the given operator. Moreover, an aggregate function may be named, using the notation `LET <name> = AGGREGATE(<operator>, <neutral element>)`.

With these constructs and given a relation `employee(name:`*string*`, salary:`*int*`, permanent:`*bool*`)`, we can sum all salaries and determine whether all employees have permanent positions.

```
SELECT AGGR(salary, +, 0) FROM employee

LET all = AGGREGATE(and, TRUE);
SELECT all(permanent) FROM employee
```

4.3.3 Overview of Data Type Operations

The design of the operations adheres to three principles: (i) Design operations as generic as possible. (ii) Achieve consistency between operations on non-temporal and temporal types. (iii) Capture the interesting phenomena.

The first principle is crucial, as the type system is quite large. To avoid a proliferation of operations, a unifying view of collections of types is mandatory. This is enabled by relating each type to either a one-dimensional or a two-dimensional space and by considering all values as either single elements or subsets of the respective space. For example, type *int* describes single elements of the one-dimensional space of integers, while *range*(*int*) describes sets of integers. Similarly, *point* describes single elements of two-dimensional space, whereas *points*, *line*, and *region* describe subsets of this space.

Next, in order to achieve consistency of operations on non-temporal and temporal types, we first define operations on non-temporal types and then systematically *lift* these operations to become temporal variants of the respective types.

Finally, to obtain a powerful query language, it is necessary to include operations that address the most important concepts from various domains (or branches of mathematics). Whereas simple set theory and first-order logic are certainly the most fundamental and best-understood parts of query languages, operations based on order relationships, topology, metric spaces, etc., are also needed. While there is no clear recipe for achieving closure of "interesting phenomena," this motivates the inclusion of concepts and operations such as distance, size of a region, and relationships of boundaries.

Section 4.3.4 develops operations on non-temporal types, based on the generic point and point set (value vs. subset of space) view of these types. Section 4.3.5 defines operations on temporal types.

4.3.4 Operations on Non-temporal Types

The classes of operations on non-temporal types are given in Table 4.3, which also lists the names of the operations on these types. Although the focus is on moving objects, and hence on temporal types, the definitions of operations on non-temporal types are essential, as these operations will later be *lifted*, to obtain operations on temporal types.

Table 4.3. Classes of operations on non-temporal types

Class	Operations
Predicates	**isempty** $=, \neq$, **intersects**, **inside** $<, \leq, \geq, >$, **before** **touches**, **attached**, **overlaps**, **on_border**, **in_interior**
Set Operations	**intersection**, **union**, **minus** **crossings**, **touch_points**, **common_border**
Aggregation	**min**, **max**, **avg**, **center**, **single**
Numeric	**no_components**, **size**, **perimeter**, **duration**, **length**, **area**
Distance and Direction	**distance**, **direction**
Base Type Specific	**and**, **or**, **not**

We take the view that we are dealing with single values and sets of these values in one- and two-dimensional space. The types can then be classified according to Table 4.4. (Remember that "temporal types" are functions of time. Types *instant* and *periods* are not temporal types in this sense.) The table contains five

Table 4.4. Classification of non-temporal types

	1D Spaces					2D Space
	discrete			continuous		
	Integer	Boolean	String	Real	Time	2D
point	*int*	*bool*	*string*	*real*	*instant*	*point*
point set	*range*(*int*)	*range*(*bool*)	*range* (*string*)	*range* (*real*)	*periods*	*points*, *line*, *region*

one-dimensional spaces, Integer, Boolean, etc., and one two-dimensional space, 2D. For example, space Integer has two types, *int* and *range*(*int*). We distinguish between 1D and 2D spaces because only the 1D spaces have a (natural) total order. The distinction between discrete and continuous one-dimensional spaces is important for certain numeric operations. To have a uniform terminology, in any of the respective spaces, we call a single element a point and a subset of the space a point set; and we classify types as point types or point set types.

Example 1. We introduce the following example relations for use within this section, representing cities, countries, rivers, and highways in Europe.

```
city(name:string, pop:int, center:point)
country(name:string, area:region)
river(name:string, route:line)
highway(name:string, route:line)
```

Notations for Signatures. The notations for signatures used when defining the data type operations next are partly based on Table 4.4. We let π and σ be type variables that range over all point and point set types of Table 4.4, respectively. If several type variables occur in a signature (e.g., for binary operations), then they are assumed to range over types of the same space. For example, in signature $\pi \times \sigma \to \alpha$ we can select one-dimensional space Integer and instantiate π to <u>*int*</u> and σ to <u>*range*(*int*)</u>; or we can select two-dimensional space 2D and then instantiate π to <u>*point*</u> and σ to either <u>*points*</u>, <u>*line*</u>, or <u>*region*</u>.

In signature $\sigma \times \sigma \to \alpha$, both arguments have to be the same type. However, in signature $\sigma_1 \times \sigma_2 \to \alpha$, type variables σ_1 and σ_2 can be instantiated independently, but must range over the same space. The notation $\alpha \otimes \beta \to \gamma$ indicates that any order of the two argument types is valid.

Some operations are restricted to certain classes of spaces, namely 1D = {Integer, Boolean, String, Real, Time}, 2D = {2D}, 1Dcont = {Real, Time}, 1Dnum = {Integer, Real, Time}, and cont = {Real, Time, 2D}. A signature is restricted to a class of spaces by putting the name of the class behind it in square brackets. For example, a signature $\alpha \to \beta$ [1D] is valid for all one-dimensional spaces.

Generic operations with generic names may have more appropriate, specific names when applied to specific types. For example, a generic **size** operation exists for point set types. For type <u>*periods*</u> the name **duration** is more appropriate. In this case, we introduce the more specific name as an *alias* with the notation **size**[**duration**].

In defining semantics, $u, v, \ldots$ denote single values of a π type, and $U, V, \ldots$ generic sets of values (point sets) of a σ type. For binary operations, u or U will refer to the first and v or V to the second argument. Furthermore, b (B) ranges over values (sets of values) of base types, and predicates are denoted by p. We use μ to range over moving objects and t (T) to range over instant values (periods).

For the definition of the semantics of operations we generally assume strict evaluation, i.e., for any function f_{op} defining the semantics of an operation *op* we assume $f_{op}(\ldots, \perp, \ldots) = \perp$. Undefined arguments are therefore not handled explicitly in definitions.

Predicates. We consider unary and binary predicates. At this abstract level, we can ask whether a single point is undefined, and whether a point set is empty. The generic predicate **isempty**[**undefined**] is used for this purpose.

The design of binary predicates is based on the following strategy. First, we consider possible relationships between two points (single values), two point sets, and a point vs. a point set in the respective space. Second, orthogonal to this, predicates are based on three different concepts, namely set theory, order relationships, and topology. This design is shown in Table 4.5. The signatures

Table 4.5. Analysis of binary predicates

	Sets	Order (1D)	Topology
point vs. point	$u = v, u \neq v$	$u < v, u \leq v$ $u \geq v, u > v$	
point set vs. point set	$U = V, U \neq V$ $U \cap V \neq \emptyset$ (**intersects**) $U \subseteq V$ (**inside**)	U **before** V	$\partial U \cap \partial V \neq \emptyset$ (**touches**) $\partial U \cap V^o \neq \emptyset$ (**attached**) $U^o \cap V^o \neq \emptyset$ (**overlaps**)
point vs. point set	$u \in U$ (**inside**)	u **before** V U **before** v	$u \in \partial U$ (**on_border**) $u \in U^o$ (**in_interior**)

and definitions for these predicates are as expected and have been omitted. Predicates related to distance or direction (e.g., "north") can be obtained via numeric evaluations (see Section 4.3.4).

Set Operations. Set operations are fundamental and are available for all point-set types. Where feasible, we also allow set operations on point types, thus allowing expressions such as u **minus** v and U **minus** u. Resulting singleton or empty sets are interpreted as point values.

Defining set operations on the combination of one- and two-dimensional point sets is more involved. This is because we are using arbitrary closed or open sets in one-dimensional space, whereas only closed point sets (*points*, *line*, and *region*) exist in two-dimensional space. This renders it necessary to apply a closure operation after applying the set operations on such entities which adds all points on the boundary of an open set.

Because there are three point set types in 2D space, an analysis of which argument type combinations make sense (return interesting results) and of what the result types are is required.

Generally, set operations may return results that intermix zero-, one-, and two-dimensional point sets, i.e., points, lines, and proper regions. Usually one is interested mainly in the result of the highest dimension. This is reflected in the concept of *regularized* set operations [32]. For example, the regularized intersection removes all lower-dimensional pieces from the corresponding intersection result. We adopt regularization in the semantics of the three "standard" 2D set operations, **union**, **minus**, and **intersection**.

The union of arguments of equal types has the usual semantics, and union on different types is not defined. Difference always results in the type of the first argument. Closure has to be applied to the result. Intersection produces results of all dimensions smaller than or equal to the dimension of the lowest-dimensional argument. For example, the intersection of a <u>*line*</u> value with a <u>*region*</u> value may result in points and lines. We define the **intersection** operator for all type combinations with regularized semantics. To make the other parts of results available, we introduce specialized operators, e.g., **common_border** and **touch_points**.

The resulting set operations are given in Table 4.6. The notation $min(\sigma_1, \sigma_2)$ refers to taking the minimum in an assumed "dimensional" order: <u>*points*</u> < <u>*line*</u> < <u>*region*</u>. The table uses predicates, e.g., $is2D$, with the obvious meaning, as well as the notations $\rho(Q)$, Q^o, and ∂Q for the closure, interior, and boundary of Q, respectively.

Table 4.6. Set operations

Operation	Signature	Semantics
intersection	$\pi \times \pi \rightarrow \pi$	if $u = v$ then u else $\bot$
minus	$\pi \times \pi \rightarrow \pi$	if $u = v$ then $\bot$ else u
intersection	$\pi \otimes \sigma \rightarrow \pi$	if $u \in V$ then u else $\bot$
minus	$\pi \times \sigma \rightarrow \pi$	if $u \in V$ then $\bot$ else u
	$\sigma \times \pi \rightarrow \sigma$	if $is2D(U)$ then $\rho(U \setminus \{v\})$ else $U \setminus \{v\}$
union	$\pi \otimes \sigma \rightarrow \sigma$	if $is1D(V)$ or $type(V) =$ <u>*points*</u> then $V \cup \{u\}$ else V
intersection, minus, union	$\sigma \times \sigma \rightarrow \sigma$ [1D]	$U \cap V, U \setminus V, U \cup V$
intersection	$\sigma_1 \times \sigma_2 \rightarrow min(\sigma_1, \sigma_2)$ [2D]	[21]
minus	$\sigma_1 \times \sigma_2 \rightarrow \sigma_1$ [2D]	$\rho(Q_1 \setminus Q_2)$
union	$\sigma \times \sigma \rightarrow \sigma$ [2D]	$Q_1 \cup Q_2$
crossings	<u>*line*</u> × <u>*line*</u> → <u>*points*</u>	[21]
touch_points	<u>*region*</u> ⊗ <u>*line*</u> → <u>*points*</u>	
	<u>*region*</u> × <u>*region*</u> → <u>*points*</u>	
common_border	<u>*region*</u> × <u>*region*</u> → <u>*line*</u>	

The following example shows how, with **union** and **intersection**, we obtain the corresponding aggregate functions over sets of objects (relations).

Example 2. "Determine the region of Europe from the regions of its countries."

```
LET sum = AGGREGATE(union, TheEmptyRegion);
LET Europe = SELECT sum(area) FROM country
```

Aggregation. Aggregation reduces sets of points to points (Table 4.7). For open and half-open intervals, we use the infimum and supremum values, i.e., the maximum and minimum of their closure, for computing minimum and maximum values. This is preferable over returning undefined values. In all domains that have addition, we can compute the average (**avg**). In 2D, the average is based

Table 4.7. Aggregate operations

Operation	Signature	Semantics
min, max	$\sigma \rightarrow \pi$ [1D]	$\min(\rho(U)), \max(\rho(U))$
avg	$\sigma \rightarrow \pi$ [1Dnum]	$\frac{1}{\vert intvls(U)\vert} \sum_{T \in intvls(U)} \frac{\sup(T)+\inf(T)}{2}$
avg[center]	$\underline{points} \rightarrow \pi$ [2D]	$\frac{1}{n} \sum_{p \in U} \vec{p}$
avg[center]	$\underline{line} \rightarrow \pi$ [2D]	$\frac{1}{\Vert U \Vert} \sum_{c \in sc(U)} \vec{c} \Vert c \Vert$
avg[center]	$\underline{region} \rightarrow \pi$ [2D]	$\frac{1}{M} \int_U \vec{p}\, dA$ where $M = \int_U dA$
single	$\sigma \rightarrow \pi$	if $\exists u : U = \{u\}$ then u else $\perp$

on vector addition and is usually called **center** (of gravity). It is often useful to have a "casting" operation available to transform a singleton set into its single value; operation **single** does this conversion.

Example 3. The query "find the point where highway A1 crosses the river Rhine" can be expressed as:

```
SELECT single(crossings(R.route, H.route))
FROM river R, highway H
WHERE R.name = "Rhine" and H.name = "A1"
  and R.route intersects H.route
```

Numeric Properties of Sets. For sets of points, a number of well-known numeric properties may be computed (Table 4.8). For example, the number of

Table 4.8. Numeric operations

Operation	Signature	Semantics
no_components	$\sigma \rightarrow \underline{int}$ [1D]	$\vert intvls(U)\vert$
no_components	$\underline{points} \rightarrow \underline{int}$	$\vert U \vert$
no_components	$\underline{line} \rightarrow \underline{int}$	$\vert blocks(U)\vert$
no_components	$\underline{region} \rightarrow \underline{int}$	$\vert faces(U)\vert$
size[duration]	$\sigma \rightarrow \underline{real}$ [1Dcont]	$\sum_{T \in intvls(U)} \sup(T) - \inf(T)$
size[length]	$\underline{line} \rightarrow \underline{real}$	$\Vert U \Vert$
size[area]	$\underline{region} \rightarrow \underline{real}$	$\int_U dA$
perimeter	$\underline{region} \rightarrow \underline{real}$	$f_{\mathbf{length}}(\partial U)$

components (**no_components**) is the number of disjoint maximal connected subsets, i.e., the number of faces for a region, connected components for a line graph, and intervals for a 1D point set. The **size** is defined for all continuous set types (i.e., for <u>range</u>(<u>real</u>), <u>periods</u>, <u>line</u>, and <u>region</u>). For 1D types, the size is the sum of the lengths of component intervals; for <u>line</u>, it is the length, and for <u>region</u>, it is the area.

Example 4. "List for each country its total size and the number of disjoint land areas."

```
SELECT name, area(area), no_components(area) FROM country
```

Distance and Direction. A distance measure exists for all continuous types. The **distance** function determines the distance between the closest pair of a point from the first and the second argument.

The direction between points is sometimes of interest. The **direction** function returns the angle of the line from the first to the second point, measured in degrees ($0 \leq angle < 360$). Hence, if q is exactly north of p then **direction**$(p, q) = 90$. If $p = q$ then the undefined value $\perp$ is returned.

Example 5. "Find all cities north of and within 200 kms of Munich!"

```
LET Munich = ELEMENT(SELECT center FROM city
  WHERE name = "Munich");
SELECT name FROM city
WHERE distance(center, Munich) < 200
  and direction(Munich, center) >= 45
  and direction(Munich, center) <= 135
```

Specific Operations for Base Types. The operations **and**, **or**, and **not** on base types are also needed, although they are not related to the point vs. point set view. We mention them because they will be subject to *lifting* described below and so become applicable to temporal types.

4.3.5 Operations on Temporal Types

Values of temporal types (i.e., types $\underline{moving}(\alpha)$) are partial functions of the form $f : A_{instant} \rightarrow \bar{A}_{\alpha}$. There are four classes of operations on such functions.

Projection to Domain and Range. For values of all $\underline{moving}$ types—which are functions—operations are provided that yield the domain and range of these functions. The domain function **deftime** : $\underline{moving}(\alpha) \rightarrow \underline{periods}$ returns the times for which a function is defined.

In 1D space, operation **rangevalues** : $\underline{moving}(\alpha) \rightarrow \underline{range}(\alpha)$ returns values assumed over time as a set of intervals. For the 2D types, operations are offered to return the parts of the projections corresponding to our data types. For example, the projection of a moving point into the plane may consist of points and lines; these can be obtained by operations **locations** and **trajectory** respectively.

For $\underline{intime}$ types, the two trivial projection operations, **inst** : $\underline{intime}(\alpha) \rightarrow \underline{instant}$ and **val** : $\underline{intime}(\alpha) \rightarrow \alpha$, are offered.

All the infinite point sets that result from domain and range projections are represented in collapsed form by the corresponding point set types. For example, a set of instants is represented as a $\underline{periods}$ value, and an infinite set of regions is represented by the union of the points of the regions, which is represented in turn as a <u>region</u> value. This finite representation is enabled by the continuity condition required for types <u>moving</u>(α) (see Section 4.3.1).

The resulting design is complete in that all projection values in domain and range can be obtained.

Example 6. For illustration of operations on temporal types, we use the following relations (a slight variation of those of Section 4.2).

```
flight(airline:string, no:int, from:string, to:string, route:mpoint)
weather(name:string, kind:string, area:mregion)
site(name:string, pos:point)
```

In the first, attributes `airline` and `no` identify a flight, and the names of the departure and destination cities and the route taken for each flight are also recorded. A route is defined only for the times the plane is in flight and not when it is on the ground. Relation `weather` records named weather phenomena. Attribute `kind` gives the type of phenomenon, such as, "snow-cloud" or "tornado," and attribute `area` provides the phenomenon's evolving extent. Relation `site` contains positions of well-known sites.

Example 7. "How far does flight LH 257 travel in French air space?"

```
LET route257 = ELEMENT(SELECT route FROM flight
  WHERE airline = "LH" and no = 257);
length(intersection(France, trajectory(route257)))
```

"What are the departure and arrival times of flight LH 257?"

```
min(deftime(route257)); max(deftime(route257))
```

Example 8. "When and at distance does flight 257 pass the Eiffel tower?"
We assume a **closest** operator with signature <u>mpoint</u> $\times$ <u>point</u> $\rightarrow$ <u>intime</u>(<u>point</u>), which returns the time and position when a moving point is closest to a given fixed point in the plane. In [21] it is shown how such an operator can be derived from others.

```
LET EiffelTower =
  ELEMENT(SELECT pos FROM site WHERE name = "Eiffel Tower");
LET pass = closest(route257, EiffelTower);
inst(pass); distance(EiffelTower, val(pass))
```

Interaction with Points and Point Sets in Domain and Range. Some operations relate the functional values of <u>moving</u> types with values either in their (time) domain or their range. For example, such functions allow us to determine whether a moving point passes a given point or region. With these, one may also restrict a moving entity to given domain or range values. As an example, one may determine the value(s) of a moving real at time t or in time interval $[t_1, t_2]$.

Table 4.9. Interaction of temporal values with values in domain and range

Operation	Signature		
atinstant	$\underline{moving}(\alpha) \times \underline{instant}$	$\rightarrow \underline{intime}(\alpha)$	
atperiods	$\underline{moving}(\alpha) \times \underline{periods}$	$\rightarrow \underline{moving}(\alpha)$	
initial	$\underline{moving}(\alpha)$	$\rightarrow \underline{intime}(\alpha)$	
final	$\underline{moving}(\alpha)$	$\rightarrow \underline{intime}(\alpha)$	
present	$\underline{moving}(\alpha) \times \underline{instant}$	$\rightarrow \underline{bool}$	
present	$\underline{moving}(\alpha) \times \underline{periods}$	$\rightarrow \underline{bool}$	
at	$\underline{moving}(\alpha) \times \alpha$	$\rightarrow \underline{moving}(\alpha)$	[1D]
at	$\underline{moving}(\alpha) \times \underline{range}(\alpha)$	$\rightarrow \underline{moving}(\alpha)$	[1D]
at	$\underline{moving}(\alpha) \times \underline{point}$	$\rightarrow \underline{mpoint}$	[2D]
at	$\underline{moving}(\alpha) \times \beta$	$\rightarrow \underline{moving}(min(\alpha, \beta))$	[2D]
atmin	$\underline{moving}(\alpha)$	$\rightarrow \underline{moving}(\alpha)$	[1D]
atmax	$\underline{moving}(\alpha)$	$\rightarrow \underline{moving}(\alpha)$	[1D]
passes	$\underline{moving}(\alpha) \times \beta$	$\rightarrow \underline{bool}$	

The first and second groups of operations in Table 4.9 concern interactions with the (time) domain and range values, respectively.

In the first group, operations **atinstant** and **atperiods** restrict a moving entity to a given instant, resulting in an (instant, value) pair, or to a given set of time intervals, respectively. The **atinstant** operation is similar to the timeslice operator found in many temporal relational algebras. Operations **initial** and **final** return the first and last (instant, value) pair, respectively. Operation **present** allows one to check whether the moving value exists at a given instant, or is ever present during a given set of time intervals.

In the second group, the purpose of **at** is again restriction (like **atinstant**, **atperiods**), here to values in the range. For 1D space, restriction to a point or a point-set value returns a value of the given moving type. For example, we can restrict a moving real to the times when its value was between 3 and 4. In 2D, the resulting moving type is obtained by taking the minimum of the two argument types α and β with respect to the order $\underline{point} < \underline{points} < \underline{line} < \underline{region}$. For example, the restriction of a $\underline{moving}(\underline{region})$ by a $\underline{point}$ will result in a $\underline{moving}(\underline{point})$. This is analogous to the definitions **intersection** in 2D in Section 4.3.4.

In one-dimensional spaces, operations **atmin** and **atmax** restrict the moving value to the times when it was minimal or maximal with respect to the total order on this space. Operation **passes** determine whether the moving value ever assumed (one of) the value(s) in the second argument.

All of these operations are of interest from a language design point of view. Some of them may also be expressed in terms of other operations in the framework. For example, we have **present**$(f, t) =$ **not**(**isempty**(**val**(**atinstant**(f, t)))).

Example 9. "When and where did flight 257 enter French air space?"

```
LET entry = initial(at(route257, France));
inst(entry); val(entry)
```

Example 10. "When was the Eiffel Tower within snow storm 'Lizzy'?"

```
LET Lizzy = ELEMENT(SELECT area FROM weather
  WHERE name = "Lizzy" and kind = "snow storm");
deftime(at(Lizzy, EiffelTower))
```

Lifting Operations to Time-Dependent Operations. Section 4.3.4 systematically defines operations on non-temporal types. This section uniformly lifts these operations to apply to the corresponding $\underline{moving}$ (temporal) types. The idea is to allow any argument of a non-temporal operation to be made temporal and to return a temporal type. More specifically, the lifted version of an operation with signature $\alpha_1 \times \ldots \times \alpha_k \rightarrow \beta$ has signatures $\alpha'_1 \times \ldots \times \alpha'_k \rightarrow \underline{moving}(\beta)$ with $\alpha'_i \in \{\alpha_i, \underline{moving}(\alpha_i)\}$.

So, each argument type may be changed into a time-dependent type, which will then transform the result type into a time-dependent type. The new operations are given the same name as the operation they originate from. As an example of lifting, the **intersection** operation with signature $\underline{region} \times \underline{point} \rightarrow \underline{point}$ is lifted to the signatures $\underline{mregion} \times \underline{point} \rightarrow \underline{mpoint}$, $\underline{region} \times \underline{mpoint} \rightarrow \underline{mpoint}$, and $\underline{mregion} \times \underline{mpoint} \rightarrow \underline{mpoint}$.

To define the semantics of lifting, we note that an operation $op : \alpha_1 \times \ldots \times \alpha_k \rightarrow \beta$ can be lifted with respect to any combination of argument types. The set of lifted parameters may be described by a set $L \subseteq \{1, \ldots, k\}$, and we define:

$$\alpha_i^L = \begin{cases} \underline{moving}(\alpha_i) & \text{if } i \in L \\ \alpha_i & \text{otherwise} \end{cases}$$

Thus, the signature of any lifted version of op can be written as $op : \alpha_1^L \times \ldots \times \alpha_k^L \rightarrow \underline{moving}(\beta)$. If f_{op} is the semantics of op, we now have to define the semantics of f_{op}^L for each possible lifting L. For this we define what it means to apply a possibly lifted value to an $\underline{instant}$-value:

$$x_i^L(t) = \begin{cases} x_i(t) & \text{if } i \in L \\ x_i & \text{otherwise} \end{cases}$$

This enables a point-wise definition of the functions f_{op}^L.

$$f_{op}^L(x_1, \ldots, x_k) = \{(t, f_{op}(x_1^L(t), \ldots, x_k^L(t))) \mid t \in A_{instant}\}$$

This lifting generalizes existing operations, which perhaps did not appear to be of great utility, to new and quite useful operations. For example, an operator that determines the intersection of a region with a point may not be of great interest, but the operation that determines the intersection between a $\underline{region}$ and an $\underline{mpoint}$ ("get the part of the $\underline{mpoint}$ within the region") is quite useful. This explains why Section 4.3.4 defined the set operations for all argument types, including single points.

Time-dependent operations enable a powerful query language. Examples follow.

Example 11. We can formulate involved queries such as "For how long did the moving point mp move along the boundary of region r?"

```
duration(deftime(at(on_border(mp, r), TRUE)))
```

Predicate **on_border** yields a result of type <u>*mbool*</u>, which is defined for all times when mp is defined and has value *TRUE* or *FALSE*. Operation **at** restricts this <u>*mbool*</u> to the times when it has value *TRUE*.

Example 12. "When did snow storm 'Lizzy' consist of exactly three separate areas."

```
deftime(at(no_components(Lizzy) = 3, TRUE))
```

Here, 'Lizzy' is of type <u>*mregion*</u>, and the lifted versions of **no_components** and equality apply.

Rate of Change. An important property of any time-dependent value is its rate of change, i.e., its **derivative**. This concept is applicable to types <u>*mreal*</u> and <u>*mpoint*</u>. For the latter, we include three operators, namely **speed**, based on Euclidean distance, **turn**, based on the direction between two points, and **velocity**, based on the vector difference (viewing points as 2D vectors). The acceleration

Table 4.10. Derivative operations

Operation	Signature	Semantics
derivative	<u>*mreal*</u> → <u>*mreal*</u>	μ' where $\mu'(t) = \lim_{\delta \to 0}(f(t+\delta) - f(t))/\delta$
speed	<u>*mpoint*</u> → <u>*mreal*</u>	μ' where $\mu'(t) = \lim_{\delta \to 0} f_{\mathbf{distance}}(f(t+\delta), f(t))/\delta$
turn	<u>*mpoint*</u> → <u>*mreal*</u>	μ' where $\mu'(t) = \lim_{\delta \to 0} f_{\mathbf{direction}}(f(t+\delta), f(t))/\delta$
velocity	<u>*mpoint*</u> → <u>*mpoint*</u>	μ' where $\mu'(t) = \lim_{\delta \to 0}(\overrightarrow{f(t+\delta)} - \overrightarrow{f(t)})/\delta$

of a moving point mp may be obtained as a number by **derivative**(**speed**(mp)) and as a vector, or moving point, by **velocity**(**velocity**(mp)).

Example 13. One can observe the growth rate of a moving region: "When did 'Lizzy' expand the most?"

```
inst(initial(atmax(derivative(area(Lizzy)))))
```

Example 14. "Show on a map the parts of the route of flight 257 when the plane's speed exceeds 800 km/h."

```
trajectory(atperiods(route257,
  deftime(at(speed(route257) > 800, TRUE))))
```

The background of the map has to be produced by a different tool or query.

4.3.6 Application Example

To illustrate the use of the data types in querying, we consider an example application that concerns forest fire analysis and which allows us to explore advanced aspects of moving point and region objects.

In a number of countries, fire is one of the main agents of forest damage. Forest fire control management mainly pursues the two goals of learning from past fires and their evolution and of preventing fires in the future, by studying weather and other factors such as cover type, elevation, slope, distance to roads, and distance to human settlements. In a very simplified manner, this example considers the first goal of learning from past fires and their evolution in space and time. We assume a database containing relations with schemas

```
forest(forestname:string, territory:mregion)
forest_fire(firename:string, extent:mregion)
fire_fighter(fightername:string, location:mpoint)
```

Relation `forest` records the extents of forests, which grow and shrink over time due to, e.g., clearing, cultivation, and destruction processes. Relation `forest_fire` captures the evolution of fires, from ignition to extinction. Relation `fire_fighter` describes the motions of fire fighters on duty, from their start at the fire station up to their return. The following four queries illustrate enhanced spatio-temporal database functionality.

Example 15. "When and where did the fire called 'The Big Fire' have its largest extent?"

```
LET TheBigFire = ELEMENT(SELECT extent FROM forest_fire
  WHERE firename = "The Big Fire");
LET max_area = initial(atmax(area(TheBigFire)));
atinstant(TheBigFire, inst(max_area));
val(max_area)
```

The second argument of **atinstant** computes the time when the area of the fire was at its maximum. The area operator is used in its lifted version.

Example 16. "Determine the size of all forest areas destroyed by 'The Big Fire'." We assume that a fire may reach several, perhaps adjacent, forests.

```
LET ever = FUN (mb:mbool) passes(mb, TRUE);
LET burnt =
  SELECT size AS area(traversed(intersection(territory, extent)))
  FROM forest_fire, forest
  WHERE firename = "The Big Fire"
    and ever(intersects(territory, extent));
SELECT SUM(size)
FROM burnt
```

The **intersects** predicate of the join condition is lifted. Since the join condition expects a Boolean value, the **ever** predicate checks whether there is at least one intersection between the two <u>mregion</u> values just considered.

Example 17. "When and where was the spread of fires larger than 500 km^2?"

```
LET big_part =
  SELECT big_area AS extent when[FUN (r:region) area(r) > 500]
  FROM forest_fire;
SELECT *
FROM big_part
WHERE not(isempty(deftime(big_area)))
```

The first subquery reduces the moving region of each fire to the parts when it was large. For some fires, this may never be the case; for them, `bigarea` may be empty (always undefined). These are eliminated in the second subquery.

Example 18. " How long was fire fighter Th. Miller enclosed by 'The Big Fire' and which distance did the fire fighter cover there?

```
SELECT time AS
    duration(deftime(intersection(location, TheBigFire))),
  distance AS
    length(trajectory(intersection(location, TheBigFire)))
FROM fire_fighter
WHERE fightername = "Th. Miller"
```

We assume that the value 'TheBigFire' has already been determined as in Example 15, and that we know that Th. Miller was in this fire (otherwise, time and distance will be returned as zero).

4.3.7 Summary

This section offers an integrated, comprehensive design of abstract data types involving base types, spatial types, time types, as well as consistent temporal and spatio-temporal versions of these. Embedding this in a DBMS query language, one obtains a query language for spatio-temporal data, and moving objects in particular.

The strong points are several. The framework emphasizes genericity, closure, and consistency. An abstract level of modeling is adopted, with the design including the first comprehensive model of spatial data types (going beyond the study of just topological relationships) formulated entirely at the abstract, infinite point-set level. To our knowledge, the framework is the first to systematically and coherently use continuous functions as values of attribute data types. Finally, the idea of defining operations over non-temporal types and then uniformly lift these to operations over temporal types seems to be a new and important concept that achieves consistency between non-temporal and temporal operations.

4.4 A Discrete Model: Data Structures for Moving Objects Databases

4.4.1 Overview

In this section, which is based on [16], we define data types that can represent values of corresponding types of the abstract model just presented in Section 4.3.

Of course, the discrete types can in general only represent a subset of the values of the corresponding abstract type.

All type constructors of the abstract model will have direct counterparts in the discrete model except for the *moving* constructor. This is, because it is impossible to introduce at the discrete level a type constructor that automatically transforms types into corresponding temporal types. The type system for the discrete model therefore looks quite the same as the abstract type system shown in Table 4.2 up to the *intime* constructor, but then introduces a number of new type constructors to implement the *moving* constructor, as shown in Table 4.11.

Table 4.11. Signature describing the discrete type system

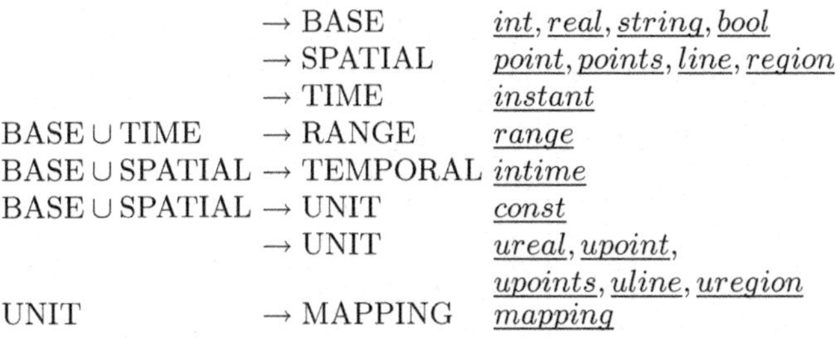

	→ BASE	*int*, *real*, *string*, *bool*
	→ SPATIAL	*point*, *points*, *line*, *region*
	→ TIME	*instant*
BASE ∪ TIME	→ RANGE	*range*
BASE ∪ SPATIAL	→ TEMPORAL	*intime*
BASE ∪ SPATIAL	→ UNIT	*const*
	→ UNIT	*ureal*, *upoint*, *upoints*, *uline*, *uregion*
UNIT	→ MAPPING	*mapping*

Let us give a brief overview of the meaning of the discrete type constructors. The base types *int*, *real*, *string*, *bool* can be implemented directly in terms of corresponding programming language types. The spatial types *point* and *points* also have direct discrete representations whereas for the types *line* and *region* linear approximations (i.e., polylines and polygons) are introduced. Type *instant* is also represented directly in terms of programming language real numbers. The *range* and *intime* types represent sets of intervals, or pairs of time instants and values, respectively. These representations are also straightforward.

The interesting part of the model is how temporal ("moving") types are represented. We here describe the so-called *sliced representation*. The basic idea is to decompose the temporal development of a value into fragments called "slices" such that within the slice this development can be described by some kind of "simple" function. This is illustrated in Figure 4.5.

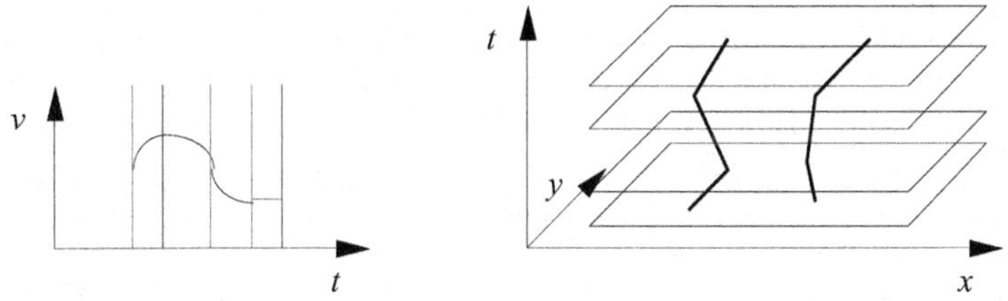

Fig. 4.5. Sliced representation of moving *real* and moving *points* value

The sliced representation is built by a type constructor *mapping* parameterized by the type describing a single slice which we call a *unit* type. A value of a unit type is a pair (i, v) where i is a time interval and v is some representation of a simple function defined within that time interval. We define unit types *ureal*, *upoint*, *upoints*, *uline*, and *uregion*. For values that can only change discretely, there is a trivial "simple" function, namely the constant function. It is provided by a *const* type constructor which produces units whose second component is just a constant of the argument type. This is in particular needed to represent moving *int*, *string*, and *bool* values. The *mapping* data structure basically just assembles a set of units and makes sure that their time intervals are disjoint.

In summary, we obtain the correspondence between abstract and discrete temporal types shown in Table 4.12.

Table 4.12. Correspondence between abstract and discrete temporal types

Abstract Type	Discrete Type
moving(*int*)	*mapping*(*const*(*int*))
moving(*string*)	*mapping*(*const*(*string*))
moving(*bool*)	*mapping*(*const*(*bool*))
moving(*real*)	*mapping*(*ureal*)
moving(*point*)	*mapping*(*upoint*)
moving(*points*)	*mapping*(*upoints*)
moving(*line*)	*mapping*(*uline*)
moving(*region*)	*mapping*(*uregion*)

In Table 4.12 we have omitted the representations *mapping*(*const*(*real*)), etc. which can be used to represent discretely changing real values and so forth, but are not so interesting for us.

In the remainder of this section we formally define the data types of the discrete model. That means, for each type we define its *domain* of values in terms of some finite representation. From an algebraic point of view, we define for each *sort* (type) a *carrier set*. For a type α we denote its carrier set as D_α.

Of course, each value in D_α is supposed to represent some value of the corresponding abstract domain, that is, the carrier set of the corresponding abstract type. For a type α of the abstract model, let A_α denote its carrier set. We can view the value $a \in A_\alpha$ that is represented by $d \in D_\alpha$ as the *semantics* of d. We will always make clear which value from A_α is meant by a value from D_α. Often this is obvious, or an informal description is sufficient. Otherwise we provide a definition of the form $\sigma(d) = a$ where σ denotes the "semantics" function.

The following Section 4.4.2 contains definitions for all non-temporal types and for the temporal types in the sliced representation. For the spatial temporal data types *moving*(*points*), *moving*(*line*), and *moving*(*region*) one can also define direct three-dimensional representations in terms of polyhedra etc.; these representations will be treated elsewhere.

In Chapter 6 of this book we will present some examples of how this high level specification translates into physical data structures and algorithms.

4.4.2 Definition of Discrete Data Types

Base Types and Time Type. The carrier sets of the *discrete base types* and the type for time rest on available programming language types. Let *Instant* = `real`.

$$D_{\underline{int}} = \texttt{int} \cup \{\bot\} \qquad D_{\underline{real}} = \texttt{real} \cup \{\bot\} \qquad D_{\underline{string}} = \texttt{string} \cup \{\bot\}$$
$$D_{\underline{bool}} = \texttt{bool} \cup \{\bot\} \quad D_{\underline{instant}} = \mathit{Instant} \cup \{\bot\}$$

The only special thing about these types is that they always include the undefined value $\bot$ as required by the abstract model. Since we are interested in continuous evolutions of values, type *instant* is defined in terms of the programming language type `real`.

We sometimes need to speak about only the defined values of some carrier set and therefore introduce a notation for it: Let $D'_\alpha = D_\alpha \setminus \{\bot\}$. We will later introduce carrier sets whose elements are sets themselves; for them we extend this notation to mean $D'_\alpha = D_\alpha \setminus \{\emptyset\}$.

Spatial Data Types. Next, we define finite representations for single points, point collections, lines, and regions in two-dimensional (2D) Euclidean space. A point is, as usual, given by a pair (x, y) of coordinates. Let *Point* = `real` × `real` and

$$D_{\underline{point}} = \mathit{Point} \cup \{\bot\}$$

The semantics of an element of $D_{\underline{point}}$ is obviously an element of $A_{\underline{point}}$. We assume lexicographical order on points, that is, given any two points $p, q \in \mathit{Point}$, we define: $p < q \Leftrightarrow (p.x < q.x) \vee (p.x = q.x \wedge p.y < q.y)$.

A value of type *points* is simply a set of points.

$$D_{\underline{points}} = 2^{\mathit{Point}}$$

Again it is clear that a value of $D_{\underline{points}}$ represents a value of the abstract domain $A_{\underline{points}}$.

The definition of discrete representations for the types *line* and *region* is based on linear approximations. A value of type *line* is essentially just a finite set of line segments in the plane. Figure 4.6 shows the correspondence between the abstract type for *line* and the discrete type. The abstract type is a set of curves in the plane which was viewed in Section 4.3 as a planar graph whose nodes are intersections of curves and whose edges are intersection-free pieces of curves. The discrete *line* type represents curves by polylines. However, one can assume a less structured view and consider the same shape to be just a collection of line segments. At the same time, any collection of line segments in the plane defines a valid collection of curves (or planar graph) of the abstract model (see Figure 4.6 (c)). Hence, modeling *line* as a set of line segments is no less expressive than the polyline view. It has the advantage that computing the projection of a (discrete representation) moving point into the plane can be done

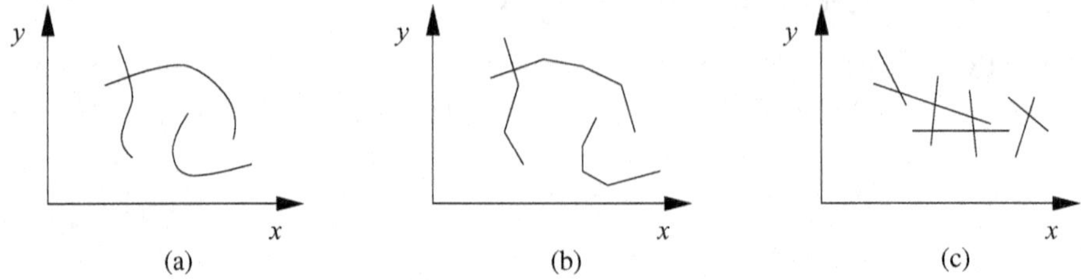

Fig. 4.6. (a) line value of the abstract model (b) line value of the discrete model (c) any set of line segments is also a line value

very efficiently as it is not necessary to compute the polyline or graph structure. Hence we prefer to use this unstructured view. Let

$$Seg = \{(u, v) \mid u, v \in Point, u < v\}$$

be the set of all line segments.

$$D_{\underline{line}} = \{S \subset Seg \mid \forall s, t \in Seg : s \neq t \land collinear(s, t) \Rightarrow disjoint(s, t)\}$$

The predicate *collinear* means that two line segments lie on the same infinite line in 2D space. Hence for a set of line segments to be a line value we only require that there are no collinear, overlapping segments. This condition ensures unique representation, as collinear overlapping segments could be merged into a single segment. The semantics of a line value is, of course, the union of the points on all of its segments.

A region value at the discrete level is essentially a collection of polygons with polygonal holes (Figure 4.7). Formal definitions are based on the notions

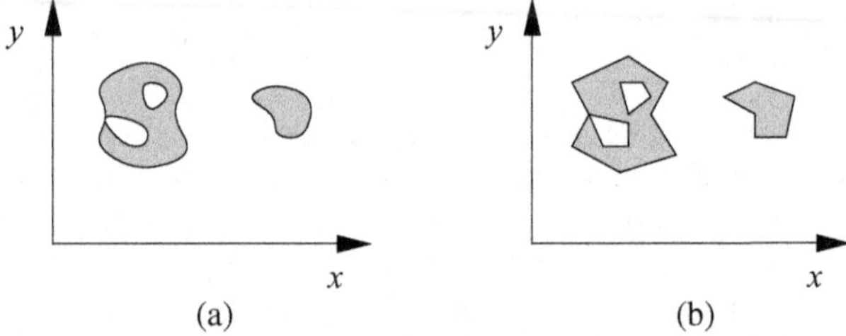

Fig. 4.7. (a) region value of the abstract model (b) region value of the discrete model

of *cycles* and *faces*. These definitions are similar to those of the ROSE algebra [22]. We need to reconsider such definitions here for two reasons: (i) They have to be modified a bit because here we have no "realm-based" [22] environment any more, and (ii) we are going to extend them to the "moving" case in the following sections.

A *cycle* is a simple polygon, defined as follows:

$$\begin{aligned}
Cycle = \{S \subset Seg \mid |S| = n, n \geq 3, \text{ such that} \\
(i)\ \forall s, t \in S : s \neq t \Rightarrow \neg p\text{-}intersect(s,t) \wedge \neg touch(s,t) \\
(ii)\ \forall p \in points(S) : card(p, S) = 2 \\
(iii)\ \exists \langle s_0, \ldots, s_{n-1} \rangle : \{s_0, \ldots, s_{n-1}\} = S \\
\wedge (\forall i \in \{0, \ldots, n-1\} : meet(s_i, s_{(i+1) \bmod n}))\}
\end{aligned}$$

Two segments *p-intersect* ("properly intersect") if they intersect in their interior (a point other than an end point); they *touch* if one end point lies in the interior of the other segment. Two segments *meet* if they have a common end point. The set $points(S)$ contains all end points of segments, hence is $points(S) = \{p \in Point \mid \exists s \in S : s = (p,q) \vee s = (q,p)\}$. The function $card(p, S)$ tells how often point p occurs in S and is defined as $card(p, S) = |\{s \in S \mid s = (p,q) \vee s = (q,p)\}|$. Hence a collection of segments is a cycle, if (i) no segments intersect properly, (ii) each end point occurs in exactly two segments, and (iii) segments can be arranged into a single cycle rather than several disjoint ones (the notation $\langle s_0, \ldots, s_{n-1} \rangle$ refers to an ordered list of segments).

A *face* is a pair consisting of an outer cycle and a possibly empty set of hole cycles.

$$\begin{aligned}
Face = \{(c, H) \mid c \in Cycle, H \subset Cycle, \text{ such that} \\
(i)\ \forall h \in H : edge\text{-}inside(h, c) \\
(ii)\ \forall h_1, h_2 \in H : h_1 \neq h_2 \Rightarrow edge\text{-}disjoint(h_1, h_2) \\
(iii)\ \text{any cycle that can be formed from the segments of } c \text{ or } H \text{ is either} \\
c \text{ or one of the cycles of } H
\end{aligned}$$

A cycle c is *edge-inside* another cycle d if its interior is a subset of the interior of d and no edges of c and d overlap. They are *edge-disjoint* if their interiors are disjoint and none of their edges overlap. Note that it is allowed that a segment of one cycle *touches* a segment of another cycle. Overlapping segments are not allowed, since then one could remove the overlapping parts entirely (e.g. two hole cycles could be merged into one hole). The last condition (iii) ensures unique representation, that is, there are no two different interpretations of a set of segments as sets of faces. This implies that a face cannot be decomposed into two or more edge-disjoint faces.

A *region* is then basically a set of disjoint faces.

$$D_{\underline{region}} = \{F \subset Face \mid f_1, f_2 \in F \wedge f_1 \neq f_2 \Rightarrow edge\text{-}disjoint(f_1, f_2)\}$$

More precisely, faces have to be *edge-disjoint*. Two faces (c_1, H_1) and (c_2, H_2) are *edge-disjoint* if either their outer cycles c_1 and c_2 are edge-disjoint, or one of the outer cycles, e.g. c_1, is *edge-inside* one of the holes of the other face (some $h \in H_2$). Hence faces may also touch each other in an isolated point, but must not have overlapping boundary segments.

The semantics of a region value should be clear: A cycle c represents all points of the plane enclosed by it as well as the points on the boundary. Given $\sigma(c)$, we have for a face $\sigma((c, H)) = closure(\sigma(c) \setminus \bigcup_{h \in H} \sigma(h))$, that is, hole areas are subtracted from the outer cycle area, but then the resulting point set is closed again in the abstract domain. The area of a region is then obviously the union of the area of its faces.

Sets of Intervals. In this subsection, we introduce the *non-constant* <u>range</u> type constructor which converts a given type $\alpha \in \text{BASE} \cup \text{TIME}$ into a type whose values are finite sets of intervals over α. Note that on all such types α a total order exists. Range types are needed, for example, to represent collections of time intervals, or the values taken by a moving real.

Let $(S, <)$ be a set with a total order. The representation of an interval over S is given by the following definition.

$$\begin{aligned} Interval(S) = \{&(s, e, lc, rc) | s, e \in S, lc, rc \in \texttt{bool}, \\ &s \leq e, (s = e) \Rightarrow (lc = rc = true)\}. \end{aligned}$$

Hence an interval is represented by its end points s and e and two flags lc and rc indicating whether it is left-closed and/or right-closed. The meaning of an interval representation (s, e, lc, rc) is

$$\sigma((s, e, lc, rc)) = \{u \in S | s < u < e\} \cup LC \cup RC$$

where the two sets LC and RC are defined as

$$LC = \begin{cases} \{s\} & \text{if } lc \\ \emptyset & \text{otherwise} \end{cases} \quad \text{and} \quad RC = \begin{cases} \{e\} & \text{if } rc \\ \emptyset & \text{otherwise} \end{cases}$$

Given an interval i, we denote with $\sigma'(i)$ the semantics expressed by $\sigma(i)$ restricted to the open part of the interval.

Whether intervals $u = (s_u, e_u, lc_u, rc_u)$ and $v = (s_v, e_v, lc_v, rc_v) \in Interval(S)$ are *disjoint* or *adjacent* is defined as follows:

$$\begin{aligned} r\text{-}disjoint(u, v) &\Leftrightarrow e_u < s_v \vee (e_u = s_v \wedge \neg(rc_u \wedge lc_v)) \\ disjoint(u, v) &\Leftrightarrow r\text{-}disjoint(u, v) \vee r\text{-}disjoint(v, u) \\ r\text{-}adjacent(u, v) &\Leftrightarrow disjoint(u, v) \wedge (e_u = s_v \wedge (rc_u \vee lc_v)) \vee \\ &\quad ((e_u < s_v \wedge rc_u \wedge lc_v) \wedge \neg(\exists w \in S \,|\, e_u < w < s_v)) \\ adjacent(u, v) &\Leftrightarrow r\text{-}adjacent(u, v) \vee r\text{-}adjacent(v, u) \end{aligned}$$

The last condition for *r-adjacent* is important for discrete domains such as <u>int</u>. Representations of finite sets of intervals over S can now be defined as

$$\begin{aligned} IntervalSet(S) = \{&V \subseteq Interval(S) \,| \\ &(u, v \in S \wedge u \neq v) \Rightarrow disjoint(u, v) \wedge \neg adjacent(u, v)\} \end{aligned}$$

The conditions ensure that a set of intervals has a unique and minimal representation. The <u>range</u> type constructor can then be defined as:

$$D_{\underline{range}(\alpha)} = IntervalSet(D'_\alpha) \quad \forall \alpha \in \text{BASE} \cup \text{TIME}$$

We also define the <u>intime</u> type constructor in this subsection which yields types whose values consist of a time instant and a value, as in the abstract model.

$$D_{\underline{intime}(\alpha)} = D_{\underline{instant}} \times D_\alpha \quad \forall \alpha \in \text{BASE} \cup \text{SPATIAL}$$

Sliced Representation for Moving Objects. In this subsection we introduce and formalize the *sliced representation* for moving objects. The sliced representation is provided by the <u>mapping</u> type constructor which represents a moving object as a set of so-called *temporal units* (*slices*). Informally speaking, a temporal unit for a moving data type α is a maximal interval of time where values taken by an instance of α can be described by a "simple" function. A temporal unit therefore records the evolution of a value v of some type α in a given time interval i, while ensuring the maintenance of type-specific constraints during such an evolution.

For a set of temporal units representing a moving object their time intervals are mutually disjoint, and if they are adjacent, their values are distinct. These requirements ensure unique and minimal representations.

Temporal units are described as a generic concept in this subsection. Their specialization to various data types is given in the next two subsections. Let S be a set. The concept of temporal unit is defined by:

$$Unit(S) = Interval(Instant) \times S$$

A pair (i, v) of $Unit(S)$ is called a *temporal unit* or simply a *unit.* Its first component is called the *unit interval,* its second component the *unit function.*

The <u>mapping</u> type constructor allows one to build sets of units with the required constraints. Let

$$\begin{aligned}
&Mapping(S) = \{U \subseteq Unit(S) \,|\, \forall (i_1, v_1) \in U, \forall (i_2, v_2) \in U : \\
&\;(i)\ i_1 = i_2 \Rightarrow v_1 = v_2 \\
&\;(ii)\ i_1 \neq i_2 \Rightarrow (disjoint(i_1, i_2) \wedge (adjacent(i_1, i_2) \Rightarrow v_1 \neq v_2))\}
\end{aligned}$$

The <u>mapping</u> type constructor is defined for any type $\alpha \in UNIT$ as:

$$D_{\underline{mapping}(\alpha)} = Mapping(D_\alpha) \quad \forall \alpha \in UNIT.$$

In the next subsections we will define the types <u>ureal</u>, <u>upoint</u>, <u>upoints</u>, <u>uline</u>, and <u>uregion</u>. Since all of them will have the structure of a unit, the just introduced type constructor <u>mapping</u>(α) can be applied to all of them.

Units describe certain simple functions of time. We will define a generic function ι on units which evaluates the unit function at a given time instant. More precisely, let α be a non-temporal type (e.g. <u>real</u>) and u_α the corresponding unit type (e.g. <u>ureal</u>) with $D_{u_\alpha} = Interval(Instant) \times S_\alpha$, where S_α is a suitably defined set. Then ι_α is a function

$$\iota_\alpha : S_\alpha \times Instant \rightarrow D_\alpha$$

Usually we will omit the index α and just denote the function by ι. Hence, ι maps a discrete representation of a unit function for a given instant of time into a discrete representation of the function value at that time. The ι function serves three purposes: (i) It allows us to express constraints on the structure of a unit in terms of constraints on the structure of the corresponding non-temporal value. (ii) It allows us to express the semantics of a unit by reusing the semantics definition of the corresponding non-temporal value. (iii) It can serve as a basis for the implementation of the **atinstant** operation on the unit.

The use of ι will become clear in the next subsections when we instantiate it for the different unit types.

Temporal Units for Base Types. For a type $\alpha \in$ BASE $\cup$ SPATIAL, we introduce the type constructor <u>*const*</u> that produces a temporal unit for α. Its carrier set is defined as:

$$D_{\underline{const}(\alpha)} = \mathit{Interval}(\mathit{Instant}) \times D'_\alpha$$

Recall that the notation D'_α refers to the carrier set of α without undefined elements or empty sets. A unit containing an undefined or empty value makes no sense as for such time intervals we can simply let no unit exist (within a <u>*mapping*</u>).

Note that, even if we introduce the type constructor <u>*const*</u> with the explicit purpose of defining temporal units for <u>*int*</u>, <u>*string*</u>, and <u>*bool*</u>, it can nevertheless be applied also to other types. This may be useful for applications where values of such types change only in discrete steps.

The trivial temporal function described by such a unit can be defined as

$$\iota(v, t) = v$$

Note that in defining ι for a specific unit type we automatically define the semantics of the unit which should be a temporal function in the abstract model. For example, for a value u of a unit type <u>*const*</u>(<u>*int*</u>) the semantics $\sigma(u)$ should be a partial function $f : A'_{\underline{instant}} \to A'_{\underline{int}}$. This is covered by a generic definition of the semantics of unit types: Let $u = (i, v)$ be a value of a unit type u_α. Then

$$\sigma(u) = f_u : A'_{\underline{instant}} \cap \sigma(i) \to A'_\alpha \quad \text{where}$$
$$f_u(t) = \sigma(\iota(v, t)) \quad \forall t \in \sigma(i)$$

Hence we reuse the semantics defined for the discrete value $\iota(v, t) \in D'_\alpha$.

This semantics definition will in most cases be sufficient. However, for some unit types (namely, <u>*uline*</u> and <u>*uregion*</u>) the discrete value obtained in the end points of the time interval by ι may be an incorrect one due to degeneracies: in such a case it has to be "cleaned up." We will below slightly extend the generic semantics definition to accommodate this. For all other units, this semantics definition suffices so that we will only define the ι function in each case.

For the representation of moving reals we introduce a unit type <u>*ureal*</u>. The "simple" function we use for the sliced representation of moving reals is either a polynomial of degree not higher than two or a square root of such a polynomial. The motivation for this choice is a trade-off between richness of the representation (e.g. square roots of degree two polynomials are needed to express time-dependent distance functions in the Euclidean metric) and simplicity of the representation of the discrete type and of its operations. With this particular choice one can implement (i.e., the discrete model is closed under) the lifted versions of **size**, **perimeter**, and **distance** operations; one cannot implement the **derivative** operation of the abstract model. The carrier set for type <u>*ureal*</u> is

$$D_{\underline{ureal}} = \mathit{Interval}(\mathit{Instant}) \times \{(a, b, c, r) \mid a, b, c \in \texttt{real}, r \in \texttt{bool}\}$$

and evaluation at time t is defined by:

$$\iota((a, b, c, r), t) = \begin{cases} at^2 + bt + c & \text{if } \neg r \\ \sqrt{at^2 + bt + c} & \text{if } r \end{cases}$$

Temporal Units for Spatial Data Types. In this subsection we specialize the concept of unit to moving instances of spatial data types.

Similar to moving reals, the temporal evolution of moving spatial objects is characterized by continuity and smoothness and can be approximated in various ways. Again we have to find the balance between richness and simplicity of representation. As indicated before, in this chapter we make the design decision to base our approximations of the temporal behavior of moving spatial objects on linear functions. Linear approximations ensure simple and efficient representations for the data types and a manageable complexity of the algorithms. Nevertheless, more complex functions like polynomials of a degree higher than one are conceivable as the basis of representation but are not considered in this paper.

Due to the concept of sliced representation, also for moving spatial objects we have to specify constraints in order to describe the permitted behavior of a value of such a type within a temporal unit. Since the end points of a time interval mark a change in the description of the data type, we require that constraints are satisfied only for the respective open interval. In the end points of the time interval a collapse of components of the moving object can happen. This is completely acceptable, since one of the reasons to introduce the sliced representation is exactly to have "simple" and "continuous" description of the moving value within each time interval and to limit "discontinuities" in the description to a finite set of instants.

Moving Points and Point Sets. The structurally simplest spatial object that can move is a single point. Hence, we start with the definition of the spatial unit type <u>*upoint*</u>. First we introduce a set *MPoint* which defines 3D lines that describe unlimited temporal evolution of 2D points.

$$\mathit{MPoint} = \{(x_0, x_1, y_0, y_1) \mid x_0, x_1, y_0, y_1 \in \texttt{real}\}$$

This describes a linearly moving point for which evaluation at time t is given by:

$$\iota((x_0, x_1, y_0, y_1), t) = (x_0 + x_1 \cdot t, y_0 + y_1 \cdot t) \quad \forall t \in \mathit{Instant}$$

The carrier set of <u>*upoint*</u> can then be very simply defined as:

$$D_{\underline{upoint}} = \mathit{Interval}(\mathit{Instant}) \times \mathit{MPoint}$$

We pass now to describe a set of moving points. The carrier set of <u>*upoints*</u> can be defined as:

$$\begin{aligned} D_{\underline{upoints}} = \{(i, M) \mid i \in \mathit{Interval}(\mathit{Instant}), M \subset \mathit{MPoint}, |M| \geq 1, \text{ and} \\ (i)\ \forall t \in \sigma'(i), \forall l, k \in M : l \neq k \Rightarrow \iota(l, t) \neq \iota(k, t) \\ (ii)\ i = (s, e, lc, rc) \wedge s = e \Rightarrow (\forall l, k \in M : l \neq k \Rightarrow \iota(l, s) \neq \iota(k, s))\} \end{aligned}$$

Here we encounter for the first time a constraint valid during the open time interval of the unit (condition (i)). Namely, a <u>*upoints*</u> unit is a collection of linearly moving points that do not intersect within the open unit interval. Condition (ii) concerns units defined only in a single time instant; for them all points have to be distinct at that instant.

For $(i, M) \in D_{\underline{upoints}}$, evaluation at time t is given by

$$\iota(M, t) = \bigcup_{m \in M} \{\iota(m)\} \quad \forall t \in \sigma(i)$$

which is clearly a set of points in $D'_{\underline{points}}$. We will generally assume that ι distributes through sets and tuples so that $\iota(M, t)$ is defined for any set M as above, and for a tuple $r = (r_1, \ldots, r_n)$, we have $\iota(r, t) = (\iota(r_1), \ldots, \iota(r_n))$.

Moving Lines. We now introduce the unit type for <u>*line*</u> called <u>*uline*</u>. Here we restrict movements of segments so that in the time interval associated to a value of <u>*uline*</u> each segment maintains its direction in the 2-dimensional space. That is, segments which rotate during their movement are not admitted. See in Figure 4.8 an example of a valid <u>*uline*</u> value. This constraint derives from the need of

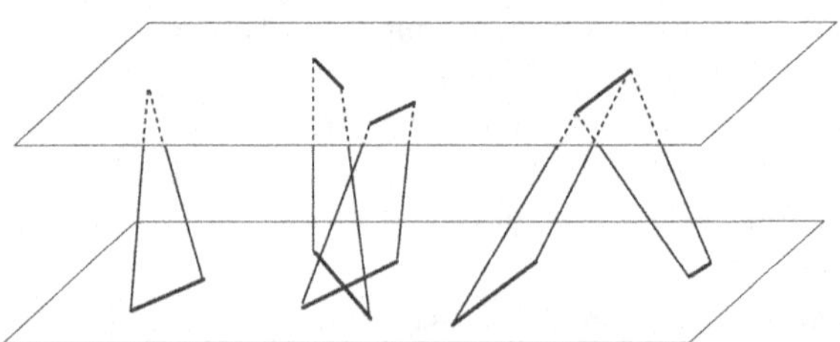

Fig. 4.8. An instance of <u>*uline*</u>

keeping a balance between ease of representation and manipulation of the data type and its expressive power. Rotating segments define curved surfaces in the 3D space that, even if they constitute a more accurate description, can always be approximated by a sequence of plane surfaces.

The carrier set of $\underline{uline}$ is therefore based on a set of moving segments with the above restriction and which never overlaps at any instant internal to the associated open time interval. Overlapping has a meaning equivalent to the one used for $\underline{line}$ values: to be collinear and to have a non-empty intersection.

To prepare the definition of $\underline{uline}$ we introduce the set of all pairs of lines in a 3D space that are coplanar, which will be used to represent moving segments:

$$MSeg = \{(s,e) \mid s, e \in MPoint, s \neq e, s \text{ is coplanar with } e\}.$$

The carrier set for $\underline{uline}$ can now be defined as:

$$\begin{aligned} D_{\underline{uline}} = \{(i, M) \mid i \in \mathit{Interval}(\mathit{Instant}), M \subset MSeg, |M| \geq 1, \text{ such that} \\ (i)\ \forall t \in \sigma'(i) : \iota(M,t) \in D'_{\underline{line}} \\ (ii)\ i = (s,e,lc,rc) \wedge s = e \Rightarrow \iota(M,s) \in D'_{\underline{line}}\} \end{aligned}$$

Here again the first condition defines constraints for the open time interval and the second treats the case of units defined only at a single instant. Note that $\iota(M,t)$ is defined due to the fact that ι distributes through sets and tuples. A $\underline{uline}$ value therefore inherits the structural conditions on $\underline{line}$ values and segments. For example, condition (i) requires that

$$(s,e) \in M \Rightarrow (\iota(s,t), \iota(e,t)) \in Seg \quad \forall t \in \sigma'(i)$$

and therefore $\iota(s,t) < \iota(e,t) \quad \forall t \in \sigma'(i)$.

The semantics defined for $\underline{uline}$ via ι according to the generic definition given earlier needs to be slightly changed to cope with degeneracies in the end points of a unit time interval, as we anticipated. In these points, in fact, moving segments can degenerate into points and different moving segments can overlap. We accommodate this by defining separate ι functions for the start time and the end time of the time interval, called ι_s and ι_e, respectively. Let $((s,e,lc,rc), M) \in D_{\underline{uline}}$. Then

$$\iota_s(M,t) = \iota_e(M,t) = \mathit{merge\text{-}segs}(\{(p,q) \in \iota(M,t) \mid p < q\}$$

This definition removes pairs of points returned by $\iota(M,t)$ that are not segments (i.e., segments degenerated into a single point); it also merges overlapping segments into maximal ones (this is the meaning of the *merge-segs* function). The generic semantics definition is then extended as follows:

$$\sigma(u) = f_u : A'_{\underline{instant}} \cap \sigma(i) \rightarrow A'_\alpha$$

where for $u = (i,v)$ and $i = (s,e,lc,rc)$

$$f_u(t) = \begin{cases} \sigma(\iota(v,t)) & \text{if } t \in \sigma'(i) \\ \sigma(\iota_s(v,t)) & \text{if } t = s \wedge lc \\ \sigma(\iota_e(v,t)) & \text{if } t = e \wedge rc \end{cases}$$

A final remark on the design decisions for the discrete type for moving lines is the following. Assume we choose instance u_1 (resp., u_2) of <u>*uline*</u> as the discrete representation at the initial (resp., final) time t_1 (t_2) of a unit for the (continuously) moving line l. Then the constraint that segments making up the discrete representation of l cannot rotate during the unit does not restrict too much the fidelity of the discrete representation. Indeed, since members of *MSeg* in a unit can be triangles, this leaves the possibility of choosing among many possible mappings between endpoints of their segments in t_1 and those in t_2, as long as the non-rotation constraint is satisfied. In Figure 4.9 an example of a discrete representation of a continuously moving line by means of an instance of <u>*uline*</u> is shown. If this approach causes a too rough approximation internally to

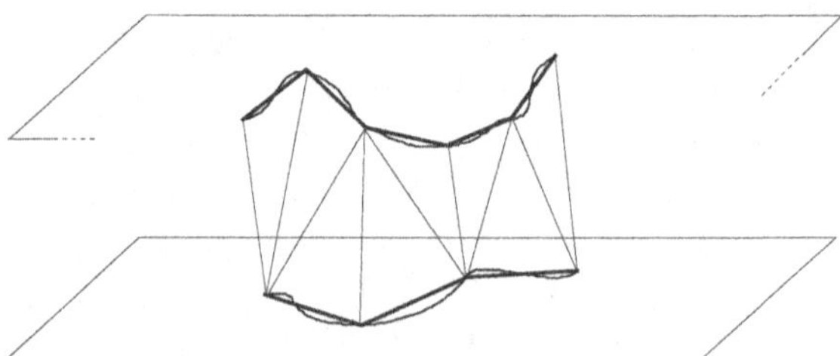

Fig. 4.9. A discrete representation of a moving line

the time unit, then possibly an additional instant, internal to the unit, has to be chosen and an additional discrete representation of l at that instant has to be introduced so that a better approximation is obtained. It can be easily seen that in the limit this sequence of discrete representations can reach an arbitrary precision in representing l.

Moving Regions. We now introduce the moving counterpart for <u>*region*</u>, namely the <u>*uregion*</u> data type. We adopt the same restriction used for moving lines, i.e., that rotation of segments in the 3-dimensional space is not admitted. We therefore base the definition of <u>*uregion*</u> on the same set of all pairs of lines in a 3D space that are coplanar, namely *MSeg*, with additional constraints ensuring that throughout the whole unit we always obtain a valid instance of the <u>*region*</u> data type. Figure 4.10 shows an example of a valid <u>*uregion*</u> value. (It also shows the degeneracies that can occur in the end points of a unit interval.)

As for a <u>*region*</u> value, we can have moving regions with (moving) holes, hence the basic building blocks are given by the concepts of *cycle* and *face* already introduced in the definition of <u>*region*</u>.

The carrier set of <u>*uregion*</u> is therefore based, informally speaking, on a set of (possibly nested) faces which never intersect at any instant internal to the associated time interval. For the formal definition of <u>*uregion*</u>, we first introduce a set intended to describe the moving version of a cycle, without restriction on

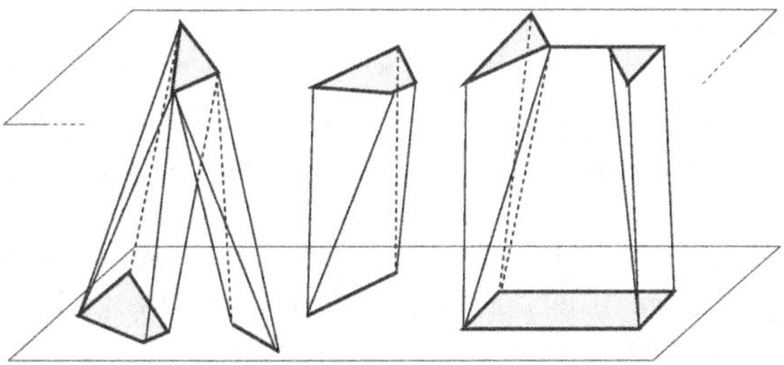

Fig. 4.10. An instance of <u>*uregion*</u>

time:

$$MCycle = \{\{s_0, \ldots, s_{n-1}\} \mid n \geq 3, \forall i \in \{0, \ldots, n-1\} : \; s_i \in MSeg\}$$

We then introduce a set for the description of the moving version of a face, without restriction on time:

$$MFace = \{(c, H) \mid c \in MCycle, H \subset MCycle\}.$$

Note that in the definitions of *MCycle* and *MFace* we have not given the constraints to impose on the sets the semantics of cycles and faces because this will be done directly in the moving region definition. The carrier set for <u>*uregion*</u> is now defined as

$$\begin{aligned} D_{\underline{uregion}} = \{(i, F) \mid i \in Interval(Instant), F \subset MFace, \text{ such that} \\ (i)\ \forall t \in \sigma'(i) : \iota(F, t) \in D'_{\underline{region}} \\ (ii)\ i = (s, e, lc, rc) \wedge s = e \Rightarrow \iota(F, s) \in D'_{\underline{region}}\} \end{aligned}$$

For the end points of the time interval again we have to provide separate functions ι_s and ι_e. Essentially these work as follows. From the pairs of points (p, q) (segments) obtained by evaluating $\iota(F, s)$ or $\iota(F, e)$, remove all pairs that are no proper segments (as for <u>*uline*</u>). Next, for all collections of overlapping segments on a single line, partition the line into fragments belonging to the same set of segments (e.g. if segment (p, q) overlaps (r, s) such that points are ordered on the line as $\langle p, r, q, s\rangle$ then there are fragments $(p, r), (r, q)$, and (q, s)). For each fragment, count the number of segments containing it. If this number is even, remove the fragment; if it is odd, put the fragment as a new segment into the result. A complete formalization of this is lengthy and omitted.

We have now concluded the formal definition of data types of the discrete model. In Chapter 6 we will show some examples of translation into physical data structures of the above specifications and we will provide some examples of algorithms implementing operations on discrete data types.

4.5 Outlook

This section presents two extensions of the approach presented so far in this chapter, and two other approaches to spatio-temporal modeling with a different focus of interest.

Section 4.5.1 addresses the problem of defining spatio-temporal predicates and their composition, in order to describe developments of relationships between (moving) objects. For example, one might want to ask in a query for moving region objects that were first disjoint, then overlapped, and finally were disjoint again. Section 4.5.2 considers time-varying partitions of the plane, for example the countries of Europe over the last centuries, and operations on such dynamic partitions.

Section 4.5.3 presents a data model based on "quanta" where space is rasterized. Consider a chess board. Atomic spatial entities or "quanta" would be all corners of fields (point quanta), horizontal or vertical edges of fields (line quanta), and fields themselves (surface quanta). Spatial data types are defined as unions of such quanta and relational algebra is extended to allow grouping ("fold") or decomposition of spatial values. By adding time intervals, this model can also describe (discretely changing) spatio-temporal data.

The final subsection Section 4.5.4 addresses the treatment of legacy databases and their applications when a "dimension", which could be a time or a space dimension, is added. For example, a static table is extended by a valid time attribute. The goal is that applications that did not know about the time attribute can run unchanged, and will yield the same results as before.

4.5.1 Spatio-temporal Predicates and Developments

Spatio-temporal predicates characterize temporal changes of relationships between spatio-temporal objects. In the following we briefly discuss some of the design issues that arise with spatio-temporal predicates.

What Are Spatio-temporal Predicates? A basic spatio-temporal predicate can be thought of as a lifted spatial predicate yielding a temporal boolean, which is aggregated by determining whether that temporal boolean was sometimes or always true. In general, a spatio-temporal predicate can be viewed as a function that aggregates the values of a (possibly changing) spatial predicate as it evolves over time. Thus, a spatio-temporal predicate is a function of type $\tau(\alpha) \times \tau(\beta) \rightarrow \mathbb{B}$ for $\alpha, \beta \in \{point, region\}$.

Consider the definition of *inside* from Section 4.3. We can define two spatio-temporal predicates *sometimes-inside* and *always-inside* that yield true if *inside* yields true at some time, respectively, at all times.

Whereas the definition for *sometimes-inside* is certainly reasonable, the definition for *always-inside* is questionable since it yields false whenever the point or the region is undefined. This is not what we would expect. For example, when the moving point has a shorter lifetime than the evolving region but is always inside the region, we would expect *always-inside* to yield true. In fact, we can distinguish different kinds of universal quantification that result from different time intervals over which aggregation can be defined to range. In the case of *inside* the expected behavior is obtained if the aggregation ranges over the lifetime of the first argument, the moving point. This is not true for all spatial predicates; actually, it depends on the nature and use of each individual predicate. For example, two spatio-temporal objects are considered as being *equal* only if they are equal on both objects' lifetimes, that is, the objects must have the same lifespans and must be always equal during these.

We denote different kinds of $\forall$-aggregation by parameterized quantifiers $\forall_\gamma$ where $\gamma \in \{\cup, \cap, \pi_1, \pi_2\}$ and where $\pi_i(x_1, \ldots, x_i, \ldots, x_n) = x_i$. These quantifiers are defined as follows.

$$\forall_\gamma p \;:=\; \lambda(S_1, S_2).\forall t \in \gamma(dom(S_1), dom(S_2)) : p(S_1(t), S_2(t))$$

This means that, for example, $\forall_{\pi_1}.inside$ denotes the spatio-temporal predicate

$$\lambda(S_1, S_2).\forall\ t \in dom(S_1).inside(S_1(t), S_2(t))$$

In general, $\lambda(x_1, x_2, \ldots).e$ denotes a function that takes arguments $x_1, x_2, \ldots$ and returns a value determined by the expression e.

With this aggregation notation we can give the definitions for the spatio-temporal versions of the eight basic spatial predicates (for two regions).

$$
\begin{array}{lcl}
Disjoint & := & \forall_\cap.disjoint \\
Meet & := & \forall_\cup.meet \\
Overlap & := & \forall_\cup.overlap \\
Equal & := & \forall_\cup.equal \\
Covers & := & \forall_{\pi_2}.covers \\
CoveredBy & := & \forall_{\pi_1}.coveredBy \\
Contains & := & \forall_{\pi_2}.contains \\
Inside & := & \forall_{\pi_1}.inside
\end{array}
$$

For a moving point and a moving region we have just the three basic predicates *Disjoint*, *Meet*, and *Inside*, which are defined as above.

The chosen aggregations (and possible variations) are motivated and discussed in great detail in [11].

Developments: Sequences of Spatio-temporal Predicates. Consider a plane entering a storm. This scenario is abstractly characterized by a moving point that initially is disjoint from an evolving region for some period of time, then touches its border, and finally remains inside of it. In other words, the described development is characterized by a sequence of spatio-temporal (and spatial) predicates: *Disjoint*, *meet*, and *Inside*. In order to define such predicate sequences we need a way of restricting the temporal scope of basic spatio-temporal predicates. This becomes possible by *predicate constrictions*: let P be a spatio-temporal predicate, and let I be a (half-) open or closed interval. Then

$$P_I \;:=\; \lambda(S_1, S_2).P(S_1|_I, S_2|_I)$$

Here $S|_I$ denotes the partial function that yields $S(t)$ for all $t \in I$ and is undefined otherwise.

When we now consider more closely how spatial situations can change over time, we observe that certain relationships can be observed only for a period of time and not for only a single time point (given that the participating objects do exist for a period of time) while other relationships can hold at instants as well as on time intervals. Predicates that can hold at time points and intervals are: *equal*, *meet*, *covers*, *coveredBy*; these are called *instant predicates*. Predicates that, in general, can only hold on intervals are: *disjoint*, *overlap*, *inside*, *contains*; these are called *period predicates*.

It is interesting to note that (in satisfiable developments for continuously moving objects) instant and period predicates always occur in alternating patterns, for example, there cannot be two spatio-temporal objects that satisfy *Inside* immediately followed by *Disjoint*. In contrast, *Inside* first followed by *meet* (or *Meet*) and then followed by *Disjoint* can be satisfied.

Next we define three operations for combining spatio-temporal and spatial predicates: $p \vdash P$ (*from*) defines a spatio-temporal predicate that for some time t checks p and then enforces P for all times after t; $P \dashv p$ (*until*) is defined dually, that is, P must hold until p is true at some time t. Finally, $P \dashv p \vdash Q$ (*then*) is true if there is some time point t when p is true so that P holds before and Q holds after t. Below we abbreviate open intervals $]t, \infty[$ and $]\infty, t[$ by simply writing $>t$ and $<t$. (Note that variable t ranges over *time*.) Let p be a spatial predicate, and let P and Q be spatio-temporal predicates. Then

$$
\begin{array}{lcl}
p \vdash P & := & \lambda(S_1, S_2).\exists t : p(S_1(t), S_2(t)) \wedge P_{>t}(S_1, S_2) \\
P \dashv p & := & \lambda(S_1, S_2).\exists t : p(S_1(t), S_2(t)) \wedge P_{<t}(S_1, S_2) \\
P \dashv p \vdash Q & := & \lambda(S_1, S_2).\exists t : p(S_1(t), S_2(t)) \wedge P_{<t}(S_1, S_2) \wedge Q_{>t}(S_1, S_2)
\end{array}
$$

These combinators obey several interesting laws; these and others are presented in [11]. In particular, the composition of predicate sequencing is associative, that is,

$$P \dashv p \vdash (Q \dashv q \vdash R) = (P \dashv p \vdash Q) \dashv q \vdash R$$

This enables us to use a succinct sequencing syntax for developments, that is, we can simply write $P \triangleright p \triangleright Q$ for $P \dashv p \vdash Q$. For example, we can define predicates for capturing the scenario of a point entering or crossing a region by:

$$
\begin{array}{lcl}
\mathit{Enter} & := & \mathit{Disjoint} \triangleright \mathit{meet} \triangleright \mathit{Inside} \\
\mathit{Cross} & := & \mathit{Disjoint} \triangleright \mathit{meet} \triangleright \mathit{Inside} \triangleright \mathit{meet} \triangleright \mathit{Disjoint}
\end{array}
$$

Sequential temporal composition is just one possibility to build new spatio-temporal predicates. Temporal alternative, negation, and reflection provide further powerful means to specify developments. These and many other combinators are defined and investigated in [11].

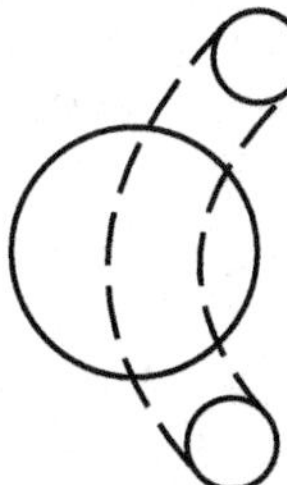

Fig. 4.11. Visual specification of the *Cross* predicate

Further Work. Spatio-temporal predicates lay the foundation for further research in spatio-temporal query languages. Two aspects have already been investigated:

First, it is, in fact, fairly simple to integrate spatio-temporal predicates into existing query languages. For example, we have shown in [10] how extending SQL by (i) the set of eight basic spatio-temporal predicates and (ii) by a macro facility to compose new predicates leads to a powerful spatio-temporal query language. Let us reconsider the example query of finding out all planes that ran into a storm. We assume having defined two relations `flights` and `weather` containing, respectively, a moving point attribute `Route` representing the flights' movements and an evolving region attribute `Extent` describing the developments of weather areas. With a predicate combinator `>>` that has the semantics of temporal composition $\triangleright$ we can formulate the query simply as:

```
SELECT id FROM flights, weather
WHERE kind = "storm"
  AND Route Disjoint>>meet>>Inside Extent
```

A second line of future work is motivated by the fact that the number of different spatio-temporal predicates is actually unlimited due to the sequencing possibility to generate new developments. Since we cannot invent names for all possible predicates we need some kind of language for specifying developments. Now an (additional) textual language for predicate specifications is not very convenient for the (end) user. Moreover, the specification of predicates can become quite longish. Hence, visual notations can be very useful to keep the specification of developments manageable by the user. Consider, for example, the predicate *Cross* which is defined for two evolving regions as follows:

$$\begin{aligned} \mathit{Cross} \;:=\; & \mathit{Disjoint} \triangleright \mathit{meet} \triangleright \mathit{Overlap} \triangleright \mathit{coveredBy} \triangleright \mathit{Inside} \triangleright \\ & \mathit{coveredBy} \triangleright \mathit{Overlap} \triangleright \mathit{meet} \triangleright \mathit{Disjoint} \end{aligned}$$

In contrast, this can be specified very easily and intuitively by a simple two-dimensional picture as shown in Figure 4.11.

The rationale behind this visual notation is described in more detail in [13,14]. The key idea is to infer from the intersections of two-dimensional traces of moving/evolving objects the temporal changes of their relationships. The visual notation is mainly intended to be used as a supplement to textual languages and can be integrated, for example, along the lines described in [8].

4.5.2 Spatio-temporal Partitions

While we have so far in this chapter dealt with the temporal evolution of *single* spatial entities, we now study the temporal evolution of *spatial partitions* as an important example of a *collection* of spatial entities satisfying specific constraints. This leads to a concept of *spatio-temporal partitions.*

Spatial Partitions. The metaphor of a *map* has turned out to be a fundamental and ubiquitous spatial concept in geography, cartography, and other related disciplines as well as in computer-assisted systems like GIS and spatial database systems. The central element of a map is a *spatial partition* which is a subdivision of the plane into pairwise disjoint *regions* where regions are separated from each other by *boundaries* and where each region is associated with an attribute or *label* having simple or even complex structure. That is, a region (possibly composed of several disconnected parts) with an attribute incorporates all points of a spatial partition having this attribute. Examples are the subdivision of the world map into countries, classification of rural areas according to their agricultural use, areas of different degrees of air pollution, etc. A spatial partition implicitly models topological relationships between the participating regions which can be regarded as integrity constraints. First, it expresses neighborhood relationships for different regions that have common boundaries. Second, different regions of a partition are always disjoint (except for common boundaries). Both topological properties are denoted as *partition constraints.* As a purely geometric structure, a map yields only a *static* description of spatial entities and required constraints between them.

A rigorous and thorough formal definition of spatial partitions and of application-specific operations defined on them has been given in [9]. The basic idea is that a spatial partition is a mapping from the Euclidean space $\mathbb{R}^2$ to some *label type*, that is, regions of a partition are assigned single labels. Adjacent regions have different labels in their interior, and a boundary is assigned the pair of labels of both adjacent regions.

Many application-specific operations on spatial partitions like *overlay*, *reclassify*, *fusion*, *cover*, *clipping*, *difference*, *superimposition*, *window*, and variations of them have been described in the literature (see for example, [1,23,29,33]). In [9] all these operations have been reduced to the three fundamental and powerful operations *intersection*, *relabel*, and *refine*. Intersecting two spatial partitions means to compute the geometric intersection of all regions and to produce a new spatial partition; each resulting region is labeled with the pair of labels of the original two intersecting regions, and the values on the boundaries are derived from these. Relabeling a spatial partition has the effect of changing the labels of its regions. This can happen by simply renaming the label of each region. Or, in particular, distinct labels of two or more regions are mapped to the same new label. If some of these regions are adjacent in the partition, the border between them disappears, and the regions are fused in the result partition. Relabeling has then a coarsening effect. Refining a partition means to look with a finer granularity on regions and to reveal and to enumerate the connected components of regions.

Spatio-temporal Partitions. Spatio-temporal partitions [12] or "temporal maps" describe the temporal development of spatial partitions and have a wide range of interesting applications. They represent collections of evolving regions satisfying the partition constraints for each time of their lifespan and maintaining these constraints over time. That is, for each time of their lifespan we obtain a stationary, two-dimensional spatial partition which changes over time due to altering shapes, sizes, or attribute values of regions. This corresponds to our temporal object view which is based on the observation that everything that changes over time can be considered as a function over time. Spatio-temporal partitions can then be viewed as functions from time to a two-dimensional spatial partition.

Temporal changes of spatial partitions can occur either in discrete and stepwise constant steps or continuously and smoothly. Examples of the first category are the reunification of West and East Germany, the splitting of Yugoslavia, the temporal development of any hierarchical decomposition of space into administrative or cadastral units like the world map into countries or districts into land parcels, or the classification of rural areas according to their agricultural use over time. A characteristic feature of these applications is that the number of discrete temporal changes is finite and that there is no change between any two subsequent *temporal change points* which is a special form of continuity.

The open issue now is what happens at temporal change points with their abrupt transition from one spatial partition to another. If we consider the time point when West and East Germany were reunified, did the spatial partition before or after the reunification belong to this time point? Since we cannot come to an objective decision, we have to decide arbitrarily and to assign one of both spatial partitions to it. This, in particular, maintains the functional character of our temporal object view. We have chosen to ascribe the temporally later spatial partition to a temporal change point. Mathematically this means that we permit a finite set of time points where the temporal function is not continuous. The application examples reveal that after a temporal change point the continuity of the temporal function proceeds for some time interval up to the next temporal change point; there are no "thin, isolated slices" containing single spatial partitions at temporal change points. Hence, we have to tighten our requirement in the sense that mathematically the temporal function has to be *upper semicontinuous* at each time.

Examples of the second category are the temporal evolution of climatic phenomena like temperature zones or high/low pressure areas, areas of air pollution with distinct degrees of intensity, or developments of forest fires in space and time. They all show a very dynamic and attribute-varying behavior over time. Application examples which have by far slower temporal changes are the increasing spread of ethnic or religious groups, the decreasing extent of mineral resources like oil fields during the course of time due to exploitation, or the subdivision of space into areas with different sets of spoken languages over time.

Application-specific spatio-temporal operations rest on the transfer of the two basic spatial partition operations *intersection* and *relabel* to the spatio-temporal case. They are temporally lifted and generalized versions of the application-specific operations on spatial partitions. The *overlay* operation is based on a spatio-temporal *intersection* operation and can be used to analyze the temporal evolution of two (or more) different attribute categories. Consider a temporal map indicating the extent of mineral resources like oil fields or coal deposits and another temporal map showing the country map over time. Then an overlay of both temporal maps can, for instance, reveal the countries that had or still have the richest mineral resources, it can show the grade of decline of mineral deposits in the different countries, and it can expose the countries which most exploited their mineral resources.

The *clipping* operation is a special case of the spatio-temporal *intersection* operation and works as a *spatio-temporal filter*. An application is a temporal map about the development of diseases. As a clipping window we use a temporal map of urban areas developing in space over time. The task is to analyze whether there is a connection between the increase or decrease of urban space and the development of certain diseases. Hence, all areas of disease outside of urban regions are excluded from consideration.

The *reclassification* operation is a special case of a spatio-temporal *relabel* operation. Consider a temporal map marking all countries of the world with their population numbers. A query can now ask for the proportion of each country's population on the world population over time, a task that can be performed by temporal relabeling. This corresponds to a reclassification of attribute categories over time without changing geometry.

The *fusion* operation is a kind of grouping operation with subsequent geometric union over time. Assume that a temporal map of districts with their land use is given. The task is to identify regions with the same land use over time. At each time neighboring districts with the same land use are replaced by a single region, that is, their common boundary line is erased. We obtain a temporal fusion operator which is based on relabeling. Reclassification and fusion are examples of *static relabeling* since the relabeling function does not change over time. We generalize this to *dynamic relabeling*. Consider the classification of income to show poor and rich areas over time. Due to the changing value of money, due to inflation, and due to social changes, the understanding of wealth and poorness varies over time. Hence, we need different and appropriate relabeling functions that are applied to distinct time intervals.

Additionally, some new operations are introduced that are more directed to the time dimension. The operation *dom* determines the domain of a temporal map, that is, all times where the map does not yield the completely undefined partition. An example is a temporal map of earthquakes and volcanic eruptions in the world as they are interesting for seismological investigations. Applying the operation *dom* on this map returns the time periods of earthquake and volcanic activity in the world.

The operation *restrict* realizes a function restriction on spatio-temporal partitions and computes a new partition. As parameters it obtains a temporal map and a set of (right half-open) time intervals describing the time periods of interest. Imagine that we have a temporal map of birth rates, and we are only interested in the birth rates between 1989 and 1991 and between 1999 and 2001 ("millennium baby"). Then we can exclude all the other time periods and compare the change of birth rates in these two time intervals.

The operation *select* allows one to scan spatio-temporal partitions over time and to check for each time whether a specified predicate is fulfilled or not. Consider a map showing the spread of fires. We could be interested in when and where the spread of fires occupied an area larger than 300 km^2.

The operation *aggregate* collects all labels of a point over time and combines them with the aid of a binary function into one label. The result is a two-dimensional spatial partition. For example, if the population numbers, the birth rates, the death rates, the population density, the average income, etc. of the countries in the world are available, we can aggregate over them and compute the maximum or minimum value each country ever had for one of these attributes.

A special kind of aggregation is realized by the *project* operation which computes the projection of a spatio-temporal partition onto the Euclidean space and which yields a spatial partition. For each point in space, all labels, except for the undefined label denoting the outside, are collected over time. That is, if a point has always had the same single label over its lifetime, this single label will appear in the resulting partition and indicate a place that has never changed. On the other hand, points of the resulting partition with a collection of labels describe places where changes occurred. An example is the projection of a temporal map illustrating the water levels of lakes onto the Euclidean plane. The result shows those parts of lakes that have always, sometimes, and never been covered with water.

For a much more detailed description and, in particular, formal definition of spatio-temporal partitions, the reader is referred to [12]. There as well as in [15] especially a concept of "spatial continuity" and a "difference measure" for regions are defined.

4.5.3 On a Spatio-temporal Relational Model Based on Quanta

In this section an outline is given of a formalism for the definition of a spatio-temporal extension to the relational model. The formalism considers *temporal* and *spatial quanta* and, based on them, defines relevant data types. This way, a series of relational algebra operations can be defined, that are closed and enable the uniform management of either conventional or spatial or spatio-temporal data.

Quanta and Data Types of Time. A *generic* data type for *time* is defined as the set $I_n = \{1, 2, \ldots, n\}, n > 0$ [25]. The elements of I_n are called *quanta of time* or (*time*) *instants.* Based on this data type, another *generic* data type is defined, PERIOD(I_n), with elements of the form $[p, q] \equiv \{i \mid i \in I_n, p \le i \le q\}$ that are called *periods* (*of time*). If the elements of I_n are replaced by n consecutive dates, then the respective data types for time are DATE and PERIOD(DATE). In a similar manner, a variety of data types can be defined like TIME and TIMESTAMP, which are supported in SQL.

Spatial Quanta and Spatial Data Types. If $I_m = \{1, 2, \ldots, m\}, m > 0$, is a subset of the integers, then I_m^2 is finite. Each element of I_m^2 is a 2D point that can be identified uniquely by an integer (see Figure 4.12 for $m = 15$). If $p \equiv (i, j)$ is such a point, then $p_N \equiv (i, j+1)$ and $p_E \equiv (i+1, j)$ are the *neighbors* of p. Points $p, p_E, p_{NE} \equiv \{(i+1, j+1)\}$ and p_N are *corner* points. In Figure 4.12, 193 and 207 are neighbors of 192, whereas 192, 193, 208 and 207 are *corner* points. Based on this terminology, the following *spatial quanta* are defined [26].

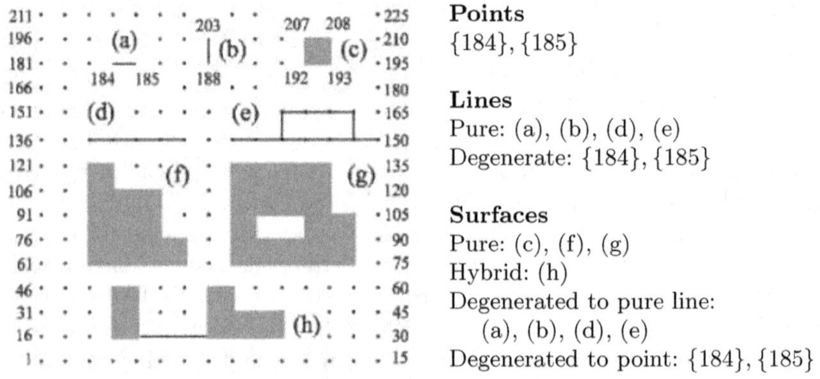

Fig. 4.12. Spatial quanta and spatial objects

Quantum Point : It is any set $\{p\}$, where p is a 2D point (see $\{192\}, \{193\}, \{208\}$ in Figure 4.12). The set of all the quantum points is denoted by Q_{POINT}.

Quantum Line : It is either a line segment $ql_{p,q}$ with edges two neighbor 2D points, p and q, or a quantum point $\{p\}$ (see $ql_{184,185}, ql_{188,203}$ and $\{184\}$ in Figure 4.12). Clearly, $ql_{p,q}$ consists of an infinite number of R^2 elements. The set of all the quantum lines is denoted by Q_{LINE}.

Quantum Surface : It is either the surface of a square $qs_{p,q,r,s}$, where p, q, r and s are corner 2D points or a quantum line (see $qs_{192,193,208,207}, ql_{184,185}$ and $\{184\}$ in Figure 4.12). Clearly, $qs_{p,q,r,s}$ consists of infinitely many elements of R^2. The set of all these surfaces is denoted by $Q_{SURFACE}$.

Assuming now that the concept of a *connected* set is known, it is defined that a non-empty connected subset $S = \bigcup_i q_i$ of R^2 is of a (2D)

- POINT data type if $q_i \in Q_{POINT}$
- LINE data type if $q_i \in Q_{LINE}$
- SURFACE data type if $q_i \in Q_{SURFACE}$

Given that $Q_{POINT} \subset Q_{LINE} \subset Q_{SURFACE}$, it follows that POINT $\subset$ LINE $\subset$ SURFACE. Examples of elements of the above data types are given in Figure 4.12. A point or line or surface element is called a *geo* or *spatial* object.

Modeling of Spatio-temporal Data. Based on the above data types, Figure 4.13 illustrates the evolution of a spatial object, Morpheas, with respect to time. As can be seen on the relevant plots, during the periods [11, 20], [21, 40] [41, 90], Morpheas was just a spring, a river and an actual lake, respectively. Relation LAKES, in the same figure, is used to record this evolution. The domain of attribute Shape is SURFACE and each of g1, g2 and g3 is a shorthand of the description of the geometry of Morpheas. In this model therefore, a *map* matches the *geometric interpretation* of the content of a relation that contains spatial data.

In [25] it has been shown that *period* is a special case of a more generic data type, interval. Functions and predicates for such data have been defined. A set of relational algebra operations has also been defined. It includes the well-known primitive operations, *Union, Except, Project, Cartesian Product, Select.* It also includes *Fold, Unfold* and some derived operations, whose formalism and functionality can also be found in [25]. Hence, only the application of *Fold* and *Unfold* on spatial data is illustrated below.

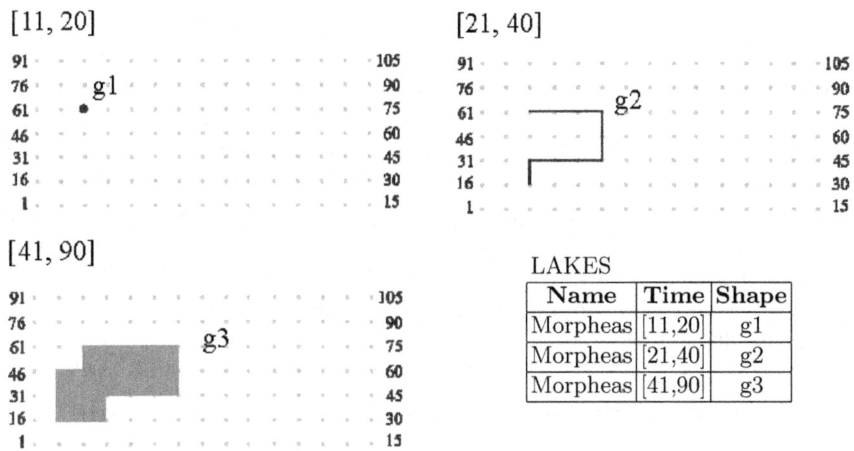

Name	Time	Shape
Morpheas	[11,20]	g1
Morpheas	[21,40]	g2
Morpheas	[41,90]	g3

Fig. 4.13. Representation of spatio-temporal data

Fold : Consider a table $R(\mathbf{A}, G)$, where $\mathbf{A}$ is a (possibly empty) list of attributes and G is an attribute of some geo data type. Assume also that $(\mathbf{a}, g_i), i = 1, 2, \ldots, n$ are all those tuples of R which satisfy the property that the *spatial union* of all g_i yields a new spatial object g. Then all these tuples result in one tuple, $(\mathbf{a}, g)$, in relation $F = Fold[G](R)$. Consider for example the plane in Figure 4.12 and assume that $R = \{(x, \{2\}), (x, l_{2,3}l_{3,4}), (x, s_{3,4,19,18}), (y, s_{6,7,22,21}\, s_{7,8,23,22}), (y, l_{6,7})\}$. Then $F = \{(x, l_{2,3}s_{3,4,19,18}), (y, s_{6,7,22,21}s_{7,8,23,22})\}$.

Unfold : Consider $R(\mathbf{A}, G)$ as before and let g be a geo object. Consider also any geo object $g_i, i = 1, 2, \ldots, n$, of one quantum, which is a subset of g. Then a tuple $(\mathbf{a}, g)$ of R yields in $U = Unfold[G](R)$ the tuples $(\mathbf{a}, g_i), i = 1, 2, \ldots, n$. Assuming for example that R is the previous relation, $U = \{(x, \{2\}), (x, \{3\}), (x, \{4\}), (x, \{18\}), (x, \{19\}), (x, l_{2,3}), (x, l_{3,4}), (x, l_{4,19}), (x, l_{19,18}), (x, l_{18,3}), (x, s_{3,4,19,18}), (y, \{6\}), (y, \{7\}), (y, \{8\}), (y, \{21\}), (y, \{22\}), (y, \{23\}), (y, l_{6,7}), (y, l_{7,22}), (y, l_{22,21}), (y, l_{21,6}), (y, l_{7,8}), (y, l_{8,23}), (y, l_{23,22}), (y, s_{6,7,22,21}), (y, s_{7,8,23,22})\}$. Based on the above two operations, a series of useful derived operations have also been defined [27], that enable the management of spatial data and, in conjunction with [25], spatio-temporal data. Regarding the management of spatial data, the functionality of these operations is relevant to that of *Spatial Union, Spatial Exception, Spatial Intersection, Overlay* etc that either have been defined by other researchers [5,22,34,35,29] or are supported by commercial Geographic Information Systems.

Conclusions. The advantages of the proposed model can be summarized as follows: All the algebraic operations are closed and, in conjunction with [25], they can be applied uniformly for the management of either spatio-temporal or spatial or temporal or conventional data. It has been identified, in particular, that certain operations, originally defined solely for the management of spatial data, are also of practical interest for the handling of temporal or conventional data. Hence, the algebra is not many-sorted and enables the uniform treatment of any of the above types of data. Regarding the case of spatial data, it has been identified that a *map* matches the geometric representation of a relation that contains such data. The model is also close to human intuition. By definition, for example, a line or a surface consists of an infinite number of 2D points, a line is treated as a degenerate surface and a point is treated as either a degenerate line or as a degenerate surface. Due to this, it is estimated that the model is also user-friendly. Finally, it is very general. This is not only because it can be applied to the previously mentioned types of data. It can also handle relations with many attributes of some *time* data type [25] and investigation results have shown that such relations may also contain n-dimensional spatial objects. Relevant research concerns implementation issues and the definition of an SQL extension.

4.5.4 Spatio-temporal Statement Modifiers

Data types and operators are generally embedded in some host language, which makes them available for use during data management. The characteristics of this language in large part determines the difficulty in migrating existing applications to a new, spatio-temporal DBMS (STDBMS). The concept of a *statement modifier* extended host language [2,31,30], largely orthogonal to the specific abstract data types offered, enables he migration of legacy applications.

This section defines technical requirements to an STDBMS that provide a foundation for making it economically feasible to migrate legacy applications to an STDBMS. It proceeds to present the design of the core of a spatio-temporal, multi-dimensional extension to SQL–92, called STSQL, that satisfies the requirements. This is achieved by offering *upward compatible*, *dimensional upward compatible*, *reducible*, and *non-reducible* query language statements.

A planning and scheduling system (termed Ecoplan), which is used for forest management [28], serves to exemplify the new concepts. A `stands` table captures regions that are homogeneous with respect to soil fertility, wood specie, and average age. An `estates` table records the IDs of estates and their owners. An estate is a legal entity covering a geographical region, possibly including one or more forests. A `plans` table captures harvest plans, with each stand being associated with one or more plans (and vise versa), an estimated harvest volume for each stand, and an optimal harvest time (a so-called ripe year) of the stand.

Migration Requirements. Let $M = (DS, QL)$ be a data model with a data structure and a query language component. For query $s \in QL$ and database $db \in DS$, we define $\langle\langle s(db) \rangle\rangle_M$ as the result of applying s to db in data model M. We use the superscripts "^{s}" and "^{d}" to indicate snapshot and dimensional entities, respectively. The dimensional slice operator, $\tau_p^{M^d, M^s}$, takes a dimensional database db^d and returns a snapshot database db^s containing all tuples that are defined at point p.

Definition 5. (UC) Model M_1 is *upward compatible* with model M_2 iff

- $\forall db_2 \in DS_2 \; (db_2 \in DS_1)$,
- $\forall s_2 \in QL_2 \; (s_2 \in QL_1)$, and
- $\forall db_2 \in DS_2 \; (\forall s_2 \in QL_2 \; (\langle\langle s_2(db_2) \rangle\rangle_{M_2} = \langle\langle s_2(db_2) \rangle\rangle_{M_1}))$.

The conditions imply that all existing databases and query expressions in the old DBMS, captured by M_2, are also legal in the new DBMS, captured by M_1 and that all existing queries compute the same results in the new and old DBMSs.

Definition 6. (DUC) Model M^d is *dimensional upward compatible* with model M^s iff

- M^d is upward compatible with M^s and
- $\forall db^s \in DS^s \; (\forall \mathcal{U} \; (\forall q^s \in QL^s \; (\langle\langle q^s(\mathcal{U}(db^s)) \rangle\rangle_{M^s} = \langle\langle q^s(\mathcal{U}(\mathcal{D}(db^s))) \rangle\rangle_{M^d})))$.

DUC ensures that legacy applications remain operational even if the database is rendered dimensional. Intuitively, a query q^s must return the same result on an associated snapshot database db^s as on the dimensional counterpart of the database, $\mathcal{D}(db^s)$ (operator $\mathcal{D}$ adds dimensions to its argument). A sequence of modification statements, $\mathcal{U}$, may not affect this. To satisfy DUC, legacy queries ignore spatial dimensions and are evaluated only on tuples with time periods that overlap *now*.

To illustrate the compatibility requirements, consider the following statements issued in an STDBMS satisfying UC and DUC with respect to SQL-92.

```
> SELECT * FROM plans;
> ALTER TABLE plans ADD harvest1 PERIOD AS VALID;
> SELECT * FROM plans;
```

The first statement is syntactically an SQL–92 query and is issued on the legacy table, plans. Due to UC, it returns the same result as in the old DBMS. The next statement alters plans, adding a valid-time dimension to indicate harvest periods of stands. The last statement, now on a dimensional table, yields the same result as the first due to DUC.

To generalize the relational model to a dimensional relational model, we adopt the view that a dimensional table is a collection of snapshot tables, each of which has an associated multi-dimensional point and contains all snapshot tuples with an associated multi-dimensional region that contains the point.

Definition 7. (SR) Data model M^d is *snapshot reducible with respect to* data model M^s iff

$$\forall q^s \in QL^s(\exists q^d \in QL^d(\forall db^d \in DS^d(\forall p(\tau_p^{M^d,M^s}(q^d(db^d)) = q^s(\tau_p^{M^d,M^s}(db^d))))))$$

In addition, it is desirable that q^d be *syntactically similar snapshot reducible* to q^s [4].

Definition 8. (SSSR) Data model M^d is a *syntactically similar snapshot-reducible extension* of model M^s iff

- data model M^d is snapshot reducible with respect to data model M^s, and
- there exist two (possibly empty) strings, S_1 and S_2, such that each query q^d in QL^d that is snapshot reducible with respect to a query q^s in QL^s is syntactically identical to $S_1 q^s S_2$.

The SSSR requirement enables the SQL–92 programmer to easily formulate spatio-temporal queries.

```
> ALTER TABLE estates ADD es_area 2D_REGION AS SPACE;
> ALTER TABLE stands ADD st_area 2D_REGION AS SPACE;
> REDUCIBLE (es_area) AS area SELECT * FROM estates;
> REDUCIBLE (es_area, st_area) AS area
  SELECT es_ID, st_ID FROM estates, stands;
```

The first two statements render estates and stands dimensional. The two queries have an SQL–92 core. The prepended string, REDUCIBLE, is a *statement modifier* that determines the handling of the dimension attributes in the queries. The presence of REDUCIBLE implies that, conceptually, the queries are computed point-by-point. More specifically, for each point in space, the legacy SQL statement following the statement modifier is evaluated on the snapshot database corresponding to that point. The results for each point in space are integrated into a single dimensional table.

Many useful dimensional queries cannot be specified as reducible generalizations of snapshot queries, and there is a need for queries where no built-in processing of the dimension attributes. The modifier NONREDUCIBLE specifies that dimension attributes are to be considered as regular attributes. Together with the predicates and functions offered by the dimensional data types, this yields full control over the dimension attributes.

```
> NONREDUCIBLE (es_area, st_area)
    SELECT s.st_ID, s.st_area FROM stands s WHERE  NOT EXISTS (
      SELECT * FROM estates e WHERE e.es_area CONTAINS s.st_area);
```

This query retrieves each stand for which no single estate exists that covers the stand's area. We consider the regions of the stands as being non-decomposable and constrain them with a spatial predicate. This contrasts the REDUCIBLE queries from before, where regions are decomposed into their constituent points.

STSQL Design. The first step in the design of STSQL is to introduce new, dimensional abstract data types. Time values are anchored time periods while spatial values are unions of regions. The corresponding data types are PERIOD, 1D_REGION, 2D_REGION and 3D_REGION, respectively [3]. The second step is to make tables *dimensional*, by enabling the designation of certain time or space valued attributes as dimensional. In STSQL, a dimension attribute is specified as either a VALID, a TRANSACTION, or a SPACE attribute.

STSQL permits a table to have any number of dimension attributes. This generality is useful for many purposes. Several VALID-type attributes may record different temporal aspects of a tuple. For example, the plans table has a VALID attribute harvest1 recording when a stand is supposed to be harvested. We add another VALID attribute harvest2 that records an alternative harvest period. The two resulting harvest attributes reflect different (possible) worlds. It is equally easy to envision uses of multiple space dimension attributes: the multiple-worlds argument applies equally well to space, and tuples may have several different kinds of spatial aspects [6].

When formulating queries on dimensional tables, it is advantageous to proceed in several steps. All dimensions are initially ignored, and the core STSQL query, typically an SQL-92 query, is formulated. Next, the query's statement modifier is specified. For each dimension of each table in the query, it must be stated how the dimension is used in the query. Specifically, each dimensions that should be evaluated with reducible semantics is identified. Each occurrence of REDUCIBLE requires the participation of exactly one dimension from each table. Following this, each dimensions to be given NONREDUCIBLE semantics is identified. This semantics is chosen if we want to formulate user-defined predicates (e.g., `CONTAINS`) on the dimension attribute or if we want to override the DUC-consistent semantics, which are given to dimension attributes not mentioned in the statement modifier.

We conclude by formulating the following query: *for each stand that is ripe in 2000, determine its harvest periods.* The `stands` and the `plans` tables are joined using a reducible join over the valid times, to associate stands with relevant plans only. Next, we are only interested in the current data on `stands`. Assuming that a transaction-time dimension `st_tt` has been added to `stands`, we want to consider only tuples that overlap *now*. This is the semantics provided by DUC, so no modifier is specified for `st_tt`. The location of a stand is not relevant and, thus, must be disregarded. This is again DUC semantics, so no modifier for `st_area` is specified. Finally, we want to retrieve (and handle) the harvest periods as regular attributes. This is achieved by specifying a non-reducible modifier for these dimensions.

```
> REDUCIBLE (st_vt,pl_vt) AS vt AND NONREDUCIBLE (harvest1, harvest2)
    SELECT st.st_ID, harvest1, harvest2
    FROM   stands st, plans pl
    WHERE  pl.st_ID = st.st_ID AND pl.ripe = 2000;
```

Conclusion and Future Research. We formulated requirements to a new dimensional DBMS aiming at addressing legacy-related concerns. The objectives are to make it possible for legacy database applications using a conventional SQL-92-based DBMS to be migrated to a dimensional DBMS without without affecting the legacy applications, while also reusing programmer expertise. A spatio-temporal extension to SQL–92, termed STSQL, that provides built-in data management support for spatio-temporal data has been designed to meet the above requirements.

Several directions for further explorations may be identified. First, we have only described the initial design of the core of STSQL, and further formalizations of the language are in order. Next, we have chosen one possible semantics for DUC statements. Other semantics are possible; further studies are needed to identify these and explore their utility.

References

1. J.K. Berry. Fundamental Operations in Computer-Assisted Map Analysis. *Int. Journal of Geographic Information Systems*, 1(2):119–136, 1987.
2. M.H. Böhlen and C. S. Jensen. Seamless Integration of Time into SQL. Technical Report R-96-49, Department of Computer Science, Aalborg University.
3. M.H. Böhlen, C.S. Jensen, and B. Skjellaug. Spatio-Temporal Database Support for Legacy Applications. In *Proceedings of the 1998 ACM Symposium on Applied Computing*, pp. 226–234. Atlanta, Georgia, February 1998.
4. M.H. Böhlen, C.S. Jensen, and R.T. Snodgrass. Evaluating the Completeness of TSQL2. In *Recent Advances in Temporal Databases, International Workshop on Temporal Databases*, pp. 153–172. Springer, Berlin, Zürich, Switzerland, September 1995.

5. E.P.F. Chan and R. Zhu. QL/G - A Query Language for Geometric Data Bases. In *Proc. 1st International Conference on GIS, Urban Regional and Environmental Planning*, pp. 271–286. Samos, Greece, 1996.
6. H. Couclelis. People Manipulate Objects (but Cultivate Fields): Beyond the Raster-Vector Debate in GIS. In *Lecture Notes in Computer Science*, Vol. 639, pp. 65–77, Springer-Verlag, 1992.
7. M. Erwig, R.H. Güting, M. Schneider, and M. Vazirgiannis. Spatio-Temporal Data Types: An Approach to Modeling and Querying Moving Objects in Databases. *GeoInformatica*, 3(3):265–291, 1999.
8. M. Erwig and B. Meyer. Heterogeneous Visual Languages – Integrating Visual and Textual Programming. In *11th IEEE Symp. on Visual Languages*, pp. 318–325, 1995.
9. M. Erwig and M. Schneider. Partition and Conquer. In *3rd Int. Conf. on Spatial Information Theory*, LNCS 1329, pp. 389–408, 1997.
10. M. Erwig and M. Schneider. Developments in Spatio-Temporal Query Languages. In *IEEE Int. Workshop on Spatio-Temporal Data Models and Languages*, pp. 441–449, 1999.
11. M. Erwig and M. Schneider. Spatio-Temporal Predicates. Technical Report 262, FernUniversität Hagen, 1999.
12. M. Erwig and M. Schneider. The Honeycomb Model of Spatio-Temporal Partitions. In *Int. Workshop on Spatio-Temporal Database Management*, LNCS 1678, pp. 39–59, 1999.
13. M. Erwig and M. Schneider. Visual Specifications of Spatio-Temporal Developments. In *15th IEEE Symp. on Visual Languages*, pp. 187–188, 1999.
14. M. Erwig and M. Schneider. Query-By-Trace: Visual Predicate Specification in Spatio-Temporal Databases. In *5th IFIP Conf. on Visual Databases*, 2000. To appear.
15. M. Erwig, M. Schneider, and R.H. Güting. Temporal Objects for Spatio-Temporal Data Models and a Comparison of Their Representations. In *Int. Workshop on Advances in Database Technologies*, LNCS 1552, pp. 454–465, 1998.
16. L. Forlizzi, R.H. Güting, E. Nardelli, and M. Schneider. A Data Model and Data Structures for Moving Objects Databases. In *Proceedings of the ACM SIGMOD International Conference on Management of Data*. Dallas, Texas, 2000.
17. S. Gaal. *Point Set Topology*. Academic Press, 1964.
18. R.H. Güting. Second-Order Signature: A Tool for Specifying Data Models, Query Processing, and Optimization. In *Proceedings of the ACM SIGMOD International Conference on Management of Data*, pp. 277–286. Washington, 1993.
19. R.H. Güting. An Introduction to Spatial Database Systems. *VLDB Journal*, 3:357–399, 1994.
20. R.H. Güting, M.H. Böhlen, M. Erwig, C.S. Jensen, N.A. Lorentzos, M. Schneider, and M.Vazirgiannis. A Foundation for Representing and Querying Moving Objects. Technical Report Informatik 238, FernUniversität Hagen, 1998. Available at `http://www.fernuni-hagen.de/inf/pi4/papers/Foundation.ps.gz`.
21. R.H. Güting, M.H. Böhlen, M. Erwig, C.S. Jensen, N.A. Lorentzos, M. Schneider, and M.Vazirgiannis. A Foundation for Representing and Querying Moving Objects. *ACM Transactions on Database Systems*, 25(1), 2000.
22. R.H. Güting and M. Schneider. Realm-Based Spatial Data Types: The ROSE Algebra. *VLDB Journal*, 4(2):100–143, 1995.
23. Z. Huang, P. Svensson, and H. Hauska. Solving Spatial Analysis Problems with GeoSAL, a Spatial Query Language. In *6th Int. Working Conf. on Scientific and Statistical Database Management*, 1992.

24. J. Loeckx, H. D. Ehrich, and M. Wolf. *Specification of Abstract Data Types.* John Wiley & Sons, Inc. and B.G. Teubner Publishers, 1996.
25. N.A. Lorentzos and Y.G. Mitsopoulos. SQL Extension for Interval Data. *IEEE Transactions on Knowledge and Data Engineering*, 9(3):480–499, 1997.
26. N.A. Lorentzos, N. Tryfona, and J.R. Rios Viqueira. Relational Algebra for Spatial Data Management. In *Proc. International Workshop Integrated Spatial Databases: Digital Images and GIS.* Portland, Maine, June 1999.
27. N.A. Lorentzos, J.R. Rios Viqueira, and N. Tryfona. Quantum-Based Spatial Data Model. Technical Report, Informatics Laboratory, Agricultural University of Athens, 2000.
28. G. Misund, B. Johansen, G. Hasle, and J. Haukland. Integration of geographical information technology and constraint reasoning — A promising approach to forest management. Technical Report STF33A 95009, SINTEF Applied Mathematics, Oslo, Norway, June 1995.
29. M. Scholl and A. Voisard. Thematic Map Modeling. In *1st Int. Symp. on Large Spatial Databases*, LNCS 409, pp. 167–190, 1989.
30. R.T. Snodgrass, M.H. Böhlen, C.S. Jensen, and A. Steiner. Adding Transaction Time to SQL/Temporal. ANSI X3H2-96-152r, ISO–ANSI SQL/Temporal Change Proposal, ISO/IEC JTC1/SC21/WG3 DBL MCI-143, May 1996.
31. R.T. Snodgrass, M.H. Böhlen, C.S. Jensen, and A. Steiner. Adding Valid Time to SQL/Temporal. ANSI X3H2-96-151r1, ISO–ANSI SQL/Temporal Change Proposal, ISO/IEC JTC1/SC21/WG3 DBL MCI-142, May 1996.
32. R.B. Tilove. Set Membership Classification: A Unified Approach to Geometric Intersection Problems. *IEEE Transactions on Computers C-29*, pp. 874–883, 1980.
33. C.D. Tomlin. *Geographic Information Systems and Cartographic Modeling.* Prentice Hall, 1990.
34. J.W. van Roessel. Conceptual Folding and Unfolding of Spatial Data for Spatial Queries. In V.B. Robinson and H. Tom, eds., *Towards SQL Database Extensions for Geographic Information Systems*, pp. 133–148, National Institute of Standards and Technology, Gaithersburg, Maryland, 1993. Report NISTIR 5258.
35. J.W. van Roessel. An Integrated Point-Attribute Model for Four Types of Areal Gis Features. In R.G. Healey T.C. Waugh, ed., *Proc. 6th International Symposium on Spatial Data Handling (SDH94)*, pp. 127–144. Edinburgh, Scotland, UK, 1994.

5 Spatio-temporal Models and Languages: An Approach Based on Constraints

Stéphane Grumbach[1], Manolis Koubarakis[2], Philippe Rigaux[3], Michel Scholl[1], and Spiros Skiadopoulos[4]

[1] INRIA, Rocquencourt, France
[2] Technical University of Crete, Chania, Crete, Greece
[3] CNAM, Paris, France
[4] National Technical University of Athens, Greece

5.1 Introduction

The introduction of spatio-temporal information in database systems presents us with an important data modelling challenge: the design of data models general and powerful enough to handle conventional thematic data, purely temporal or spatial concepts and spatio-temporal concepts.

General purpose DBMS (e.g., relational) are not appropriate for storing and manipulating spatio-temporal data because of the complex structure of temporal and geometric information and the intricate temporal and spatial relationships among sets of related objects. Moreover, the costly operations involved in temporal and spatial object management seem at first glance to prevent the logical-level approach based on simple data structures (e.g., relations) manipulated through a limited set of simple operations (e.g., relational algebra).

In the temporal world the main research trend has therefore been to introduce a variety of temporal data models with *new temporal operations*, and then worry about the semantics and appropriateness of these operations [45,44]. Similarly, in the spatial world the main approach has been to introduce extensions of conventional database models with abstract spatial data types encapsulating geometric structures and operations (see for example [39,35,40,19,20]). Chapter 4 of this book follows this school of thought for the modeling and querying of spatio-temporal information. Similarly, in the commercial world, DBMS such as Oracle [21] and Illustra [47] provide in their latest version separate modules devoted to temporal and spatial data.

In this chapter we take an approach that differs significantly from the trends outlined above. We notice that the abstract data type approach *lacks uniformity* for the representation of data. To avoid this, we adopt the constraint data model, first introduced by Kanellakis, Kuper, and Revesz [25,33], and show that it is a very successful paradigm for the representation of spatio-temporal data in a *unified* framework. The constraint data model allows a uniform representation of all kinds of information present in spatio-temporal applications, and supports *declarative query languages* well-suited for complex spatio-temporal queries. Contrary to the temporal and spatial data models mentioned above, it also enables the straightforward modeling of *indefinite information*, a feature particularly useful in many spatio-temporal applications.

T. Sellis et al. (Eds.): Spatio-temporal Databases, LNCS 2520, pp. 177–201, 2003.
© Springer-Verlag Berlin Heidelberg 2003

The basic idea of the constraint data model is to represent temporal and spatial objects as infinite collections of points satisfying first-order formulae. For example, an interval is defined by the conjunction of two order constraints. Similarly, a non-convex polygon, which is the intersection of a set of half-planes, is defined by the conjunction of the inequalities defining each half-plane. Finally, a non-convex polygon is defined by the union (logical disjunction) of a set of convex polygons.

In this chapter we use first-order formulae with linear (and in some cases polynomial) constraints for the definition of spatio-temporal objects. Linear constraints over the rational numbers have been shown to be a flexible and powerful way to represent many kinds of temporal data e.g., absolute, relative, periodic and indefinite [24,1,26,27,37,46]. Utilising the power of linear constraints, the user can *avoid worrying about the nasty details of special temporal operations* (as in many of the models in [45,44]), and *instead use standard relational algebra* for expressing temporal queries. There is also the added benefit of having a uniform way to represent infinite and indefinite information [24,26] something which is not possible in other temporal models.

Similarly, linear constraints over rational numbers have been shown to be appropriate for the symbolic representation of spatial and spatio-temporal data [16,32,33]. In this paradigm spatio-temporal objects can be seen as infinite sets of points at the abstract level, and can be represented by quantifier-free formulas with linear constraints at the symbolic level. This approach contrasts with the representation of objects by their boundary, which leads to cumbersome data models with ad-hoc operations, and no standard emerging. The fundamental benefits of the constraint approach is a *uniform representation* of any kind of spatio-temporal object, *no limitation in the dimension*, and *independence from the physical level* and its algorithms (i.e., potential for query optimization). Most importantly, it is also possible to manipulate spatio-temporal objects through the *standard languages* of relational calculus and algebra.

This chapter is organized as follows. In the following section, we introduce the original constraint database model of [25] and present examples of representing and querying definite spatio-temporal information. In Section 5.3 we extend this model to account for indefinite spatio-temporal information following the proposal of [28,27]. Section 5.4 points out some shortcomings of the original constraint database model when used to represent spatio-temporal data. Then it introduces the data model and algebra of the system DEDALE that has been designed to overcome these shortcomings. Section 5.5 presents the user query language for DEDALE and shows it in action in a real-life spatio-temporal application. Finally, Section 5.6 summarizes the chapter and discusses related research.

5.2 Representing Spatio-temporal Information Using Constraints

We introduce here the framework of *constraint databases* [25,33] which extends the classical paradigm of the relational model, and demonstrate that spatio-temporal data can be represented gracefully using constraints.

In this framework spatio-temporal data are modeled as infinite sets in the rational space. For example, a convex interval on the rational line is seen as an infinite set of points bounded by two endpoints. A polygon on the plane is seen as the infinite set of points of $\mathbb{Q}^2$ inside its frontier, and a 3-dimensional pyramid is seen as the infinite set of points of $\mathbb{Q}^3$ inside its facets. Similarly, the trajectory of a moving object is seen as an infinite set of points representing the various positions of the object during successive intervals of time (see Example 1).

Following the terminology of Chapter 4 of this book, our *abstract model* of spatial data consists of infinite relations over the universe of the rational numbers $\mathbb{Q}$. These infinite relations can only be described and manipulated through a finite representation. Following the trends of constraint databases [25], we use first-order logic to represent the relations of interest, and define a *symbolic level* of representation (a discrete model in the terminology of Chapter 4). We distinguish clearly between the *intensional* representation of a relation at the symbolic level, and its *extensional* interpretation at the abstract level.

We consider linear constraints in the first-order language $\mathcal{L} = \{\leqslant, +\} \cup \mathbb{Q}$ over the structure $\mathcal{Q} = \langle \mathbb{Q}, \leqslant, +, (q)_{q \in \mathbb{Q}} \rangle$ of the linearly ordered set of the rational numbers with rational constants and addition. Constraints are linear equations and inequalities of the form: $\sum_{i=1}^{p} a_i\, x_i \Theta a_0$, where Θ is a predicate among $=$ or $\leqslant$, the x_i's denote variables and the a_i's are integer constants. Note that rational constants can always be avoided in linear equations and inequalities. The multiplication symbol is used as an abbreviation, $a_i x_i$ stands for $x_i + \cdots + x_i$ (a_i times).

Let $\sigma = \{R_1, ..., R_n\}$ be a database schema such that $\mathcal{L} \cap \sigma = \varnothing$, where $R_1, ..., R_n$ are relation symbols. We distinguish between *logical predicates* (e.g., $=, \leqslant$) in $\mathcal{L}$ and *relations* in σ. The basic concept of our abstract model is the linear constraint relation which is defined as follows.

Definition 1. Let $S \subseteq \mathbb{Q}^k$ be a k-ary relation. The relation S is a *linear constraint relation* if there exists a formula $\varphi(x_1, ..., x_k)$ in $\mathcal{L}$ with k distinct free variables $x_1, ..., x_k$ such that:

$$\mathcal{Q} \models \forall x_1 \cdots x_k (S(x_1, ..., x_k) \leftrightarrow \varphi(x_1, ..., x_k))$$

Formula φ is called a *representation* of S.

We denote by *LCR*($\mathbb{Q}^k$) the class of linear constraint relations over $\mathbb{Q}^k$. This class constitutes a rather drastic restriction over the class of all infinite relations over $\mathbb{Q}^k$. The restriction is necessary to ensure reasonable query complexity and is sufficient to encompass spatial data in computational geometry, GIS, etc. as it has already been widely demonstrated in the literature [33].

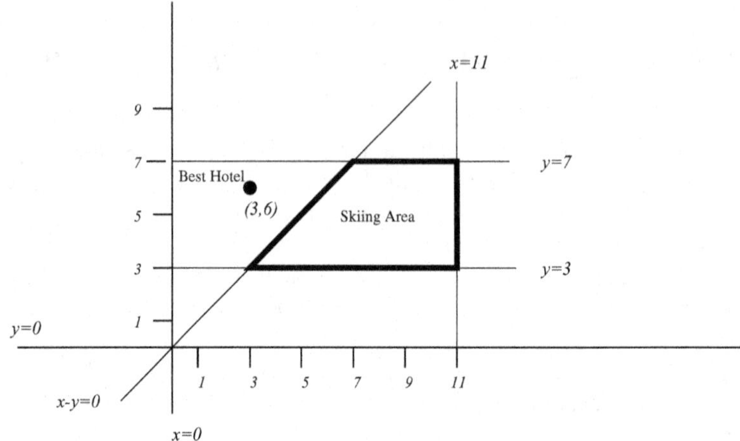

Fig. 5.1. A very simplified view of a ski resort

The basic concepts of our symbolic representation are the generalized tuple and the generalized relation (originally introduced in [25]).

Definition 2. A *generalized tuple* is a conjunction of linear constraints from the first-order language $\mathcal{L}$. A *generalized relation* is a finite set of generalized tuples (i.e., a generalized relation corresponds to a formula of $\mathcal{L}$ in disjunctive normal form or DNF).

Thus our symbolic representation offers a finite representation for infinite sets of points in d-dimensional space $\mathbb{Q}^d$ with no limitation on d.

The concept of a linear constraint relation defined above can be extended to cover relations that combine an uninterpreted domain D (used for the modelling of thematic data) with the interpreted one $\mathbb{Q}$. The concept extends easily to such relations, with representation formulae in $\mathcal{L} \cup D$, with two sorts of constraints, (i) equality constraints over objects of D, and linear constraints over objects of $\mathbb{Q}$. We will denote by $LCR(D, \mathbb{Q}^2)$, the set of linear constraint relations over $D \times \mathbb{Q}^2$. The concepts of our symbolic representation (generalized tuple and relation) can be similarly extended with equality and inequality constraints over the domain D.

Let us now illustrate the use of linear constraint relations in a real life application provided by the French laboratory LAMA, Grenoble [8].

Example 1. We consider a ski resort and study the spatio-temporal behavior of several types of actors. The resort is known as a set of buildings and areas with well-defined utilities, e.g. hotels, night clubs, various kinds of stores and of course the skiing area. The typical behavior of human actors (tourists for instance) is statistically modeled with respect to their socio-economical category (age, income, nationality, etc.) and represented as a sequence of activities and positions during a typical vacation day. For instance the tourist category will be described as the following succession of activities: sleeping, walking, skiing, eating, reading, watching movies, etc., each activity being associated with a time interval. The trajectory of each tourist is described by partitioning the day in time slices and associating with each slice the position(s) where a representative of the category is likely to be found during this time slice. The ultimate goal of the application is to improve the organization of the ski resort by detecting places where no one ever goes, equipment which is underutilized, categories which share the same behavior, spatial distribution of tourists in the resort with respect to time and so on.

The following generalized relation *People* represents the known information about the activities of tourist John from midnight ($t = 0$) to noon ($t = 12$) of a typical day. The location of John's hotel together with other geometric information is depicted in Figure 5.1). We have used simple linear constraints in this example although the full power of the language *caL* is available.

People

Name	Category	Activity	Trajectory
John	Tourist	Sleeping	$0 \leqslant t < 8 \wedge x = 3 \wedge y = 6$
John	Tourist	Eating	$8 \leqslant t < 9 \wedge x = 3 \wedge y = 6$
John	Tourist	Skiing	$9 \leqslant t < 12 \wedge x - y \geqslant 0 \wedge y \geqslant 3 \wedge y \leqslant 7 \wedge x \leqslant 11$

The column to notice is *Trajectory* (the rest of the columns are as in standard relational databases). The attribute values for column *Trajectory* form an infinite set of triples (t, x, y) which is represented in a finite way by linear constraints. *Trajectory* essentially represents three attributes in the sense of the relational model (time plus two coordinates), each of them with an infinite set of values. We choose not to show these attributes explicitly and refer to them by using the corresponding variables t, x and y. We will follow the same practice in the rest of the paper. When we give examples of generalized relations we often blur the distinction between the columns or attributes, and the corresponding variable names.

The above definitions of linear constraint relations and generalized relations are easily extended to definitions for *linear constraint databases* and *generalized databases.* The details are omitted for brevity.

5.2.1 An Algebra for Relations with Constraints

Spatio-temporal data represented by generalized relations can be queried using first-order logic (relational calculus) or relational algebra [25]. In this section we will give detailed definitions of relational algebra over linear constraint relations. The definitions for relational calculus are as in the standard relational case and are omitted. Relational calculus and algebra are equivalent over linear constraint relations (this has been shown in [18,28,27,36]).

The algebra for generalized relations consists of *union*, $\cup$, *cartesian product*, $\times$, *difference*, $-$, *selection*, σ_F, where F is an atomic constraint, and *projection*, π. These operations are defined as follows.

Let R_1 and R_2 be two relations, and respectively e_1 and e_2 be sets of generalized tuples defining them.

1. $R_1 \cup R_2 = e_1 \cup e_2$.
2. $R_1 \times R_2 = \{t_1 \wedge t_2 \mid t_1 \in e_1, t_2 \in e_2\}$.
3. $R_1 - R_2 = \{t_1 \wedge t_2 \mid t_1 \in e_1, t_2 \in (e_2)^c\}$, where e^c is the set of tuples or disjuncts of a DNF formula corresponding to $\neg e$.
4. $\sigma_F(R) = R \times \{F\}$.
5. $\pi_{\overline{x}}\, R_1 = \{\pi_{\overline{x}}\, t \mid t \in e_1\}$, where

$$\pi_{\overline{x}}\, t = \bigwedge_{1 \leqslant k \leqslant K, 1 \leqslant \ell \leqslant L} b^k \overline{x} - b_0^k \leqslant a_0^\ell - a^\ell \overline{x} \ \wedge \bigwedge_{1 \leqslant i \leqslant I} c^i \overline{x} \leqslant c_0^i.$$

is given by the Fourier-Motzkin Elimination method [42] from a tuple t defining a polyhedron $P(\overline{x}, y) \subseteq \mathbb{Q}^{n+1}$ described by the inequalities (once the coefficients of y have been normalized):

$$\begin{cases} a^\ell \overline{x} + y \leqslant a_0^\ell & \text{for } \ell = 1, \ldots, L \\ b^k \overline{x} - y \leqslant b_0^k & \text{for } k = 1, \ldots, K \\ c^i \overline{x} \leqslant c_0^i & \text{for } i = 1, \ldots, I \end{cases}$$

where $\overline{x}$ ranges over $\mathbb{Q}^n$, and y over $\mathbb{Q}$.

Using the above operations *join* can easily be defined using cartesian product and selection. Also, *intersection* of two generalized relations over a common set of attributes is easily seen to be equal to the join of these relations.

The semantics of the symbolic operators applied to sets of generalized tuples simulates relational operators applied to infinite relations, and provides a correct mathematical representation of the result that complies with the constraint representation. For selection and cross product, this is done in a somehow lazy way, by just concatenating the input(s). The result might be inconsistent or redundant: a semantic evaluation, denoted *simplification*, must be carried out at some step of the query execution process in order to eliminate redundancies and to detect inconsistencies. The algorithms and complexities for the constraints manipulation can be found in [15,16]. For the purposes of our discussion let us just consider an example.

Example 2. Let us consider the database of Example 1 and the query "Where is John between 10 and 12?". In relational calculus, this query can be expressed by the formula

$$\mathit{name}, x, y: \ (\exists t)(\mathit{People}(\mathit{name}, t, x, y) \wedge \mathit{name} = \mathit{John} \wedge 10 \leqslant t < 12)$$

and its answer is the following relation:

Name	Place
John	$x - y \geqslant 0 \wedge y \geqslant 3 \wedge y \leqslant 7 \wedge x \leqslant 11$

In the algebra for generalized relations, the same query can be expressed as follows:

$$\pi_{Name,x,y}(\sigma_{Name=John \wedge 10 \leqslant t < 12}(People))$$

To evaluate this algebraic query we first apply selection to each tuple of the relation *People*. The result of this operation is the tuple

$$(John,\ Tourist,\ Ski,\ 10 \leqslant t < 12 \wedge x - y \geqslant 0 \wedge y \geqslant 3 \wedge y \leqslant 7 \wedge x \leqslant 11)$$

This tuple comes from the third tuple of this relation (the first two tuples do not survive the selection because they contain constraints that contradict the selection condition $12 \leqslant t < 14$). Finally, the projection operation is executed (it is straightforward for this example!) to arrive at the result.

5.3 Indefinite Information in Spatio-temporal Databases

The ideas of the previous section can be extended to accommodate indefinite information [26,27,28,29,30,31,32]. The resulting *indefinite constraint database scheme* is very powerful and it can be used to model many new spatio-temporal applications. This scheme is also important from the point of view of Artificial Intelligence because it essentially unifies the representational capabilities of *constraint networks* [12] with these of relational databases.

To motivate the need to represent indefinite information consider Figure 5.2 which exhibits a situation that might be represented in the database of the Ministry of the Environment and Natural Resources of some European country. We assume that there is some species inhabiting the rectangular area A. Now assume that there is some form of atmospheric pollution which has been generated due to some industrial accident at point $(5,5)$. The extent of the pollution is not known precisely at this time. All we know is that the minimum area polluted is given by rectangle B and the maximum by rectangle C.[1]

In this example our information about the species occupying area A is *definite* (i.e., we know *precisely* the location and extent of the species habitat). Unfortunately we have *indefinite* or *imprecise* information about the polluted area. Due to our lack of precise information about the polluted area, the situation in the real world is compatible with many possibilities or *possible worlds* as they have been called by philosophers and logicians [22]. In one possible world the polluted area might be enclosed exactly by rectangle B. In another the polluted area might be enclosed by the rectangle defined by points $(1,1)$ and $(7,7)$. In fact, there is an infinite number of such possible worlds and in each one of them the polluted area is defined by a rectangle enclosing rectangle B and enclosed by rectangle C.

At the abstract level, each one of these possible worlds corresponds to two infinite sets of points defining the species habitat and the polluted area. The

[1] We do not address here the question whether a rectangle is an appropriate geometric means for capturing this kind of data.

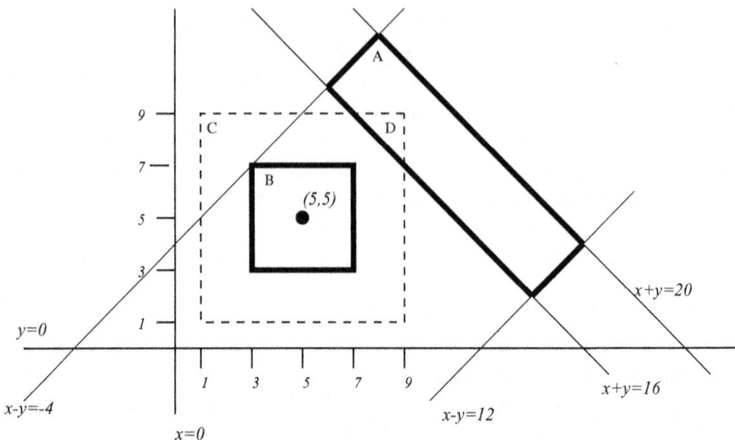

Fig. 5.2. Species (A) and pollution (B, C)

ideas of Section 5.2 now imply that, at the symbolic level, each possible world can be represented by a generalized database. For example, if we assume that the polluted area is enclosed exactly by rectangle B, the information in the above example can be captured by the following generalized relations:

Species

Area
$x + y \leqslant 20 \wedge x + y \geqslant 16 \wedge x - y \leqslant 12 \wedge x - y \geqslant -4$

Polluted

Area
$x \geqslant 3 \wedge x \leqslant 7 \wedge y \geqslant 3 \wedge y \leqslant 7$

The above example motivates us to consider *sets of possible worlds* i.e., *sets of linear constraint relations* as the appropriate formal concept for the representation of indefinite spatio-temporal information in our abstract model. This is captured by the following definitions.

Definition 3. A *possible world* is a linear constraint relation.

Possible worlds can be used for the representation of definite information as explained above. When we have indefinite information, the concept of indefinite linear constraint relation is required.

Definition 4. Let $\mathbf{S} \subseteq \wp(\mathbb{Q}^k)$ be a set of possible worlds (the notation $\wp(\cdot)$ denotes powerset). The set of possible worlds $\mathbf{S}$ is an *indefinite linear constraint relation* if there exist formulae $CS(\omega_1, ..., \omega_n)$ and $\varphi(x_1, ..., x_k, \omega_1, ..., \omega_n)$ in $\mathcal{L}$ with distinct free variables $\omega_1, ..., \omega_n$ and $x_1, ..., x_k$ such that:

$$\mathbf{S} = \{\ W \subseteq \mathbb{Q}^k :\ (x_0^1, \ldots, x_0^k) \in W \text{ iff }$$
$$\mathcal{Q} \models \exists \omega_1 \cdots \omega_n (CS(\omega_1, ..., \omega_n) \wedge \varphi(x_0^1, \ldots, x_0^k, \omega_1, ..., \omega_n))\ \}$$

We denote by *ILCR*($\mathbb{Q}^k$) the set of indefinite linear constraint relations over $\mathbb{Q}^k$. As in the previous section we also consider relations that combine the uninterpreted domain D with the interpreted one $\mathbb{Q}$. In the past indefinite relations over an uninterpreted domain had been studied in the literature on *null values* ([23,14] are the most outstanding works in this area).

It is straightforward to extend the above definition to cover the concept of indefinite linear constraint databases as sets of indefinite linear constraint relations [28,27]. The details are omitted for brevity.

At the symbolic level, indefinite linear constraint relations (or databases) will be represented by indefinite generalized relations (or databases). These concepts are defined below.

Definition 5. An *indefinite generalized relation* is a set of generalized tuples (as defined in Definition 2). These tuples are also allowed to contain Skolem constants. *Skolem constants*, usually denoted by $\omega_1, ..., \omega_n$, are essentially equivalent to the existentially quantified variables $\omega_1, ..., \omega_n$ of the formula

$$\exists \omega_1 \cdots \omega_n (CS(\omega_1, ..., \omega_n) \wedge \varphi(x_0^1, \ldots, x_0^k, \omega_1, ..., \omega_n))$$

of Definition 4.

Definition 6. An *indefinite generalized database* is a set of indefinite generalized relations with an associated *constraint store* $CS(\omega_1, ..., \omega_n)$. The constraint store is a quantifier free formula of $\mathcal{L}$ which is used to constrain the Skolem constants of all generalized relations.

Example 3. The following indefinite generalized database represents the information in Figure 5.2.

Species

Area
$x + y \leqslant 20 \wedge x + y \geqslant 16 \wedge x - y \leqslant 12 \wedge x - y \geqslant -4$

Polluted

Area
$5 - \omega \leqslant x \wedge x \leqslant 5 + \omega \wedge 5 - \omega \leqslant y \wedge y \leqslant 5 + \omega$

$$CS(\omega): \quad \omega \geqslant 2 \wedge \omega \leqslant 4$$

The following example modelled after [43] deals with moving objects.

Example 4. Let us consider an object moving on a straight line in $\mathbb{Q}^2$ with motion vector $d = (d_1, d_2)$ and speed v. Let us also assume that its initial position at time t_0 is $(x(t_0), y(t_0))$. The position $(x(t), y(t))$ of the object at future time t can be computed using the following equations:

$$x(t) = x(t_0) + v(t - t_0)d_1 \text{ and } y(t) = y(t_0) + v(t - t_0)d_2$$

In this example we go beyond linear constraints and consider *indefinite polynomial constraint databases.* For the definite case, polynomial constraint databases have been studied in various papers [25,33]. We do not give detailed definitions for this case, since the required definitions are similar to the ones above, and follow immediately from the general scheme of indefinite constraint databases given in [27].

We consider a concrete example of moving objects and assume that we have an indefinite polynomial constraint database containing information about the moving object $Car1$:

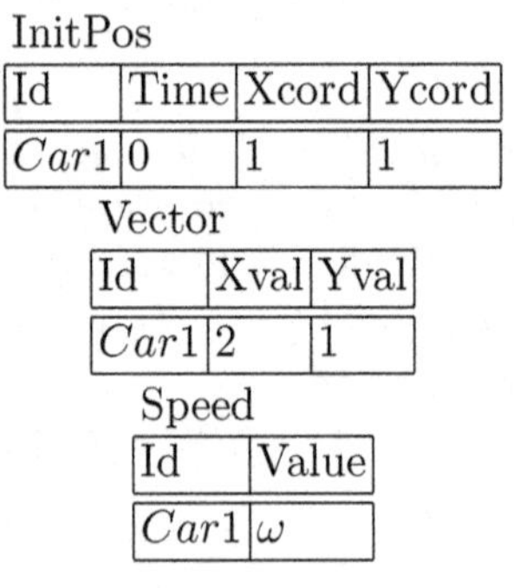

InitPos

Id	Time	Xcord	Ycord
$Car1$	0	1	1

Vector

Id	Xval	Yval
$Car1$	2	1

Speed

Id	Value
$Car1$	ω

$$CS(\omega):\ 40 \leqslant \omega \leqslant 50$$

$$\begin{aligned} FuturePos(o, t, x_0 + v(t - t_0)d_1, y_0 + v(t - t_0)d_2) \leftarrow\ & InitPos(o, t_0, x_0, y_0), \\ & Vector(o, d_1, d_2), \\ & Speed(o, v) \end{aligned}$$

The only generalized relation with truly indefinite information is *Speed*. The speed of $Car1$ is not known precisely: all we know is that it is between 40 and 50 miles per time unit. The future position of any object is computed using the rule which defines the generalized relation *FuturePos* with schema (*Id*, *Time*, *X*, *Y*). We do not give a detailed semantics for such rules since they follow immediately from relevant literature [25,28].

In the next section, we will pose queries to the above database to compute the position of object $Car1$ at times other than 0.

5.3.1 Querying Indefinite Information

An algebra and calculus for indefinite constraint databases has been defined in [26,27,28,29,30]. The calculus is an extension of standard relational calculus with the *modal* operators $\Diamond$ and $\Box$ for expressing *possibility* and *certainty* queries. The motivation for introducing modal operators for the querying of databases with indefinite information has been given in many earlier papers [23,14].

This modal calculus and the corresponding algebra are equivalent over indefinite linear constraint databases [27,28]. We omit detailed definitions for the calculus and instead concentrate on the algebra (but we show examples of calculus queries below).

The algebra for indefinite generalized relations extends the one discussed in Section 5.2 as follows:

1. The definitions of union, cartesian product, difference, selection and projection remain unchanged. These operations do not interfere with Skolem constants or the constraint store.
2. If only these operations are contained in a query, the answer relation might contain Skolem constants. Thus for an answer relation to be meaningful it must be accompanied by a *constraint store* which gives meaning to the Skolem constants.
   If $CS(\omega_1, ..., \omega_n)$ is the constraint store of the input database, the constraint store of the answer relation is $\pi_{\omega_{i1},...,\omega_{im}}(CS(\omega_1, ..., \omega_n))$ where $\omega_{i1}, ..., \omega_{im}$ are these Skolem constants among $\omega_1, ..., \omega_n$ that appear in the answer relation.
3. Two new *modal* operators $POSS$ (for *possibility*) and $CERT$ (for *certainty*) are introduced. These operators correspond to the natural language expressions "possibly" and "certainly" that might appear in a query over a database with indefinite information.
   $POSS$ and $CERT$ take into account the Skolem constants and the constraints in the constraint store. Applying these operators on an input indefinite relation, results in a *definite relation.* In other words, the information in the database might be indefinite but we can *definitely* compute what is possible (or certain) given the information in the database.

Before we proceed to define the new operators $POSS$ and $CERT$, let us give an example of a query which does not involve these operators.

Example 5. Consider the database of Example 3 and the query "Compute the intersection of the area inhabited by the species with the polluted area".

In the calculus for indefinite linear constraint relations this query can be expressed as follows:

$$x, y : \ Species(x, y) \wedge Polluted(x, y)$$

In the algebra for indefinite generalized relations the query can be expressed by *Species* ∩ *Polluted*. Evaluating this expression is straightforward and gives the following result:

SpeciesInDanger

$Area$
$x + y \leqslant 20 \wedge x + y \geqslant 16 \wedge x - y \leqslant 12 \wedge x - y \geqslant -4 \wedge$ $5 - \omega \leqslant x \wedge x \leqslant 5 + \omega \wedge 5 - \omega \leqslant y \wedge y \leqslant 5 + \omega$

$$CS(\omega) : \ \omega \geqslant 2 \wedge \omega \leqslant 4$$

Notice that the Skolem constant ω is contained in the constraints of the answer relation. Thus for the answer relation to be meaningful we include the projection of $CS(\omega)$ on ω.

An answer such as the one above might be hard to understand even with the provision of formula $CS(\omega)$. Towards the end of this section we propose ways around this problem.

Let us now define operation $POSS$. Let R be an indefinite linear constraint relation, e the indefinite generalized relation representing R, and $CS(\omega_1, ..., \omega_n)$ the constraint store associated with R. Then $POSS(R)$ is a new *linear constraint relation* over the same schema consisting of all tuples in *any* of the possible worlds of R.

A more operational definition of $POSS(R)$ is as follows:

$$POSS(R) = \{\ \pi_{var(t)}(t \wedge CS(\omega_1, ..., \omega_n)) :\ t \in e \text{ and}$$

$$var(t) \text{ is the set of variables of tuple } t\ \}$$

In other words, each tuple in the generalized relation representing $POSS(R)$ is obtained by eliminating all Skolem constants from the conjunction of constraints $CS(\omega_1, ..., \omega_n) \wedge t$ where t is a generalized tuple of the symbolic representation of R.

The following queries use operation $POSS$.

Example 6. Consider the database of Example 3 and the query "Compute the area inhabited by the species which has possibly been polluted by the accident".

In the calculus for indefinite linear constraint relations this query can be expressed as follows:

$$x, y :\ \Diamond(Species(x, y) \wedge Polluted(x, y))$$

In the algebra for indefinite generalized relations the query can be expressed by $POSS(Species \cap Polluted)$. Evaluating this expression involves computing the intersection $Species \cap Polluted$ as above and then calculating $\pi_{x,y}(t)$ for all tuples t of $Species \cap Polluted$.

The calculation of

$$\pi_{x,y}(x + y \leqslant 20 \wedge x + y \geqslant 16 \wedge x - y \leqslant 12 \wedge x - y \geqslant -4 \wedge 5 - \omega \leqslant x \wedge$$

$$x \leqslant 5 + \omega \wedge 5 - \omega \leqslant y \wedge y \leqslant 5 + \omega \wedge$$

$$\omega \geqslant 2 \wedge \omega \leqslant 4)$$

results in

$$x \leqslant 9 \wedge y \leqslant 9 \wedge x + y \geqslant 16$$

and corresponds to the area D of Figure 5.2. Thus the answer to the query is the following:

SpeciesPossiblyInDanger

Area
$x \leqslant 9 \wedge y \leqslant 9 \wedge x + y \geqslant 16$

$$CS(\omega): \quad true$$

Notice that the returned constraint store is empty (contains only the trivial constraint *true*) because the answer relation is definite.

Example 7. Consider the indefinite constraint database of Example 4 and the query "What are the possible future positions of object $Car1$ at time 5".

In the calculus for indefinite linear constraint relations this query can be expressed as follows:

$$x, y: \ \Diamond FuturePos(Car1, 5, x, y)$$

In the algebra for indefinite generalized relations this query can be expressed by

$$\pi_{X,Y}(POSS(\sigma_{Id=Car1 \wedge Time=5}(FuturePos)))$$

and has the following answer:

PosAt5

Area
$401 \leqslant x \leqslant 501 \wedge y = \frac{1}{2}x + \frac{1}{2}$

$$CS: \ true$$

Because the database does not contain definite information about the speed of $Car1$, a line segment in $\mathbb{Q}^2$ is returned as the answer to the query.

Let us now define operation $CERT$. Let R be an indefinite linear constraint relation, e the indefinite generalized relation representing R, and $CS(\omega_1, ..., \omega_n)$ the constraint store associated with R. Then $CERT(R)$ is a new *linear constraint relation* over the same schema consisting of only these tuples that are contained in *every* possible world of R.

A more operational definition of $CERT$ is as follows:

$$CERT(R) = \{ \ \pi_{var(t)}(CS(\omega_1, ..., \omega_n) \wedge t) : \ t \in e^c \text{ and }$$

$$var(t) \text{ is the set of variables of tuple } t \ \}^c$$

where the application of the operator ^{c} can be understood as follows. If A is a set of generalized tuples then A^c is the set of tuples or disjuncts of a DNF formula corresponding to formula $\neg\psi$ where ψ is the formula in DNF corresponding to A.

As witnessed by its definition, $CERT$ is a very expensive operation [30,31,32]. The following example uses operation $CERT$.

Example 8. Let us consider the database of Example 3 and the query "Compute the area occupied by the species which has certainly been polluted by the accident".

In the calculus for indefinite linear constraint relations this query can be expressed as follows:

$$x, y : \Box(Species(x, y) \wedge Polluted(x, y))$$

In the algebra for indefinite generalized relations the query can be expressed by $CERT(Species \cap Polluted)$. If we evaluate this query, we get the empty relation (as it might be expected by examining Figure 5.2).

The queries of Example 6 and 8 should now be compared with the query of Example 5. We have already mentioned that the answer to the query of Example 5 is rather hard to understand even with the provision of the constraint $CS(\omega)$. The operators $POSS$ and $CERT$ can assist in comprehending the answer in such cases because they provide *upper* and *lower bounds* to the set of tuples contained in the answer to the original query. With the provision of a graphical user interface, the user can then be shown the geometric objects corresponding to these upper and lower bounds. This will greatly enhance the comprehension of the returned answers.

5.4 Beyond Flat Constraint Relations: The DEDALE Approach

As we have shown in Sections 5.2 and 5.3, the original constraint database model of [25] (and its extension to the indefinite case [28]) gives us a powerful framework for representing and querying spatio-temporal data. However, it has been observed [3,15,16,34] that this version of the constraint database model has two shortcomings:

- Constraint query languages like the ones of Sections 5.2 and 5.3 do not always offer natural means for expressing spatio-temporal queries. This shortcoming is due to the fact that spatio-temporal objects must be "broken" into component tuples in the original constraint database model (for instance, the relation in Example 1 has 3 tuples for the same object).
  This observation has led to the development of constraint database models based on non-first normal form relations [2]. Our own data model for the system DEDALE [15,16] is discussed below. Similar models have been developed independently by [3,34].
- Often the chosen constraint language is not expressive enough. For example, the language $\mathcal{L}$ of linear constraints used in Sections 5.2 and 5.3 cannot express several interesting spatial functions e.g., distance or convex hull [3,15,16].
  This problem can be solved in two ways. We can go to a more powerful constraint language (e.g., polynomial constraints, as we did in Example 4)[2]

[2] But this is not always satisfactory because all the implementation advantages of linear constraints can be lost [3].

or we can carefully introduce appropriate additional primitives in the query language. The latter approach has been followed by [3,15,16,34].

Let us now present the data model of DEDALE which is a carefully designed formalism for addressing the above considerations [15,16]. This data model extends the original constraint-database model presented in Section 5.2 thus it does *not* support indefinite information. The addition of functionality for the support of indefinite information in DEDALE is currently an open question. This section concentrates only on the data model and languages of DEDALE while the system itself is discussed in Chapter 7.

In the DEDALE data model spatio-temporal data is treated as (possibly infinite) sets of points in $\mathbb{Q}^d$ space with no limitation on the dimension d. These sets are then used as values in tuples as it is the case in nested relational models [2]. The interesting thing to notice here is that *nesting can be limited to one level* (this suffices for spatio-temporal data). In addition, we allow to construct sets of points in the uninterpreted domain as well as in the rational domain, in order in particular to represent non-geometric time-evolving attributes.

We next introduce the data types of our model. In contrast to the approach of Chapter 4 our datatypes only distinguish between spatio-temporal and thematic data (i.e., we do *not* have a special data for each kind of spatio-temporal object).

We assume the existence of two *atomic types* $\mathcal{Q}$ and U called the *rational* and *uninterpreted type* respectively. The domains of these types are $\mathbb{Q}$ and D. In practice we might need many uninterpreted types to cater for different kinds of thematic data; we ignore this issue in the rest of this paper.

We also have the following complex types: *set type, tuple type* and *relation type*. These types are defined as follows:

Definition 7. 1. If $A_1, \ldots, A_{k_0}, A_{l_1}, \ldots, A_{l_i}$ are attribute names and $k_0, k_1, \ldots, k_i$ are positive natural numbers then

$$\{[A_1 : U, \ldots, A_{k_0} : U, A_{l_1} : \mathcal{Q}^{k_1}, \ldots, A_{l_i} : \mathcal{Q}^{k_i}]\}$$

denotes a *set type* with domain

$$LCR(D^{k_0}, \mathbb{Q}^{k_1}, \ldots, \mathbb{Q}^{k_i}).$$

2. If $T_1, ..., T_n$ are atomic or set types, and $A_1, ..., A_n$ are attribute names then

$$[A_1 : T_1, ..., A_n : T_n]$$

is a *tuple type* with domain

$$dom([A_1 : T_1, ..., A_n : T_n]) = \{[A_1 : a_1, ..., A_n : a_n] \mid a_i \in dom(T_i)\}.$$

3. If T is a tuple type then $\{T\}$ is a *relation type* with domain

$$dom(\{T\}) = \wp_f(dom(T))$$

where $\wp_f(S)$ denotes the set of finite subsets of S.

A *relation schema* is a relation type. A *database schema* is a finite collection of relation types. An *instance* of a relation schema is defined as usual. In the sequel of this chapter the word *relation* is used for an object of relation type, and we distinguish *relations* from *sets* of set type.

Example 9. Let us consider again the ski-resort application of Example 1. The following is a relation schema for the actors of the application (considered as moving objects with time-varying activities).

$$\begin{aligned} People = \{[\ & \text{Name} : string, \\ & \text{Category} : string, \\ & \text{Activity} : \{[\text{Name} : string, \text{Time} : \mathcal{Q}]\}, \\ & \text{Trajectory} : \{[\text{Space} : \mathcal{Q}^2, \text{Time} : \mathcal{Q}]\} \\ ]\}\ & \end{aligned}$$

The following table shows how the instance of relation *People* from Example 1 can now be expressed in our nested model.

Name	Category	Activity	Trajectory
John	Tourist	(`name=''Sleeping''` $\wedge\ 0 \leqslant t < 8$) $\vee$ (`name=''Eating''` $\wedge\ 8 \leqslant t < 9$) $\vee$ (*name*= "*Skiing*" $\wedge\ 9 \leqslant t < 12$)	$(x = 3 \wedge y = 6 \wedge 0 \leqslant t < 9)$ $\vee$ $(x - y \geqslant 0 \wedge\ y \geqslant 3 \wedge y \leqslant 7$ $\wedge x \leqslant 11 \wedge\ 9 \leqslant t < 12)$

In the above example the atomic type *string* is used as an uninterpreted type. Attributes *Name* and *Category* are represented as classical atomic values, while the nested attributes *Activity* and *Trajectory* are relations in respectively $LCR(D, \mathbb{Q})$ and $LCR(\mathbb{Q}^2, \mathbb{Q})$ which are represented as FO formulas. For example, in *Trajectory*, a constraint c bounds either the first two coordinates (interpreted as x and y) or the last one (interpreted as $Time$). The trajectory can be seen as a sequence of time intervals associated with the point set where the object can be found during this interval. Figure 5.3 shows graphically another example of a similar spatio-temporal object.

5.4.1 The DEDALE Algebra

We now consider an algebraic query language to manipulate the complex relations introduced above. The DEDALE algebra was originally defined in [15,16] and its basic operations are as follows:

- *Set operations*: *union*, $\cup$, *intersection*, $\cap$, and *set difference*, $-$, apply to pairs of inputs of the same *set* or *relation* type.
- *Selection*, σ_F, applies to inputs of *relation* or *set* type. F is an atomic constraint over variables corresponding to the attributes given by name or position. This constraint is either of linear form (e.g., $4X + 3Y = 2$), or a set membership constraint (e.g., $X \in S$).

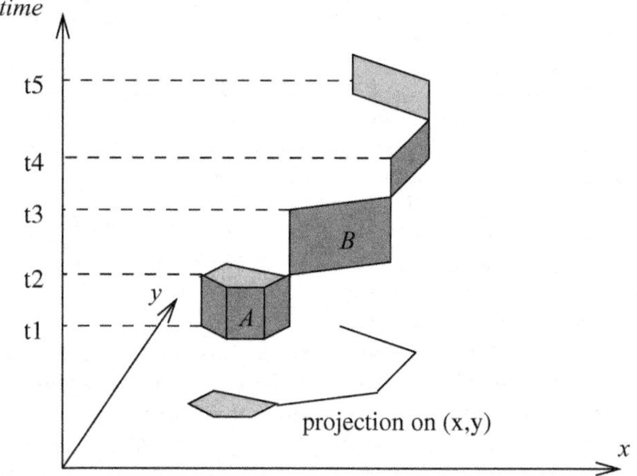

Fig. 5.3. A spatio-temporal object

- *Projection*, π, applies to inputs of *set* or *relation* type.
  For objects t of tuple type, $t.i$ denotes the i'th attribute when relevant.
- *Cartesian product*, $\times$, applies to pairs of inputs of either both *set* types or both *relation* types.
- *Restructuring*, MAP applies to inputs of *relation* type. If $E(X)$ is an algebraic expression of tuple type T', with a tuple variable $X : T$, and R is a relation of type $\{T\}$, then $MAP_{\lambda X.E(X)}(R)$ defines the relation of type $\{T'\}$, in which each tuple t of R, has been replaced by $E(t)$.

The semantics of the above operations on inputs of *relation* type corresponds to the semantics of classical relational algebra over finite relations, while the interpretation of the operations on inputs of *set* type corresponds to the semantics of the operations on linear constraint relations as presented in Section 5.2.1.

Example 10. Let us consider the following *clipping query* over the relation *People* of Example 9: "Give those people who crossed the area specified by rectangle *Rect* and their associated trajectory". Assuming that *Rect* is represented by the constraints

$$\{\ 9 \leqslant x \leqslant 13,\ 5 \leqslant y \leqslant 9\ \}$$

the algebra expression for this query is:

$$MAP_{\lambda X.[X.name,(Rect \times X.trajectory)]}(People)$$

Since *Rect* and *People* are defined partly on the same variables, the expression *Rect* $\times$ *X.trajectory* is a (natural) join on x and y. As in the relational case, its semantics is the triplets (x, y, t) in the trajectory of *People* such that (x, y) can be also found in *Rect*. Since both extensions are infinite, this result is represented by linear constraints.

For the instance of Example 9 the above query yields the following result:

Name	Trajectory
John	$(x = 3 \wedge y = 6 \wedge 0 \leqslant t < 9$ $\wedge\ 9 \leqslant x \leqslant 13 \wedge 5 \leqslant y \leqslant 9)$ $\vee$ $(x - y \geqslant 0 \wedge y \geqslant 3 \wedge y \leqslant 7 \wedge x \leqslant 11$ $\wedge\ 9 \leqslant t < 12$ $\wedge\ 9 \leqslant x \leqslant 13 \wedge 5 \leqslant y \leqslant 9)$

The result features a new formula for *trajectory* which represents exactly those points with coordinates satisfying both the formula that represented the initial trajectory and the formula representing the rectangle: in other words the intersection on x and y. Note that the first disjunct in *Trajectory* is unsatisfiable and can be deleted. Similarly the constraints in the second disjunct can be simplified to arrive at the following relation:

Name	*Trajectory*
John	$x \geqslant 9 \wedge x \leqslant 11 \wedge y \geqslant 5 \wedge y \leqslant 7$ $\wedge\ 9 \leqslant t < 12$

In summary, any relational algebraic expression can be applied through the *MAP* operator to point sets: this allows to express easily intersections (spatial joins), (geometric) projections, differences, complex selections, etc. The language is abstract and general: it is uniform for alphanumerical and spatial data, and does not limit the dimension or the geometric type of objects.

The DEDALE algebra also contains the primitive operations *unnest* and *unionest*, and some other operations definable using the primitive operations [15,16]. The algebra also includes new operators for *axis*, *median*, *dist* (for distance) and *connect?* (for connectivity). These operators have been added in order to be able to express interesting spatial queries such as convex hull, Voronoi diagrams, metric properties and topological queries. Linear constraint relations are closed over these operators and the queries expressed remain tractable (in the data complexity measure). More details can be found in [15,16].

5.5 The User Query Language of DEDALE

This section presents the user query language of DEDALE, along with some examples of queries. This language meets two requirements. First it hides as much as possible the underlying complexity of the data model. In particular the rules related to typing restrictions and to the nested structure are transparent to the end-user. Second the language has a direct and easy translation towards the DEDALE algebra, thereby allowing the design of an optimizer generating efficient evaluations.

5.5.1 The Syntax

Our language relies on an SQL-like syntax. It contains the following components:

1. Algebraic operations such as **join, union** etc. that operate on sets (i.e. spatial objects).
2. The limited set of primitives which have been introduced in the DEDALE algebra to augment its expressive power (**axis, median, dist**, and **connect?**).
3. Macro-operations which simplify the expression of the most common spatial data manipulations.

Note that macro-operations are simple shortcuts for complex algebraic expressions, while primitives may be considered as external functions with respect to the algebraic operators. The details of the language can be found in [16]. The rest of this section concentrates only on the algebraic operations and their use in a spatio-temporal application.

The query language follows the SQL syntax to construct algebraic expressions on nested attributes in the **select** clause as well as boolean expressions on nested attributes in the **where** clause. If p_1 and p_2 denote two pointset attributes (which can be either nested attributes or the results of algebraic expressions), then:

1. p_1 **join** p_2 denotes a *natural* join on the components common to p_1 and p_2.
2. p_1 **union** p_2 denotes the union of p_1 and p_2.
3. p_1 **except** p_2 denotes the difference of p_1 and p_2.
4. $p_1.c[.i]$ projects p_1 on the ith coordinate of component c. When there is only one component, its name can be omitted.
5. $p_1.c.i$ **as** j renames the ith coordinate of c as j.

join is the most general operation: depending on the existence of common components in p_1 and p_2, one obtains a cartesian product (no common components), a join (some common components), or an intersection (same components). In the latter case the equivalent **inter** keyword can be used for the sake of clarity.

The renaming of components or variables is obtained with **as** and allows to control the semantics of the **join**. Observe that the (spatial) selection is obtained as (p_1 **join** *cst*) where *cst* is any formula that represents a point set. For clarity, we sometimes use the equivalent syntax **restrict** p_1 **with** *cst*.

We can use the boolean counterpart of an operator **op**, denoted **op?**, in the **where** clause. It returns **true** if the resulting set is non empty and **false** otherwise. Therefore the general form of a query is

select $att_1, \ldots, att_n, E_1, \ldots E_m$
from $R_1[, \ldots]$
where $B_1, \ldots B_l$

where att_i denotes atomic attributes, E_j algebraic expressions over nested attributes, and B_k boolean algebraic expressions.

5.5.2 Example Queries

Let us now illustrate the use of our nested relational model and SQL query language in the real-life application of Example 1 [8]. The data modeling of this application is quite simple. It consists of two collections of objects:

1. A classical map (2-dimensional spatial objects) that describes the buildings and areas of interest in the ski resort.
2. A set of moving objects, each of which represents some typical socio-economical behavior. For simplicity, we will consider only two such objects named John and Monica.

Note that moving objects contain two time-varying attributes: their activity (pure relational information) and their geometry (shape and position). Observe also that while the shape here is irrelevant since we represent each object by a single point, nothing in the model prevents us from describing complex shaped moving objects. We give below the DEDALE schema for this application:

Relation Resort
(name: **string**,
geom: (space: **float(2)**)
)

Relation People
(name: **string**
activity: (alpha: **string**,
            time: **float(1)**)
traj :   (space: **float(2)**,
            time: **float(1)**)
)

Although extremely simple, this schema allows for a wide range of queries illustrating various combinations of spatial, temporal and relational criteria. We give a sample list of these queries in the sequel, along with some comments on either query design or its underlying evaluation in DEDALE. The same application gives rise to complex temporal relationships and queries which have been considered in the context of the TEMPOS project (see [13,41]).

All queries below run in the current implementation of DEDALE.

1. *Where is John between 10 and 12?*

   select (**restrict** traj **with** '10 < time < 12').space
   from People
   where name = 'John'

   A temporal query with spatial output: the constraint '10 < time < 12' is added to the trajectory of John, and tuples in the resulting *traj* relation which are still satisfiable (if any) are projected on the *space* component.
2. *When does Monica stay at the bar's terrace?*

   select (p.traj **join** r.geom).time
   from Resort r, People p
   where p.name = 'Monica'
   and r.name = 'Terrace'

A spatial query with temporal output. The spatial join on the *space* component common to *t.traj* and *s.geom* yields the part of Monica's trajectory inside the terrace's geometry: its projection on *time* is the result.

3. *Where is John while Monica is at the bar's terrace?*

   select (p2.traj **join** (p1.traj **join** r.geom).time).space
   from Resort r, People p1, People p2
   where p1.name = 'Monica'
   and p2.name = 'John'
   and r.name = 'Terrace'

   This query is a composition of a spatial join and a temporal join. The internal join retrieves time t during which Monica is at the terrace; t is the input argument to the temporal join with John's trajectory.

4. *Show the places where Monica sleeps*

   select ((**restrict** activity **with** "alpha = 'Sleeping' ") **join** traj).space
   from People
   where name = 'Monica'

   Here, a pure temporal query based on alphanumerical criteria ("retrieve the periods that correspond to a given activity") is composed with a temporal join. Note that the temporal join is "internal" to a single object since it involves the two nested relations (*activity* and *traj*) of Monica.

5. *Where did Monica and John meet?*

   select (p1.traj **inter** p2.traj).space
   from People p1, People p2
   where p1.name = 'Monica'
   and p2.name = 'John'

   The join involves simultaneously time and space.

6. *When were Monica and John at the same place?*

   select (p1.traj **join** r.geom).time **inter** (p2.traj **join** r.geom).time
   from Resort r, People p1, People p2
   where p1.name = 'Monica'
   and p2.name = 'John'

   Each object in the ski resort map is intersected with the trajectories of John and Monica. This yields the places where both of them spent some time during a typical day. A further join on *time* restricts the result to the places where they both were at the same instant.

7. *Who ate in the skiing area, and when?*

   select ((**restrict** p.activity **with** "alpha = 'Eating' " **join** p.traj)
   join r.geom).time, p.name
   from Resort r, People p
   where r.name = 'Skiing Area'

A first internal temporal join between the two time-varying attributes of a moving actor gives places where this actor eats. The following spatial join checks whether these places intersect the skiing area.

8. *What did John between the ski and his dinner?*

```
selectect  name,
        (((activity join "alpha = 'Ski' ").[time as t1]
        join activity
        join (activity join "alpha = 'Eat' ").[time as t2]
        ) join "t1.1 ⩽ time.1 ⩽ t2.1").alpha
from   People
where   name = 'John'
```

The comparison between three time components entails both a blow-up in dimension due to the cross-products and an unrestricted selection that links variables coming from different components. This yields an intermediate result whose global dimension is 9. Fortunately, the evaluation techniques over dimension-restricted queries developed in [17] allow to apply only operations on time intervals. In that case, the evaluation is as follows: $t1.1$ should be less than $Max(time.1)$ and $t2.1$ should be greater than $Min(time.1)$. The result is correct, as long as a projection of some component (in that case *alpha*) is made as a last operation, as required for dimension-restricted queries. The description of the evaluation techniques for such queries is beyond the scope of this chapter (the interested reader is referred to [17]).

5.6 Conclusions

This chapter presented the constraint-based approach to representing and querying spatio-temporal data. Three different models were discussed: the original constraint database model of [25], the indefinite constraint database model of [28,27] and the DEDALE model [16]. In all cases we used linear (and in one case polynomial) constraints to represent spatio-temporal data in a uniform way. This can be contrasted with the approach of Chapter 4 where an abstract data type is devised for each useful spatio-temporal concept.

Similar views to the ones developed in this chapter have also been expressed by other constraint-database researchers. Data models and query languages similar to the ones developed for DEDALE have also been presented in [3,4] and [34]. DEDALE is currently the only implemented system based on these ideas.

Spatio-temporal data have also been studied in [9,11,7] where a spatio-temporal model based on parametric rectangles is proposed. The same group of researchers has developed the constraint database system MLPQ-GIS [38] which is able to deal with spatio-temporal applications. The main difference of this system from DEDALE is that it is based on flat relations and the original constraint database model and uses a query language based on DATALOG [38]. In a related paper the issue of interoperability of spatio-temporal data is discussed [10]. This is an interesting topic not covered in this chapter.

We should also mention the constraint database system CCUBE which is based on an object-oriented model and also uses linear constraints for the representation of spatio-temporal objects [5,6].

References

1. M. Baudinet, J. Chomicki, and P. Wolper. Temporal Deductive Databases. In Tansel, A., et al., editor, *Temporal Databases: Theory, Design, and Implementation*, chapter 13. Benjamin/Cummings Pub. Co., 1993.
2. C. Beeri. Data Models and Languages for Databases. In *Proceedings of 2nd International Conference on Database Theory (LNCS 326)*, pages 58–67, 1989.
3. A. Belussi, E. Bertino, and B. Catania. Manipulating Spatial Data in Constraint Databases. In M. Scholl and A. Voisard, editors, *Proc. of the Fifth Int. Symp. on Spatial Databases*, number 1262 in Lecture Notes in Computer Science, pages 115–140, Berlin, Germany, July 1997. Springer Verlag, Berlin.
4. A. Belussi, E. Bertino, and B. Catania. An Extended Algebra for Constraint Databases. *IEEE Transactions on Knowledge and Data Engineering*, 10(5):686–705, 1998.
5. A. Brodsky and V.E. Segal. The CCUBE Constraint Object-Oriented Database System: an Overview. *Constraints*, 2(3-4):245–277, 1997.
6. A. Brodsky, V.E. Segal, J. Chen, and P.A Exarkhopoulo. The CCUBE Constraint Object-Oriented Database System. In *Proceedings of SIGMOD 1999*, pages 577–579, 2000.
7. M. Cai, D. Keshwani, and P.Z. Revesz. Parametric rectangles: A model for querying and animation of spatio-temporal databases. In *Proceedings of EDBT 2000*, pages 430–444, 2000.
8. S. Chardonnel and P. Dumolard. Personal communication.
9. J. Chomicki, Y. Liu, and P.Z. Revesz. Animating spatiotemporal constraint databases. In *Spatio-Temporal Database Management (Proceedings of the International Workshop STDBM'99)*, volume 1678 of *LNCS*, pages 224–241. Springer, 1999.
10. J. Chomicki and P.Z. Revesz. Constraint-based Interoperability of Spatiotemporal Databases. *GeoInformatica*, 3(3):211–243, 1999.
11. J. Chomicki and P.Z. Revesz. A geometric framework for specifying spatiotemporal objects. In *Proceedings of TIME'99*, pages 41–46, 1999.
12. R. Dechter, I. Meiri, and J. Pearl. Temporal Constraint Networks. *Artificial Intelligence*, 49(1-3):61–95, 1991. Special Volume on Knowledge Representation.
13. M. Dumas, M.-C. Fauvet, and P.-C. Scholl. Handling Temporal Grouping and Pattern-Matching Queries in a Temporal Object Model. In *Proc. Intl. Conf. on Information and Knowledge Management*, pages 424–431, 1998.
14. G. Grahne. The Problem of Incomplete Information in Relational Databases. Technical Report Report A-1989-1, Department of Computer Science, University of Helsinki, Finland, 1989. Also published as Lecture Notes in Computer Science 554, Springer Verlag, 1991.
15. S. Grumbach, P. Rigaux, M. Scholl, and L. Segoufin. DEDALE: A Spatial Constraint Database. In *Proc. Intl. Workshop on Database Programming Languages*, pages 38–59, 1997.
16. S. Grumbach, P. Rigaux, and L. Segoufin. The DEDALE System for Complex Spatial Queries. In *Proc. ACM SIGMOD Symp. on the Management of Data*, pages 213–224, 1998.

17. S. Grumbach, P. Rigaux, and L. Segoufin. On the Orthographic Dimension of Constraint Databases. In *Proc. Intl. Conf. on Database Theory*, pages 199–216, 1999.
18. S. Grumbach, J. Su, and C. Tollu. Linear Constraint Query Languages: Expressive Power and Complexity. In D. Leivant, editor, *Logic and Computational Complexity*, Indianapolis, 1994. Springer Verlag. LNCS 960.
19. R.H. Güting. Gral: An Extensible Relational Database System for Geometric Applications. In *Proc. Intl. Conf. on Very Large Data Bases (VLDB)*, 1989.
20. R.H. Güting and M. Schneider. Realm-Based Spatial Data Types: The ROSE Algebra. *The VLDB Journal*, 4(3):243–286, 1995.
21. J. Herring. The ORACLE 7 Spatial Data Option. Technical report, ORACLE Corp., 1996.
22. G.E. Hughes and M.J. Cresswell. *An Introduction to Modal Logic.* Methuen, 1968.
23. T. Imielinski and W. Lipski. Incomplete Information in Relational Databases. *Journal of ACM*, 31(4):761–791, 1984.
24. F. Kabanza, J.-M. Stevenne, and P. Wolper. Handling Infinite Temporal Data. In *Proceedings of ACM SIGACT-SIGMOD-SIGART Symposium on Principles of Database Systems*, pages 392–403, 1990. Full version appears in JCSS, Vol. 51, No. 1., pages 3-17, 1995.
25. P.C. Kanellakis, G.M. Kuper, and P.Z. Revesz. Constraint query languages. In *Proc. ACM Symp. on Principles of Database Systems*, pages 299–313, 1990. A longer version appears in JCSS, Vol. 51, No. 1, 1995.
26. M. Koubarakis. Representation and Querying in Temporal Databases: the Power of Temporal Constraints. In *Proceedings of the 9th International Conference on Data Engineering*, pages 327–334, April 1993.
27. M. Koubarakis. Database Models for Infinite and Indefinite Temporal Information. *Information Systems*, 19(2):141–173, March 1994.
28. M. Koubarakis. Foundations of Indefinite Constraint Databases. In A. Borning, editor, *Proceedings of the 2nd International Workshop on the Principles and Practice of Constraint Programming (PPCP'94)*, volume 874 of *Lecture Notes in Computer Science*, pages 266–280. Springer Verlag, 1994.
29. M. Koubarakis. Databases and Temporal Constraints: Semantics and Complexity. In J. Clifford and A. Tuzhilin, editors, *Recent Advances in Temporal Databases (Proceedings of the International Workshop on Temporal Databases, Zürich, Switzerland, September 1995)*, Workshops in Computing, pages 93–109. Springer, 1995.
30. M. Koubarakis. The Complexity of Query Evaluation in Indefinite Temporal Constraint Databases. *Theoretical Computer Science*, 171:25–60, January 1997. Special Issue on Uncertainty in Databases and Deductive Systems, Editor: L.V.S. Lakshmanan.
31. M. Koubarakis and S. Skiadopoulos. Querying Temporal Constraint Networks in PTIME. In *Proceedings of AAAI-99*, pages 745–750, 1999.
32. M. Koubarakis and S. Skiadopoulos. Tractable Query Answering in Indefinite Constraint Databases: Basic Results and Applications to Querying Spatio-Temporal Information. In *Spatio-Temporal Database Management (Proceedings of the International Workshop STDBM'99)*, volume 1678 of *LNCS*, pages 204–223. Springer, 1999.
33. G. Kuper, L. Libkin, and J. Paredaens, editors. *Constraint Databases.* Springer-Verlag, 2000.
34. G. Kuper, S. Ramaswamy, K. Shim, and J. Su. A Constraint-Based Spatial Extension to SQL. In *Proc. Intl. Symp. on Geographic Information Systems*, 1998.

35. J. Orenstein and F. Manola. PROBE: Spatial Data Modeling and Query Processing in an Image Database Application. *IEEE Transactions on Software Engineering*, 14(5):611–628, 1988.
36. J. Paredaens, J. Van den Bussche, and D. Van Gucht. Towards a Theory of Spatial Database Queries. In *Proc. 13th ACM Symp. on Principles of Database Systems*, pages 279–288, 1994.
37. P.Z. Revesz. A Closed Form Evaluation for Datalog Queries with Integer (Gap)-Order Constraints. *Theoretical Computer Science*, 116(1):117–149, 1993.
38. P.Z. Revesz, R. Chen, P. Kanjamala, Y. Li, Y. Liu, and Y. Wang. The MLPQ/GIS Constraint Database. In *Proceedings of SIGMOD 2000*, 2000.
39. N. Roussopoulos, C. Faloutsos, and T. Sellis. An Efficient Pictorial Database System for PSQL. *IEEE Transactions on Software Engineering*, 14(5):639–650, 1988.
40. M. Scholl and A. Voisard. Thematic Map Modeling. In *Proc. Intl. Symp. on Large Spatial Databases (SSD)*, LNCS No. 409, pages 167–192. Springer-Verlag, 1989.
41. P.-C. Scholl, M.-C. Fauvet, and J.-F. Canavaggio. Un Modèle d'Historique pour un SGBD Temporel. *TSI*, 17(3), mars 1998.
42. A. Schrijver. *Theory of Linear and Integer Programming.* Wiley, 1986.
43. A.P. Sistla, O. Wolfson, S. Chamberlain, and S. Dao. Modeling and Querying Moving Objects. In *Proceedings of ICDE-97*, 1997.
44. R.T. Snodgrass, editor. *The TSQL2 Temporal Query Language.* Kluwer Academic Publishers, 1995.
45. A. Tansel, J. Clifford, S. Gadia, S. Jajodia, A. Segev, and R. Snodgrass, editors. *Temporal Databases: Theory, Design, and Implementation.* Database Systems and Applications Series. Benjamin/Cummings Pub. Co., 1993.
46. D. Toman, J. Chomicki, and D.S. Rogers. Datalog with Integer Periodicity Constraints. In *Proceedings of the International Symposium on Logic Programming*, pages 189–203, 1994.
47. M. Ubell. The Montage Extensible DataBlade Architecture. In *Proc. ACM SIGMOD Intl. Conference on Management of Data*, 1994.

6 Access Methods and Query Processing Techniques

Adriano Di Pasquale[3], Luca Forlizzi[3], Christian S. Jensen[1], Yannis Manolopoulos[2], Enrico Nardelli[3], Dieter Pfoser[1], Guido Proietti[3,4], Simonas Šaltenis[1], Yannis Theodoridis[5], Theodoros Tzouramanis[2], and Michael Vassilakopoulos[2]

[1] Aalborg University, Denmark
[2] Aristotle University, Thessaloniki, Greece
[3] Universita Degli Studi di L'Aquila, Italy
[4] National Research Council, Roma, Italy
[5] University of Piraeus, Greece

6.1 Introduction

The performance of a database management system (DBMS) is fundamentally dependent on the access methods and query processing techniques available to the system. Traditionally, relational DBMSs have relied on well-known access methods, such as the ubiquitous B^+-tree, hashing with chaining, and, in some cases, linear hashing [52]. Object-oriented and object-relational systems have also adopted these structures to a great extend.

During the past decade, new applications of database technology — with requirements for non-standard data types and novel update and querying capabilities — have emerged that motivate a re-examination of a host of issues related to access methods and query processing techniques.

As an example, a range of applications, like cadastral, utilities, shortest path finding, etc., involve geographic, or spatial, data, which are not supported well by existing technology, making it not only desirable, but plain necessary to examine access methods and query processing techniques afresh.

Specifically, Oracle's Spatial Data Engine uses Linear Quadtrees [2] at a conceptual level, but uses B^+-trees as the storage mechanism for the quadtrees at the implementation level. As another example, R-trees [34] have been implemented by Oracle. However, R-trees are mapped to B^+-trees [75], in order to not change other system components, such as the transaction manager, the recovery manager, the buffer manager, etc. Thus, R-trees do not support deletions physically, but only logically, because this is the practice of B^{link}-trees [77], which is the brand of B-trees implemented in commercial systems.

When, as we have seen, DBMSs supporting only the traditional access methods fall short in supporting spatial data, it is not surprising that new access methods and query processing techniques are needed if DBMSs are to support the many and diverse emerging applications that call for the management of spatio-temporal data. While access methods and techniques exist that support time and, as discussed above, space, existing proposals are unable to simultaneously support time and space, either efficiently or at all.

T. Sellis et al. (Eds.): Spatio-temporal Databases, LNCS 2520, pp. 203–261, 2003.
© Springer-Verlag Berlin Heidelberg 2003

This chapter introduces a number of spatio-temporal access methods (STAMs) and query processing techniques related to spatio-temporal applications, such as bitemporal spatial applications, trajectory monitoring, and the processing of evolving raster images, such as thematic layers or satellite images. These structures are divided into two categories, according to the spatial methods they extend: R-tree-based methods versus quadtree-based methods.

Early R-tree-based approaches for indexing spatio-temporal data either treat time as a spatial dimension or, conceptually, accommodate time by maintaining a time-indexed collection of R-trees. The chapter presents more R-tree-based methods that are customized more comprehensively to accommodate the special properties of the different kinds of spatio-temporal data considered, and that as a result generally demonstrate better performance.

Briefly, the quadtree-based methods described in this chapter enhance previous methods based on Linear Quadtrees by appropriately embedding additional structuring for linking evolving raster images.

The chapter also examines other issues related to the physical database level, namely benchmarking, data generation, distributed indexing techniques, and query optimization. Finally, relevant work by other researchers is covered, and an epilog concludes the chapter.

6.2 R-Tree-Based Methods

6.2.1 Preliminary Approaches

The most straightforward way to index spatio-temporal data is to consider time simply as an additional spatial dimension, along with the other spatial dimensions. As a result, a two-dimensional rectangle (x_1, y_1, x_2, y_2) with an associated time interval $[t_1, t_2)$ is viewed as a three-dimensional box $(x_1, y_1, x_2, y_2, t_1, t_2)$. Viewing time as another dimension is attractive because several tools for handling the resulting multi-dimensional data are already available [30]. The approach of treating time as just another dimension may have the drawback of excessive dead space [96].

The technique of overlapping offers an alternative solution. In general, overlapping has been used at a number of occasions, where successive data snapshots are similar. For example, it has been used as a technique to compress similar text files [8], B-trees and B^+-trees [11,50,85], as well as main-memory quadtrees [98,99]. In the context of STAMs, the technique of overlapping has been adopted in the cases presented in the sequel.

Thus, most of the proposed STAMs can be characterized as belonging to one of the following two categories:

- the "time is an extra dimension" approach, and
- the "overlapping trees" approach.

We proceed to present several access methods that follow these approaches and were developed in the CHOROCHRONOS framework; in Section 6.7, we introduce methods proposed by other researchers.

3D R-Tree

The 3D R-tree proposed in [96] exactly considers time as an extra dimension and represents two-dimensional rectangles with time intervals as three-dimensional boxes. Figure 6.1 illustrates an example 3D R-tree storing five boxes (A, B, C, D, and E) organized in two nodes (R1 and R2). This tree can be the original R-tree [34] or any of its variants.

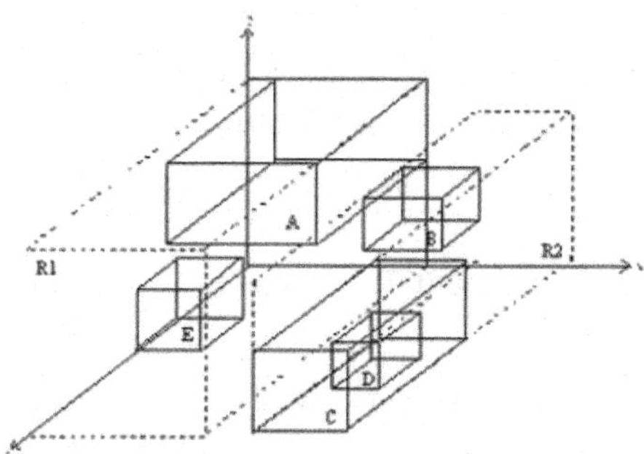

Fig. 6.1. The 3D R-tree

The 3D R-tree approach assumes that both ends of the interval $[t_1, t_2)$ of each rectangle are known and fixed. If the end time t_2 is not known, this approach does not work well. For instance, in Figure 6.1, assume that an object extends from some fixed time until the current time, *now* (refer to [16] for a thorough discussion on the notion of *now*). One approach is to represent *now* by a time instant sufficiently far in the future. But this leads to excessive boxes and consequent poor performance. Standard spatial access methods, such as the R-tree and its variants, are not well suited to handle such "open" and expanding objects. One special case where this problem can be overcome is when all movements are known a priori. This would cause only "closed" objects to be entries of the R-tree.

The 3D R-tree was implemented and evaluated analytically and experimentally [96,100], and it was compared with the alternative solution of maintaining two separate indices: a spatial (e.g., a 2D R-tree) and a temporal one (e.g., a 1D R-tree or a segment tree). Synthetic (uniform-like) datasets were used, and the retrieval costs for pure temporal (during, before), pure spatial (overlap, above), and spatio-temporal operators (the four combinations) were measured. The results suggest that the unified scheme of a single 3D R-tree is obviously superior when spatio-temporal queries are posed, whereas for mixed workloads, the decision depends on the selectivity of the operators.

2+3 R-Tree

One possible solution to the problem of "open" geometries is to maintain a pair of two R-trees [63]:

- a 2D R-tree that stores two-dimensional entries that represent current (spatial) information about data, and
- a 3D R-tree that stores three-dimensional entries that represent past (spatio-temporal) information; hence the name 2+3 R-tree.

The 2+3 R-tree approach is a variation of an original idea proposed in [41,42] in the context of bitemporal databases, and which was later generalized to accommodate more general bitemporal data [10].

As long as the end time (t_2) of an object interval is unknown, it is indexed by the (2D) *front* R-tree, keeping the start time (t_1) of its position along with its object identifier. When t_2 becomes known, then:

- the associated entry is migrated from the front R-tree to the (3D) *back* R-tree, and
- a new entry storing the updated current location is inserted into the front R-tree.

Should one know all object movements a priori, the front R-tree would not be used at all, and the 2+3 R-tree reduces to the 3D R-tree presented earlier. It is also important to note that both trees may need to be searched, depending on the time instant with respect to which the queries are posed.

HR-Tree

Historical R-trees (HR-trees, for short) have been proposed in [61] and implemented and evaluated in [63]. This STAM is based on the overlapping technique. In the HR-tree, conceptually a new R-tree is created each time an update occurs. Obviously, it is not practical to physically keep an entire R-tree for each update. Because an update is localized, most of the indexed data and thus the index remain unchanged across an update. Consequently, an R-tree and its successor are likely to have many identical nodes. The HR-tree exploits this and represents all R-trees only logically. As such, the HR-tree can be viewed as an acyclic graph, rather than as a collection of independent tree structures.

Figure 6.2 illustrates overlapping trees for successive time instants t_0 and t_1, where two subtrees from t_0 remain unchanged at t_1. With the aid of an array pointing to the root of the underlying R-trees, one can easily access the desired R-tree when performing a timeslice query. In fact, once the root node of the desired R-tree for the time instant specified in the query is obtained, the query processing cost is the same as if all R-trees where kept physically.

The concept of overlapping trees is simple to understand and implement. Moreover, when the number of objects that change location in space is relatively small, this approach is space efficient. However, if the number of moving objects from one time instant to another is large, this approach degenerates to independent tree structures, since no common paths are likely to be found.

Recently, Nascimento et al. [63] implemented the HR-tree and the 2+3 R-tree and presented a performance comparison, also including a 3D R-tree implementation, using synthetic datasets generated by GSTD (the scenarios illustrated in Section 6.5). They assumed spatio-temporal data specified as follows.

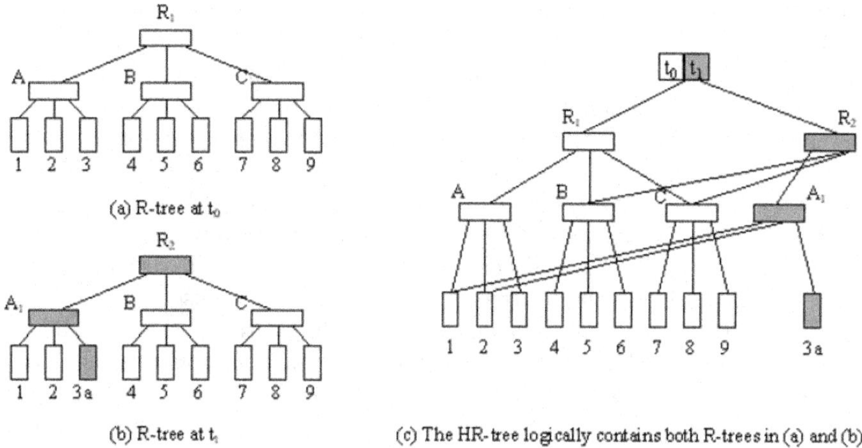

Fig. 6.2. HR-tree example

- the data set consisted of two-dimensional points, which were moving in a discrete manner within the unit square;
- updates were allowed only in the current state of the (hence, chronological) database;
- the timestamp of each point version grew monotonically following a transaction time pattern, and
- the cardinality of the data set remained fixed as time evolved.

The HR-tree was found to be more efficient than the other two methods for timeslice queries, whereas the reverse was true for time interval queries. Also, the HR-tree usually led to a rather large structure.

6.2.2 The Spatio-bitemporal R-Tree

The R^{ST}-tree proposed by Šaltenis and Jensen in [80] is capable of indexing spatio-bitemporal data with discretely changing spatial extents. In contrast to the indexing structures described previously, the R^{ST}-tree supports data that has two temporal dimensions and two spatial dimensions. The *valid time* of data is the time(s)—past, present, or future—when the data is true in the modeled reality, while the *transaction time* of data is the time(s) when the data was or is current in the database [36,76]. Data for which both valid and transaction time is captured is termed *bitemporal.*

As mentioned earlier, most of the previously proposed spatio-temporal indices [96,62] assume only one time dimension and use either the technique of overlapping index structures or add time as another dimension to an existing spatial index.

The former approaches do not generalize well to two time dimensions, and treating time as a spatial dimension has certain limitations. In particular, time intervals associated with data objects can be *now-relative*, meaning that their end points track the progressing current time. Consider the recording of addresses. The time a person resides at a given address may often extend from a known start time (the valid-time interval begin) to some unknown future time, which is captured by letting the valid-time interval extend to the progressing current time. The same applies to transaction time; the time a data object is inserted into the database is known, but it is unknown when the tuple will be deleted. This notion of *now* is peculiar to time and has no counterpart in space.

In order to support these aspects of time together with spatial dimensions, the R^{ST}-tree is based on the R*-tree [12] and attempts to reuse ideas presented in the GR-tree [9]. This latter index also extends the R*-tree and is arguably the best index for general bitemporal data, which encompasses now-relative data.

Since the index is based on the R*-tree, the spatial value of an object may be a point or may have extent. Examples of discretely changing spatio-temporal point data include demographic data that captures the changing locations of peoples' residences. Also, cadastral systems exemplify data with spatial extents. Here, the shapes and locations of land parcels, approximated by rectangles for indexing purposes, are recorded together with the histories of their change. Such histories may contain now-relative time intervals. We proceed to characterize now-relative data.

The Data and Queries Supported

We adopt the standard four-timestamp format for capturing valid and transaction time [83], where each tuple is timestamped with four time attributes: $VT^{\vdash}$ and $VT^{\dashv}$ for valid time; $TT^{\vdash}$ and $TT^{\dashv}$ for transaction time. To represent now-relative time intervals, $VT^{\dashv}$ can be set to *now* and $TT^{\dashv}$ can be set to UC (*until changed*).

Consider the example relation in Table 6.1. Tuple 1 records that the information "John lived at Pos1" was true from 3/97 to 5/97 and that this was recorded during 4/97 and is still current. Tuple 3 records that "Jane lives at Pos3" from

Table 6.1. The demographic relation

	Person	Position	$TT^{\vdash}$	$TT^{\dashv}$	$VT^{\vdash}$	$VT^{\dashv}$
(1)	John	Pos1	4/97	UC	3/97	5/97
(2)	Tom	Pos2	3/97	7/97	6/97	8/97
(3)	Jane	Pos3	5/97	UC	5/97	*now*
(4)	Julie	Pos4	3/97	7/97	3/97	*now*
(5)	Julie	Pos4	8/97	UC	3/97	7/97
(6)	Ann	Pos5	5/97	UC	3/97	$now + 1$
(7)	Scott	Pos6	4/97	UC	5/97	$now - 2$

5/97 until the current time, that we recorded this belief on 5/97, and that this remains part of the current state. In the case of Tuple 3, the valid-time end being equal to *now* means that we currently do not believe that Jane will live at Pos3 next month (on 10/97). This assumption can be too pessimistic. For example, there can exist a restriction that a person can only move with a month notice. We would then believe Jane to live at Pos3 next month as well. To record this type of knowledge, Clifford et al. [16] proposed to use $now + \Delta$ in the valid-time end attribute. The offset Δ can be any integer, positive or negative. The latter is useful when information about changes in positions is delayed. Tuples 6 and 7 exemplify the usage of positive and negative offsets.

In a bitemporal database, tuples are never physically deleted. Instead, they are removed from the current state, by changing the $\mathrm{TT}^{\dashv}$ value UC to the fixed value CT–1[1] (e.g., Tuple 2). A modification is modeled as a deletion followed by an insertion (e.g., an update led to Tuples 4 and 5).

The temporal aspects of a tuple can be represented by a two-dimensional *bitemporal region* in the space spanned by transaction time and valid time [36] (see Figure 6.3). A now-relative transaction-time interval yields a rectangle that "grows" in the transaction-time direction as time passes (Tuple 1). Having both

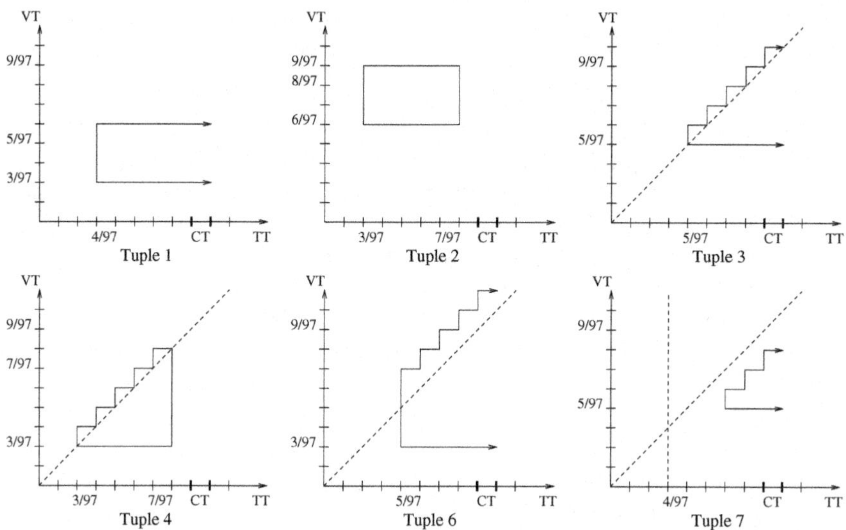

Fig. 6.3. Bitemporal regions of tuples from Table 6.1

transaction-time and valid-time intervals being now-relative yields a stair-shaped region growing in both the transaction time and valid time as time passes (Tuple 3).

This representation of the bitemporal extents of tuples suggests the use of some spatial index as the basis for a bitemporal index, which then also facili-

[1] We use closed intervals and let $[\mathrm{TT}^{\vdash}, \mathrm{TT}^{\dashv}]$ denote the interval that includes $\mathrm{TT}^{\vdash}$ and $\mathrm{TT}^{\dashv}$.

tates the incorporation of spatial dimensions into the resulting spatio-bitemporal index.

The queries supported are the well-known intersection queries, where data with a spatio-bitemporal extent that overlaps with a specified query extent, which is also spatio-bitemporal.

Index Structure and Algorithms

Index Structure
The new index has the same overall structure as the R-tree (and the R*-tree) [34]. As for the R-tree, each internal node is a record of index entries, each of which is a pair of a pointer to a node at the next level in the tree and a region that encloses all regions in the node pointed to. As something new, the leaves of the R^{ST}-tree record the exact bitemporal geometries of the spatio-bitemporal regions indexed and allow regions that grow. The same types of regions are also used as bounding regions in the non-leaf nodes (see Figure 6.4). The following format is used for index entries.

$$(\mathrm{TT}^{\vdash}, \mathrm{TT}^{\dashv}, \mathrm{VT}^{\vdash}, \mathrm{VT}^{\dashv}/\Delta, \textit{now-flag}, \langle \textit{spatial part} \rangle, \langle \textit{pointer} \rangle)$$

The first three components were introduced in the previous section and may obtain the same values as described there. Variable UC is represented as a special, reserved value from the domain of timestamps. The fourth and fifth components compactly encode the values of the $\mathrm{VT}^{\dashv}$ attribute. A value of the form $now + \Delta$ is captured by setting the *now-flag* and storing Δ in $\mathrm{VT}^{\dashv}/\Delta$; other values are stored in this attribute, without the *now-flag* set.

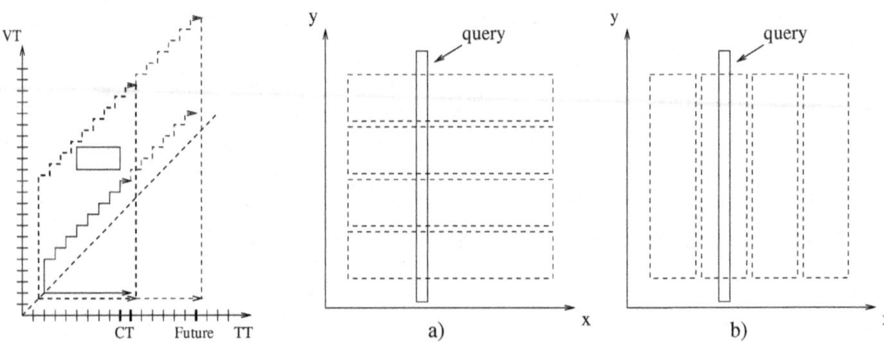

Fig. 6.4. A sample stair-shaped MBR

Fig. 6.5. Different geometries of MBRs

Index Algorithms
Since the R^{ST}-tree structure is the same as that of the R*-tree, the R*-tree search, deletion, and insertion algorithms can be re-used in the new index, provided that they employ a suite of new lower-level algorithms that manipulate

the new kinds of regions described above. These new algorithms include an algorithm that determines whether a pair of regions overlap and algorithms that compute the volume and margin of a region, the intersection of a pair of regions, and the minimum bounding region of a node. It should also be noted that a logical deletion is implemented as a physical deletion of an old region followed by an insertion of a new one with a fixed $TT^{\dashv}$.

The insertion algorithm is crucial, because it is responsible for maintaining the tree in an efficient way. The R*-tree insertion algorithm is based on heuristics that minimize the volumes of bounding regions, the overlap among bounding regions (the volume of their intersection), and the margin of bounding regions.

In the R^{ST}-tree, the quantities of volume, overlap, and margin are functions of time, and the insertion algorithm should consider not only the current values of these, but also how they evolve. This is achieved in a relatively straightforward and flexible manner, by introducing a *time parameter* p in the tree insertion algorithm, which then computes the areas (and margins) of regions as of p time units into the future. Other than this, the insertion algorithm follows that of the R*-tree, with only some differences in the splitting of overfull nodes [80]. When using a sufficiently large time parameter, a prioritization of the types of regions is obtained. Non-growing regions are naturally preferred over growing, rectangular regions, and these are preferred over growing, stair-shaped regions. Relaxed prioritizations are achieved by using smaller time parameter values. Experimental studies show that the choice of an appropriate time parameter value in an index is not very sensitive to differences in the data and query workloads.

Prioritizing Space versus Time

The heuristics used in the R*-tree are based on the assumption that intersection queries are square on average, i.e., all the dimensions are constrained by intervals of approximately the same length.

Due to the quite different semantics of the temporal and spatial dimensions, this may not always be a good assumption for the R^{ST}-tree. In some applications, most queries can be much more restrictive in the spatial dimensions than on the temporal dimensions. For example, queries in a cadastral system may retrieve the current knowledge of the full history of ownership of some piece of land. In other applications, queries can be most restrictive in the temporal dimensions. Specifically, timeslice queries, which specify time points in the temporal dimensions, have very natural semantics and are often important. Non-square queries may also be due to the use of different units of measurement in the spatial and temporal dimensions.

In order to obtain a versatile spatio-temporal index that supports well the full spectrum queries, it is desirable to introduce a mechanism that allows the R^{ST}-tree to be tuned to support better either spatially or temporally restrictive queries.

In any R-tree-based index, one dimension can be prioritized over the others by intentionally favoring minimum bounding rectangles that are narrow in this dimension and long in the other dimensions. In Figure 6.5, a two-dimensional space is considered. The two sets of minimum bounding rectangles cover the same areas and do not overlap. Scenario (b) favors queries restrictive in the x dimension and not in the y dimension.

[80] proposed a simple way to prioritize the dimensions in an R-tree-based index, which works with the existing tree algorithms. For each n-dimensional rectangle, weighted extents $((\Delta x_1)^{\alpha_1}, (\Delta x_2)^{\alpha_2}, \ldots, (\Delta x_n)^{\alpha_n})$ are used, instead of simply using the extents $(\Delta x_1, \Delta x_2, \ldots, \Delta x_n)$. If all α_i are equal to one, none of the dimensions are prioritized. The priority of dimension i is increased by setting α_i to a value greater than 1, and the priority of dimension i lowered by setting α_i to a value smaller than 1. Setting α_i to 0 makes the algorithms disregard the dimension.

Following this scheme the RST-tree uses a single parameter $\alpha \in [-1, 1]$. The volume of a four-dimensional region r is then computed as follows.

$$volume(r) = \begin{cases} bitemporal\_area(r)^{1+\alpha} \cdot spatial\_area(r) & \text{if } \alpha \leq 0 \\ bitemporal\_area(r) \cdot spatial\_area(r)^{1-\alpha} & \text{otherwise,} \end{cases}$$

where *bitemporal_area* is the area of the region time-parameterized bitemporal extent and *spatial_area* is the area of its spatial extent.

According to the criteria for classification of STAMs as proposed by Theodoridis et al. [90], the RST-tree supports two-dimensional points and regions; it is bitemporal; supports *now*-relative time intervals in both time dimensions; both the cardinality and the positions of the spatial objects may change over time; the index is dynamic; and spatial, temporal, and spatio-temporal containment queries are supported with the ability to adapt the index to spatially or temporally restrictive queries.

6.2.3 The Time-Parameterized R-Tree

So far, we have mainly considered the indexing of discretely moving spatial objects. In this section, we proceed to explore the indexing of continuously moving objects. The rapid and continued advances in positioning systems, e.g., GPS, wireless communication technologies, and electronics in general promise to render it increasingly feasible to track and record the changing positions of objects capable of continuous movement.

Continuous movement poses new challenges to database technology. In conventional databases, data is assumed to remain constant unless it is explicitly modified. Capturing continuous movement with this assumption would entail either performing very frequent updates or recording outdated, inaccurate data, neither of which are attractive alternatives.

A different tack must be adopted. The continuous movement should be captured directly, so that the mere advance of time does not necessitate explicit updates [101]. In other words, rather than storing simple positions, functions of time that express the objects' positions should be stored. Updates are then necessary only when the parameters of the functions change. We use a linear function for each object, with the parameters being the position and velocity vector of the object at the time the function is reported to the database.

The *Time-parameterized R-tree* (TPR-tree, for short) efficiently indexes the current and anticipated future positions of moving point objects (or "moving points", for short). The technique has been proposed by Šaltenis et al. in [81] and extends the R*-tree [12].

Different views of the indexed space distinguish different possible approaches to the indexing of the future linear trajectories of moving objects. Assuming the objects move in n-dimensional space (n=1,2,3), their future trajectories can be indexed as lines in (n+1)-dimensional space, where one dimension is time [91]. As an alternative, one may map the trajectories to points in a $2n$ dimensional space, which are then indexed [39]. Queries must subsequently also be transformed to counter the data transformation. Yet another alternative is to index data in its native, n-dimensional space, which is possible by parameterizing the index structure using velocity vectors and thus enabling the index to be "viewed" as of any future time. The TPR-tree adopts this latter alternative. This absence of transformations yields a quite intuitive indexing technique.

The Data and Queries Supported

We represent the linear trajectory of a moving point object by two parameters: a vector of the coordinates of a reference position at some specified time t_{ref}, $\bar{x}(t_{ref})$, and a velocity vector $\bar{v}$. Then, object positions at times not before the current time are given by $\bar{x}(t) = \bar{x}(t_{ref}) + \bar{v}(t - t_{ref})$. As will be explained shortly, the same two vectors of values, the reference position and the velocity, are used in the bounding rectangles in the TPR-tree nodes.

The TPR-tree supports queries on the future trajectories of points. A query retrieves all points with trajectories that cross the specified query region in ($\bar{x}$, t)-space. We distinguish between three kinds of queries, based on the regions they specify. Let R, R_1, and R_2 be three n-dimensional rectangles and t, $t^{\vdash} < t^{\dashv}$, three time values that are not less than the current time.

Type 1 timeslice query: $Q=(R, t)$ specifies a hyper-rectangle R located at time point t.

Type 2 window query: $Q=(R, t^{\vdash}, t^{\dashv})$ specifies a (n+1)-dimensional hyper-rectangle that has spatial coordinates specified by R and that spans in time from $t^{\vdash}$ to $t^{\dashv}$.

Type 3 moving query: $Q=(R_1, R_2, t^{\vdash}, t^{\dashv})$ specifies the (n+1)-dimensional trapezoid obtained by connecting R_1 at time $t^{\vdash}$ to R_2 at time $t^{\dashv}$.

The second type of query generalizes the first, and is itself a special case of the third type. To illustrate, consider the one-dimensional data set in Figure 6.6, which may represent temperatures measured at different locations. Here, queries $Q0$ and $Q1$ are timeslice queries, $Q2$ is a window query, and $Q3$ is a moving query.

Let $iss(Q)$ denote the time when a query Q is issued. The two parameters, reference position and velocity vector, of an object as seen by a query Q depend on $iss(Q)$, because objects update their parameters as time goes. Consider object $o1$: its movement is described by one trajectory for queries with $iss(Q) < 1$, another trajectory for queries with $1 \leq iss(Q) < 3$, and a third trajectory for

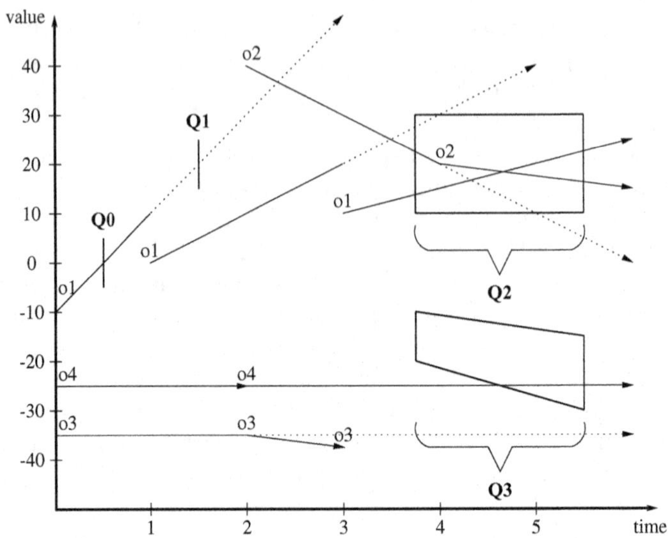

Fig. 6.6. Query examples for one-dimensional data

queries with $3 \leq iss(Q)$. For example, the answer to query $Q1$ is $o1$ if $iss(Q1) < 1$, and no object qualifies for this query if $iss(Q1) \geq 1$.

This example illustrates that queries far in the future are likely to yield answers that are of little use, because the positions predicted at query time will be less and less accurate as queries move into the future, and because updates not known at query time may occur in the meantime. Therefore, real-world applications can be expected to issue queries that are concentrated in some limited time window, termed the *querying window* (W), that extends from the current time. We assume that $iss(Q) \leq t \leq iss(Q) + W$ for Type 1 queries, and $iss(Q) \leq t^{\vdash} \leq t^{\dashv} \leq iss(Q) + W$ for queries of Types 2 and 3.

Index Structure and Algorithms

Index Structure

The TPR-tree is a balanced, multi-way tree with the structure of an R-tree. Entries in leaves are pairs of the position of a moving-point object and a pointer to the moving-point object, and entries in internal nodes are pairs of a pointer to a subtree and a rectangle that bounds the positions of all moving objects or other bounding rectangles in that subtree.

In an entry of a leaf, the position of a moving point is represented by a reference position at time t_{ref} and a corresponding velocity vector. We choose t_{ref} to be equal to the index load time, t_l. Other possibilities include setting t_{ref} to some constant value, e.g., 0, or using different t_{ref} values in different nodes.

To bound a group of n-dimensional moving points, n-dimensional bounding hyper-rectangles ("rectangles", for short) are used that are also time-parameterized, i.e., their coordinates are functions of time. A time-parameterized bounding rectangle bounds all enclosed points or rectangles at all times not earlier than the current time.

A tradeoff exists between how tightly a bounding rectangle bounds the enclosed moving points or rectangles across time and the storage needed to capture the bounding rectangles. It would be ideal to employ time-parameterized bounding rectangles that are *always minimum*, but the storage cost appears to be excessive.

Instead of using such always minimum bounding rectangles, the TPR-tree employs "conservative" bounding rectangles, which are minimum at some time point, but possibly (and most likely!) not at later times. Following the representation of moving points, we let $t_{ref}{=}t_l$ and capture a time-parameterized bounding rectangle as a tuple $([x_1^{\vdash}, x_1^{\dashv}], [x_2^{\vdash}, x_2^{\dashv}], \ldots, [x_d^{\vdash}, x_d^{\dashv}], [v_1^{\vdash}, v_1^{\dashv}], [v_2^{\vdash}, v_2^{\dashv}], \ldots, [v_d^{\vdash}, v_d^{\dashv}])$ that contains a minimum bounding rectangle of all the enclosed points or rectangles at time t_l and the minimums and maximums of the coordinates of velocity vectors of the enclosed objects. The bounding rectangle at time t is then given as follows: $[x_i^{\vdash}(t), x_i^{\dashv}(t)] = [x_i^{\vdash} + v_i^{\vdash}(t - t_l), x_i^{\dashv} + v_i^{\dashv}(t - t_l)]$, where $i = 1, \ldots, d$.

Figure 6.7 illustrates conservative bounding intervals. The begin of the conservative interval in the figure starts at the position of object A at time 0 and moves left at the speed of object B, and the end of the interval starts at object B at time 0 and moves right at the speed of object A. It is worth noting that conservative bounding intervals never shrink. In the best case, when all of the enclosed points have the same velocity vector, a conservative bounding interval has constant size, although it may move.

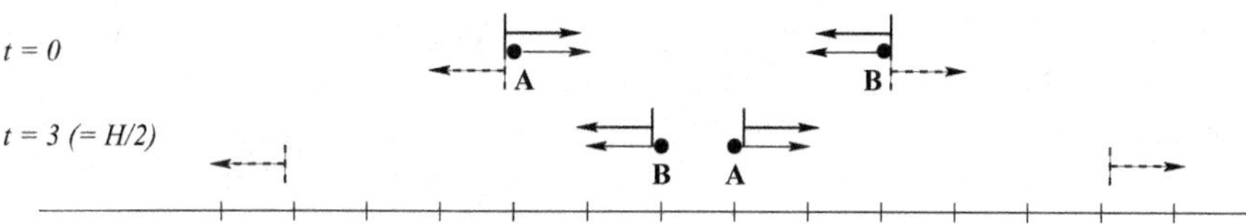

Fig. 6.7. Conservative (dashed) vs. always minimum (solid) bounding intervals

Such bounding rectangles are termed load-time bounding rectangles because they are minimal at t_l and bounding for all times not before t_l. Because they never shrink, but are likely to grow too much, it is desirable to be able to adjust them occasionally. As the index is only queried for times greater or equal to the current time, it follows that it is attractive to adjust the bounding rectangles every time any of the moving points or rectangles that they bound are updated.

We call the resulting rectangles update-time bounding rectangles. Each update-time bounding rectangle is minimal at the time of the last update that "touched" it, but all bounding rectangles are stored according to the same reference time (t_l). Figure 6.8 illustrates load time and update time bounding intervals.

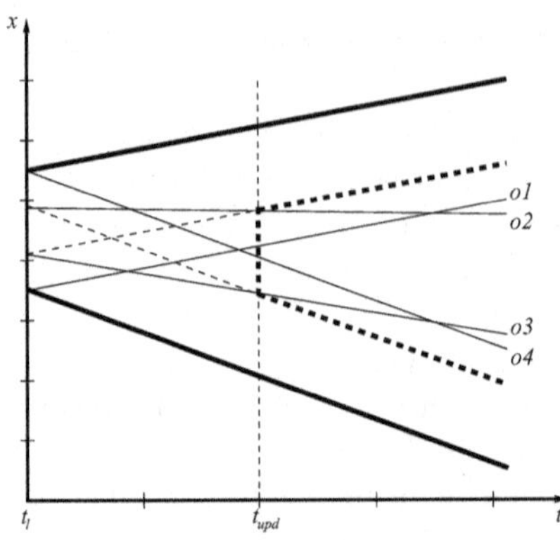

Fig. 6.8. Load-time (bold) and update-time (dashed) bounding Intervals for four moving points

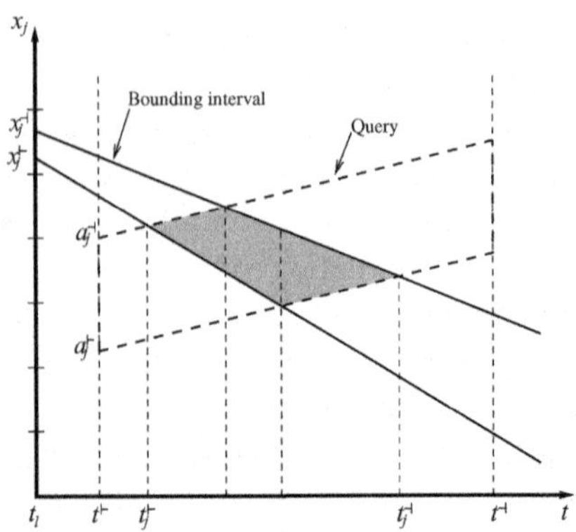

Fig. 6.9. Intersection of a bounding interval and a query

Algorithms for Querying

Answering a timeslice query using the TPR-tree proceeds as for the regular R-tree, the only difference being that all bounding rectangles are computed for the time t^q specified in the query before intersection is checked. Thus, a bounding interval specified by $(x^{\vdash}, x^{\dashv}, v^{\vdash}, v^{\dashv})$ satisfies a query $(([a^{\vdash}, a^{\dashv}]), t^q)$, if and only if $a^{\vdash} \leq x^{\dashv} + v^{\dashv}(t^q - t_l) \wedge a^{\dashv} \geq x^{\vdash} + v^{\vdash}(t^q - t_l)$.

To answer window and moving queries, we need to be able to check if, in the $(\bar{x}, t)$-space, the trapezoid of a query (see Figure 6.9) intersects with the trapezoid formed by the part of the trajectory of a bounding rectangle that is between the start and end times of the query. With one spatial dimension, this is relatively simple. For more dimensions, generic polyhedron-polyhedron intersection tests can be used [35], but due to the restricted nature of this problem, a simpler and more efficient algorithm was devised for the TPR-tree [81].

Algorithms for Insertion and Bulk Loading

The insertion and bulk loading algorithms of the R*-tree aim to minimize objective functions such as the volumes of the bounding rectangles, their margins (perimeters), and the overlap among the bounding rectangles (the volume of their intersection). In our context, these functions are time-dependent, and we consider their evolution in the time interval $[t_l, t_l + H]$, where H is the average length of the time periods when queries "see" a newly formed bounding rectangle. In particular, given an objective function $A(t)$, the following integral is minimized:

$$\int_{t_l}^{t_l+H} A(t)\ dt$$

If $A(t)$ is volume, the integral computes the volume of the trapezoid that represents part of the trajectory of a bounding rectangle in $(\bar{x}, t)$-space (see Figure 6.9). Parameter H is larger than the querying window W, and $H - W$ is inversely proportional to the average index update rate. The more frequently that object trajectories are updated, the shorter a bounding rectangle lives before it is recomputed.

The TPR-tree insertion algorithm is the same as that of the R*-tree, except that, instead of measures of volume, intersection volume, margin, and distance, integrals of these functions are used as in above formula. In addition, the algorithm for splitting overfull nodes chooses possible distributions of entries into two new nodes is adjusted.

The TPR-tree bulk loading algorithm attempts to minimize the volume integrals of the tree time-parameterized bounding rectangles across $[t_l, t_l + H]$. This is achieved by choosing an appropriate trade off between packing the moving points according to their velocities and packing them according to their reference positions. The former favors relatively large values of H, while the latter is more suitable for small values of H. The algorithm, described in detail in [81], is based on the STR algorithm [80].

6.2.4 Trajectory Bundle

Similarly to the TPR-tree, the Trajectory Bundle [69] is an R-tree-based access method, but while the TPR-tree indexes the current and anticipated future positions of moving objects, the Trajectory Bundle indexes the past trajectories of point objects capable of continuous movement. The trajectory of an object moving in two-dimensional space is similar to a "string" in three-dimensional space, where the third dimension is time. More specifically, the position of an object is sampled, which leads to a polyline representation of continuous movement (see Figure 6.10).

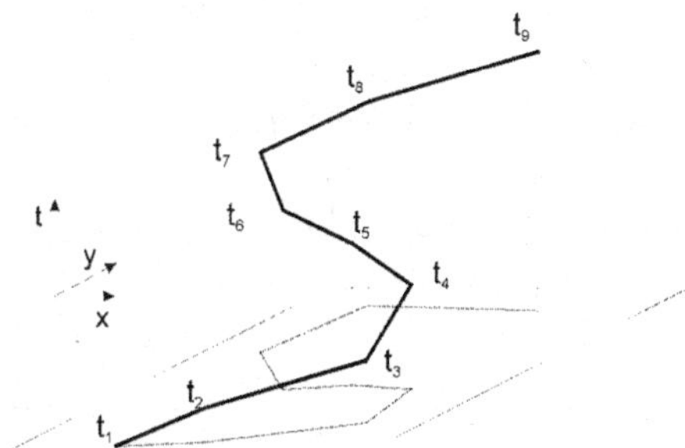

Fig. 6.10. Moving object trajectory

Several approaches to the indexing of historical spatio-temporal data exist. However, most assume that the spatial data changes discretely over time and do not address continuous movement. Although the time dimension of this spatio-temporal space can be perceived as a spatial dimension, its semantics are different. In particular, the presence of a time dimension leads to derived information, e.g., speed, acceleration, traveled distance, etc., which the access method must contend with. Next, a successful access method must recognize that the individual line segments it indexes are parts of larger constructs, namely trajectories.

The Data and Queries Supported

The Trajectory Bundle indexes the past trajectories of point objects, which are assumed to be represented as polylines in the three-dimensional space spanned by valid time and two spatial dimensions—see Figure 6.10.

The aspects of spatio-temporal data mentioned above result in new types of queries that an access method must satisfy. We distinguish between two types.

- *coordinate-based* queries, such as point, range, and nearest-neighbor queries in the resulting three-dimensional space, and
- *semantics-based queries*, usually involving trajectory metadata, such as speed and heading of objects.

Coordinate-based queries are inherited from spatial and temporal databases. The semantics-based queries are classified in *trajectory* queries, which rely on parts of trajectories that go beyond individual segments, and *navigational* queries.

Trajectory queries stem from *spatio-temporal topology* and involve predicates such as "enters," "leaves," "crosses," or "bypasses" [24]. For example, whether an object enters a given area can be determined only after examining more than one segment of its trajectory. An object entered an area with respect to a given time interval if the start point of its least recent segment is outside the area and the end point of its most recent segment is inside the given area. Similar definitions hold for the other predicates.

Navigational queries relate to information *derived* from the trajectory information and include "speed," "heading," "area covered," etc. Such quantities depend on the time interval considered, e.g., the heading of an object in the last ten minutes may have been strictly East, but considering the last hour, it may have been Northeast.

Further, *combined queries* are important. Such queries extract information related to partial trajectories by identifying the relevant trajectories and then the relevant parts of the trajectories. Trajectories can be selected based on their object identifier. Alternatively, they can be identified via a spatio-temporal range predicate, by a trajectory query, or by a query using derived information. The relevant parts of the identified trajectories are again delimited by a spatio-temporal range, a trajectory query, or derived properties. The example query "What were the trajectories of objects that left Tucson between 7 a.m. and 8 a.m. today during the first hour after they left?" uses the range "between 7 a.m. and 8 a.m." to identify the trajectories, while the temporal range "during the first hour after they left" delimits the relevant parts of the trajectories. Along these lines, a variety of combined queries can be constructed.

Index Structure and Algorithms

Unlike all previous access methods the Trajectory Bundle strictly preserves trajectories. Each leaf contains only segments belonging to one single trajectory.

This clustering of line segments based on their trajectory membership comes at a cost. Specifically, the R-tree attempts to place segments that are spatially close in the same leaf. In the Trajectory Bundle, such segments can be in different nodes. This tends to increase overlap among sibling nodes, which affects the query performance for conventional range queries. However, as we shall see, the trajectory preservation is important for answering certain spatio-temporal query types.

Insertion Algorithms

The Trajectory Bundle uses the R-tree structure and algorithms as its basis. It effectively "cuts" trajectories into pieces consisting of (up to) M line segments, where M is the number of segments that fit in a leaf. Figure 6.11 illustrates insertion of a new line segment, which is divided into six steps. First, the leaf that contains its predecessor segment is found. This node is found by traversing the tree from the root, stepping into every child node that overlaps with the minimum bounding box of the new line segment (stage 1 in Figure 6.11). In case the leaf is full, a new leaf node must be introduced. To obtain trajectory preservation, the new leaf is placed at the "end" of the tree, meaning that the new leaf is inserted at the right-most parent node. In Figure 6.11, we step up the tree until we find a non-full parent node (stages 2 through 4). We choose the right-most path (stage 5) to insert the new node. If the parent node has space (stage 6), we insert the new leaf as shown in the figure. In case it is full, we split it by creating a new node at intermediate level 1 that has the new leaf as its only descendent. If necessary, the split is propagated upwards, i.e., a new "right most" branch is created.

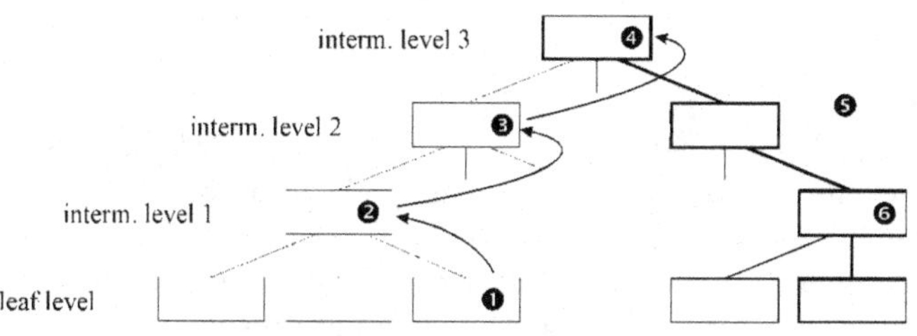

Fig. 6.11. Insertion

It can be argued that this insertion strategy leads a high degree of overlap in the index. This would certainly be the case if we were indexing arbitrary three-dimensional data. However, when indexing spatio-temporal data, we only "neglect" the two spatial dimensions, with respect to space discrimination. The temporal dimension offers some discrimination because data is inserted with increasing time coordinates. The insertion strategy makes use of this property.

Trajectory Preservation
The leaves in the Trajectory Bundle each contain a partial trajectory, and a trajectory is contained in a set of "disconnected" leaves. We shall see shortly that it is necessary to be able to retrieve segments based on their trajectory identity. To enable this, doubly linked lists are introduced that connect leaves belonging to the same trajectory. Figure 6.12 shows a partial index structure and a trajectory that illustrate this approach. For clarity, the trajectory is drawn as a band. The trajectory symbolized by the gray band is fragmented across six leaves, c1, c3, etc., which form a doubly linked list as symbolized by the two-way arrows. Upon arriving at an arbitrary leaf, these links allow retrieval of partial trajectories at minimal cost.

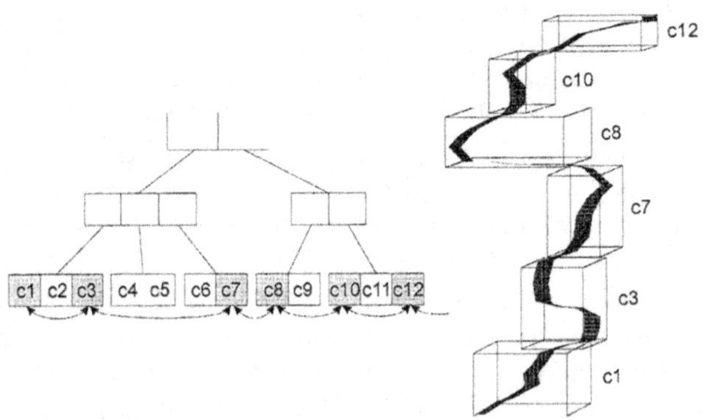

Fig. 6.12. Trajectory bundle tree structure

Algorithms for Querying
The Trajectory Bundle can be used for processing *coordinate-based*, *semantics-based*, and *combined* queries. The former are well-known and are not discussed further. Semantics-based queries comprise trajectory and navigational queries. We will show how to reduce trajectory queries to ordinary range queries. Navigational queries are special in that they compute results that are not explicitly stored. Since these computations are not based on indexing, we omit discussion of them. Algorithms for combined queries are different in that not only a spatial, but also a semantic search, is performed, i.e., they not only involve range queries, but also retrieve other segments belonging to trajectories identified in range queries.

Algorithms for Trajectory Queries
Trajectory queries involve relationships such as enters, leaves, crosses, and bypasses. These relationships can be computed using the algorithm for range queries.

Consider the query "Which taxis left Tucson between 7 a.m. and 8 a.m. today?" This query specifies a spatio-temporal range, namely "Tucson between

7 a.m. and 8 a.m. today." The cube in Figure 6.13 represents a spatio-temporal range. Trajectory t_2 belongs to an object leaving the range, and trajectories t_1, t_3, and t_4 are entering, crossing, and bypassing the range. To detect a trajectory leaving a range, we have to examine the segments of the trajectories intersecting the four sides of the spatio-temporal range as shown in Figure 6.13. If a trajectory leaves or enters a range, we will only find one segment, and it will be directed inwards or outward depending on whether it enters or leaves the range. In case a trajectory crosses (t_3) we will find two or more segments. In the case it bypasses (t_4), we will not find any segment. Thus, we can use modified range queries to evaluate the topological predicates of trajectory queries.

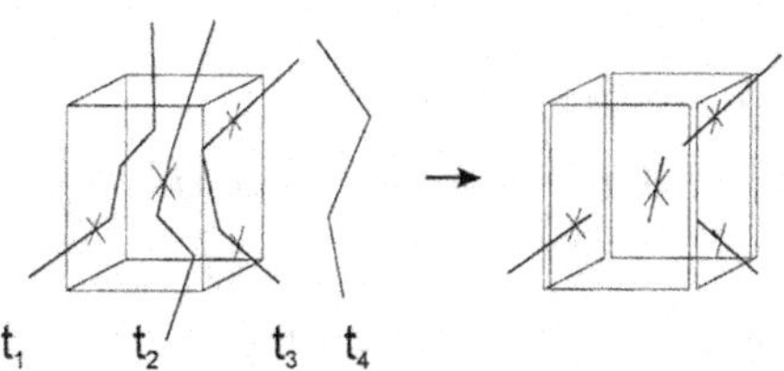

Fig. 6.13. Trajectory queries

Algorithms for Combined Queries
The first step in processing combined queries is to retrieve an initial set of segments based on some spatio-temporal range. We can apply the standard range-search algorithm used in the R-tree. In Figure 6.14, we search the tree using the cube c_1 and retrieve two segments of trajectory t_2 (labeled 1 and 2), and four segments of trajectory t_1 (labeled 3 through 6). The six segments are shown in dark gray and are contained in cube c_1. In the second step, we make use of the doubly linked lists in the Trajectory Bundle and retrieve for segments 1 and 2 of t_2 and segments 3 through 6 of t_1 the partial trajectories contained in range c_2. We have two possibilities: a connected segment can be in the same leaf or in another leaf. If it is in the same, finding it is trivial. If it is in another node, the doubly linked list is used.

When retrieving partial trajectories, care must be taken not to retrieve the same trajectory more than once. Once a partial trajectory is retrieved, we store its object identifier, and, for each retrieved trajectory, we check whether it was retrieved already.

6.3 Quadtree-Based Methods

All methods of this section assume binary images of $S \times S$ unit squares termed *pixels*, where a pixel associated with the foreground (background) is assumed to be black (respectively, white). Without loss of generality, let $S=2^m$, where m is an integer used to decompose the image. More specifically, at level m, which is

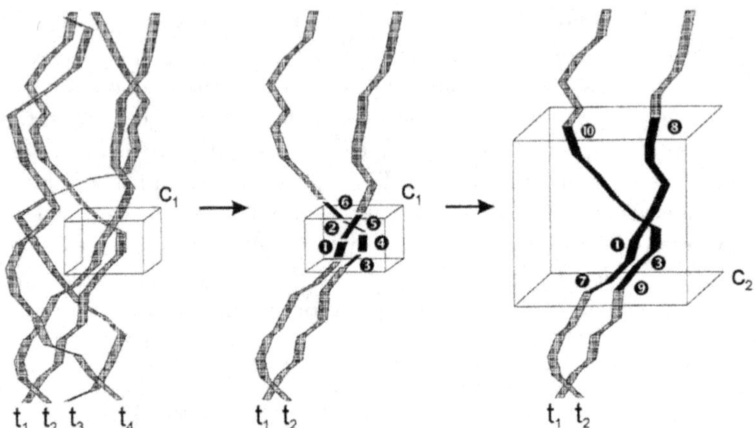

Fig. 6.14. Stages in processing combined spatio-temporal queries

stage 0 of the decomposition, there is the whole image, of side length S. At the first stage of decomposition, the image consists of four quadrants of side length $S/2$. At the second stage, each quadrant is then subdivided into four quadrants of side length $S/2^2$. The decomposition stops when a quadrant is wholly black or wholly white. The decomposition can be represented as a tree of outdegree 4, where the root (at level m) corresponds to the whole image, and each node (at level ℓ, where $1 \leq \ell \leq m$) corresponds to a quadrant of side length $S/2^{m-\ell}$. The sons of a node are labeled from left to right NW (North-West), NE (North-East), SW (South-West) and SE (South-East). Leaves are black or white (also termed *homogeneous*), and non-leaf nodes are gray (termed *non-homogeneous*).

6.3.1 The MOF-Tree

Description of the MOF-Tree

The MOF-tree has been proposed in [57,51] for indexing *multiple overlapping features*, but it can also be used as a STAM. The MOF-tree is based on a recursive space decomposition into four quadrants of equal size, in the same way as in quadtrees. Assume that a set $\langle \mathcal{I}_1, \mathcal{I}_2, \ldots, \mathcal{I}_N \rangle$ of images of size $S \times S$ is given. Image $\mathcal{I}_j$ corresponds to the j-th layer of the image $\mathcal{I} = \bigcup_{j=1,N} \mathcal{I}_j$. In the following, we will use the term *feature* to refer to a specific layer of $\mathcal{I}$. The space decomposition stops only when a quadrant is either fully covered by *all* the features in it, or is completely uncovered. In both cases, the quadrant is homogeneous.

The decomposition can be represented as a tree of outdegree 4, as described in the introductory paragraph above. Internal nodes are associated with non-homogeneous quadrants, while homogeneous quadrants correspond to leaves. In Figure 6.15, an example of a MOF-tree in a $2^2 \times 2^2$ image space representing two images is given, in which features partially covering a quadrant are depicted inside a circle, while features totally covering a quadrant are depicted inside a square. Note that internal nodes can be fully covered by some feature and partially covered by others (for example, see the SE son of the root, which is fully covered by the vertical feature and partially covered by the horizontal feature).

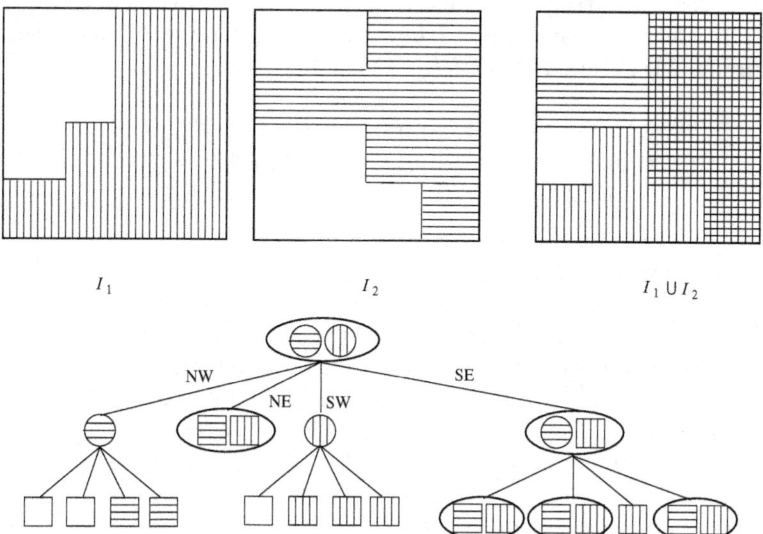

Fig. 6.15. Two overlapping features and the corresponding MOF-tree

Implementing a MOF-Tree in Secondary Memory

A linear MOF-tree version can be obtained by coding all its nodes (except the leaves associated with empty quadrants) by means of a base-5 *locational key* (l-key, for short) of length m, that can be recursively defined as follows: let the root of the MOF-tree have l-key 0; a node q' at level ℓ whose father q has l-key $L(q)$ will have l-key [28]:

$$L(q') \;=\; L(q) + s \cdot 5^{\ell}$$

where $s = 1, 2, 3, 4$ if q' is a NW, NE, SW, SE child of q, respectively. This allows the use of a conventional index such as the B$^+$-tree to efficiently support random access to every MOF-tree node [2].

For a large class of operations to be executed on the MOF-tree, we need to know, when accessing an internal node, which features are contained in and fully cover the associated quadrant. Hence, an internal node is represented by means of a record containing an N-bit vector *FEATURE*, whose j-th bit is set to 1 if and only if the j-th feature is contained in the associated quadrant, and an N-bit vector *COVER*, whose j-th bit is set to 1 if and only if the j-th feature fully covers the associated quadrant. Records associated with leaves lack the *COVER* vector. To distinguish between the two kinds of records, we also associate with each of them a *LEAF* bit, whose value is 1 if and only if the corresponding node is a leaf.

The obtained list of records can be sorted according to increasing values of the l-keys. For example, the MOF-tree in Figure 6.15 has the following sequence, where the structure of internal records is (*LEAF*, l-key, *FEATURE*, *COVER*), while for leaf records (shown in italics for readability), we have a structure (*LEAF*, l-key, *FEATURE*):

(0,00,11,00), (0,10,01,00), *(1,13,01)*, *(1,14,01)*, *(1,20,11)*,
(0,30,10,00), *(1,32,10)*, *(1,33,10)*, *(1,34,10)*, (0,40,11,10),
(1,41,11), *(1,42,11)*, *(1,43,10)*, *(1,44,11)*.

Spatio-temporal Queries

Traditional spatial queries for two-dimensional spatial data can be naturally extended to spatio-temporal queries over the set $\mathcal{I}$ of two-dimensional images. Let $\mathcal{I}_j(x,y)$ denote the pixel in $\mathcal{I}_j$ having coordinates (x,y), and let assume $\mathcal{I}_j(x,y) = 1$ if the pixel is black, $\mathcal{I}_j(x,y) = 0$ otherwise. Here, we consider only the *exist*($P(x_0, y_0, t_0), \Delta_x, \Delta_y, \Delta_t$) query, which must return `true` if and only if there exists at least an image $\mathcal{I}_j \in \mathcal{I}$, with $t_0 \le j < t_0 + \Delta_t$ such that $\mathcal{I}_j(x,y) = 1$ and $x_0 \le x < x_0 + \Delta_x, y_0 \le y < y_0 + \Delta_y$.

Let $w = \{(x,y) \in \mathbb{N}^2 | x_0 \le x < x_0 + \Delta_x, y_0 \le y < y_0 + \Delta_y\}$ denote the so-called *window query*, that is, the image space spanned by the given query. As shown by Proietti in [72], the query is solved by initially decomposing in $O(\Delta_x + \Delta_y)$ time the window into its constituting *maximal blocks*, that is, the black blocks that would be generated by applying the quadtree decomposition process to the window. Dyer has shown that the number of maximal quadtree squares inside w is $O(\Delta_x + \Delta_y)$ [23]. Afterwards, we associate with each maximal block p its respective l-key $L(p)$ and we sort these l-keys in increasing order. For each maximal block p, we have to know whether or not p contains at least one black pixel associated with $\mathcal{I}_j, t_0 \le j < t_0 + \Delta_t$. To do that, we search the B$^+$-tree with the l-key $L(p)$. One of the following can happen:

1. $L(p)$ appears in the B$^+$-tree. Then, we check whether the *FEATURE* vector has at least one bit in the interval $[t_0, t_0 + \Delta_t - 1]$ equal to 1: if so, we return `true`; otherwise, we examine the next maximal block, if any.
2. $L(p)$ does not appear in the B$^+$-tree. In this case, we check the record associated with the l-key immediately smaller than $L(p)$. Let x be such l-key. Two cases are possible:
   2.1. x is associated with a quadrant p' containing p. This happens if and only if the string obtained from x by discarding all the zeros after the rightmost non-zero digit is a prefix of $L(p)$. In this case we say that p' is an *ancestor* of p and that p is a *descendant* of p'. Then, we check whether the *FEATURE* vector of x has at least one bit in the interval $[t_0, t_0 + \Delta_t - 1]$ equal to 1: if so, we return `true`; otherwise, we examine the next maximal block, if any.

2.2. Otherwise, x is associated with a block p' disjoint from p. In this case, p is fully uncovered, and we examine the next maximal block, if any.

The following theorem is from [51].

Theorem 1. *Let us consider a sequence of two-dimensional binary images* $\langle \mathcal{I}_1, \mathcal{I}_2, \ldots, \mathcal{I}_N \rangle$ *in a* $S \times S$ *image space, represented by using a MOF-tree stored in a* B^+*-tree of order* r*. Then, an* $exist(P(x_0, y_0, t_0), \Delta_x, \Delta_y, \Delta_t)$ *query can be solved with* $O((\Delta_x + \Delta_y) \cdot \log_r n)$ *accesses to secondary memory, where* n *is the number of elements in the MOF-tree.*

Proof. Knuth [40] showed that the height $h(r, n)$ of a B$^+$-tree of order r with n elements is:

$$h(r, n) \leq 1 + \log_r \left(\frac{n+1}{2} \right).$$

From the algorithm described above, it follows that for each maximal block $O(\log_r n)$ accesses to secondary memory occur in case (1). Concerning case (2), we note that, apart from the initial descent of the B$^+$-tree, which has a cost of $O(\log_r n)$ accesses, we can at most execute an additional access to secondary memory to check the elements preceding the one accessed. As the number of maximal blocks is $O(\Delta_x + \Delta_y)$, the theorem follows. ■

Note that by using a multiple Linear Quadtrees representation, the query can be performed by applying the algorithm proposed by Nardelli et al. in [58] Δ_t times, thus obtaining an $O\left((\Delta_x + \Delta_y) \cdot \sum_{j=t_0}^{t_0+\Delta_t-1} \log_r n_j\right) = O(\Delta_t \cdot (\Delta_x + \Delta_y) \cdot \log_r n)$ time complexity, where n_j is the number of elements in the MOF-tree associated with $\mathcal{I}_j$. Hence, the improvement obtained by using an MOF-tree is linear in the number of queried features.

6.3.2 The MOF$^+$-Tree

Description of the MOF$^+$-Tree

An interesting variant of the MOF-tree, named the *MOF*$^+$*-tree*, was proposed by Proietti in [71]. This variant can be obtained by means of a particular coding technique of the *FEATURE* vector, which eliminates the need for the *COVER* vector.

This coding technique depends on both the feature distribution and the refinement process. To illustrate the coding, we analyze how the pointer version of the MOF$^+$-tree is built. The underlying idea in the building process is that we can describe the distribution of a given feature by simply marking the changing from a non-homogeneous to a homogeneous state. More precisely, let us focus on a specific feature $\mathcal{I}_j$ of $\mathcal{I}$, associated with the j-th bit of the *FEATURE* vector. At the root level, if $\mathcal{I}_j$ covers (either partially or fully) the image space, we set *FEATURE*[j]=1; otherwise, we set *FEATURE*[j]=0. In this latter case, the j-th bits of the *FEATURE* vectors of all the root descendants down to the leaf level are set to 0 as well. In the former case, two situations are possible:

1. The image space is fully covered by $\mathcal{I}_j$: in this case, the j-th bits of the *FEATURE* vectors of all the root descendants down to the leaf level are set to 1;
2. Otherwise, the j-th bits of the *FEATURE* vectors of all the root children are set to 0, thus introducing an *alternation* in the j-th bit values. The process goes on recursively with alternations in the j-th bit value until a quadrant q homogeneous (i.e., either fully covered or uncovered) with respect to $\mathcal{I}_j$ is reached: in this case, the j-th bit of the *FEATURE* vector of q is copied into the corresponding j-th bits of all its non-leaf children[2], thus introducing a *persistence* in the j-th bit values. Afterwards, the j-th bits of the *FEATURE* vectors of all the descendants of q down to the leaf level are set to 1 if $\mathcal{I}_j$ covers q, and otherwise to 0.

Since the sequence of values of the j-th bits of the *FEATURE* vectors along any path from the root to a leaf is tied to the refinement process, the three possible states (i.e., absence, partial presence, or total presence of a given feature) can be represented using only one bit. More precisely, suppose we want to know the distribution of $\mathcal{I}_j$ with respect to a non-leaf node q. For the sake of generality, suppose that q has a non-leaf child q', having in turn a non-leaf child q''. This is the most general case, since if q does not have any non-leaf nephew, then we can establish the distribution of $\mathcal{I}_j$ in q by simply looking to all its (at most 16) leaf descendants. Depending on the values of the j-th bits of the *FEATURE* vectors of q, q', and q'', the cases reported in Table 6.2 are possible.

Table 6.2. Distribution of $\mathcal{I}_j$ with respect to q

q q' q''	Distribution of $\mathcal{I}_j$ with respect to q
0 0 0	fully uncovered
0 0 1	fully covered
0 1 0	partially covered
0 1 1	if the parent of q has the j-th bit of the *FEATURE* vector set to 0, then q is fully covered; otherwise, it is partially covered (if q is the root, this configuration is not admissible)
1 0 0	if the parent of q has the j-th bit of the *FEATURE* vector set to 0, then q is partially covered; otherwise, it is fully uncovered (if q is the root, then it is partially covered)
1 0 1	partially covered
1 1 0	fully uncovered
1 1 1	fully covered

Implementing a MOF$^+$-Tree in Secondary Memory

Differently from the MOF-tree, a linear MOF$^+$-tree version can be obtained by making use of a unique record structure, that is (*LEAF*, l-key, *FEATURE*). This

[2] Notice that, by definition, a leaf q has the j-th bit of the *FEATURE* vector set to 1 if and only if $\mathcal{I}_j$ covers q.

allows up to 30% space savings [71]. For example, for the image in Figure 6.15, we have the following sequence (leaf records are shown in italics):

(0,00,11), (0,10,00), *(1,13,01)*, *(1,14,01)*, *(1,20,11)*, (0,30,00), *(1,32,10)*, *(1,33,10)*, *(1,34,10)*, (0,40,00), *(1,41,11)*, *(1,42,11)*, *(1,43,10)*, *(1,44,11)*.

It is easy to verify that by using a MOF$^+$-tree, any spatio-temporal query can be solved with at most a constant number of additional accesses on secondary storage with respect to a MOF-tree. The exist query can be solved by making use of the algorithm described in the previous section, with case (1) rewritten as follows:

1.1. $L(p)$ appears in the B$^+$-tree and the *LEAF* bit is set to 1. Then, we check whether or not the *FEATURE* vector has at least one bit in the interval $[t_0, t_0 + \Delta_t - 1]$ equal to 1. If so, we return `true`; otherwise, we examine the next maximal block, if any.
1.2. $L(p)$ appears in the B$^+$-tree and the *LEAF* bit is set to 0. Then, to determine whether at least one feature in the interval $[t_0, t_0 + \Delta_t - 1]$ covers p, we check the children and the nephews of p, and we apply the rules from Table 6.2, which entails at most a constant number of additional descents of the index.

Therefore, it can be proved that the $exist(P(x_0, y_0, t_0), \Delta_x, \Delta_y, \Delta_t)$ query can be solved with $O((\Delta_x + \Delta_y) \cdot \log_r n)$ accesses to secondary memory, where n is the number of elements in the MOF$^+$-tree.

6.3.3 Overlapping Linear Quadtrees

Description and Implementation of OLQ

The *Overlapping Linear Quadtrees* (OLQ, for short), proposed by Tzouramanis et al. [92,93], is a structure suitable for storing consecutive binary raster images according to transaction time. This corresponds to a database of evolving images (e.g., satellite meteorological images). This structure saves considerable space without sacrificing query performance in accessing every single image. Moreover, it can be used for answering efficiently window queries for consecutive images (spatio-temporal queries).

If a sequence of N images has to be stored in a Linear Quadtree, each image having a unique timestamp t_i (for i=1, 2, ..., N), then updates will overwrite old versions, and only the most recently inserted images are retained. However, in applications where spatial queries refer to the past, all past versions also need to be accessible. OLQ converts the ephemeral Linear Quadtree to a *persistent* data structure, where past states are also maintained [22].

In this subsection, we present this structure and five temporal window queries: strict containment, border intersect, general border intersect, cover, and fuzzy cover. Experiments with the OLQ based on synthetic pairs of evolving images (random images with specified aggregation) have shown [85] considerable storage savings in comparison to a group of independent Linear Quadtrees. Moreover, the I/O performance of different queries has been studied based on the same synthetic data as well as on real images [93]. It has been shown that using algorithms that take advantage of the special properties of OLQ, in comparison to straightforward algorithms, leads to significantly better I/O performance.

In the sequel, we present how overlapping is applied to Linear Region Quadtrees (the most widely used variations of Region Quadtrees for secondary memory). A Linear Quadtree representation consists of a list of values where there is one value for each black node of the pointer-based quadtree. The value of a node is an address describing the position and size of the corresponding block in the image. These addresses can be stored in an efficient structure for secondary memory (such as a B-tree or any of its variations). The most popular linear implementations are the FL (Fixed Length), the FD (Fixed length – Depth) and the VL (Variable Length) linear implementations [78]. As justified in [92], the FD implementation was the most appropriate choice for OLQs. In this implementation, the address of a black quadtree node has two fixed size parts: the first part denotes the path (directional code) to this node (starting from the root) and the second part the depth of this node. The right part of Figure 6.16 presents a quadtree which corresponds to the binary image shown on the left of the same figure. In the left part of the figure, also, the directional code of each black node of the depicted tree can be seen.

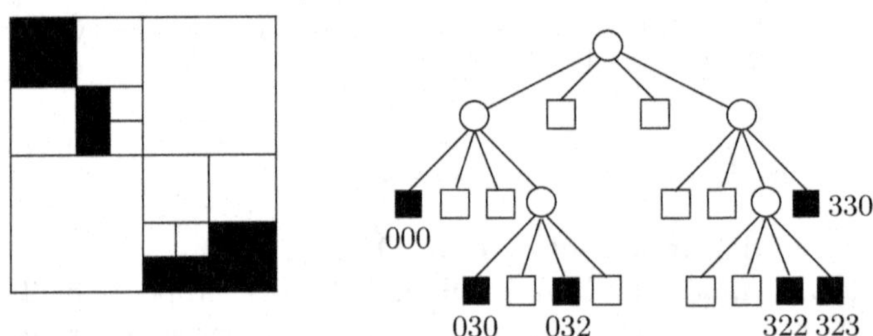

Fig. 6.16. An image, its quadtree and the linear codes of black nodes

Each quadtree, in a sequence of quadtrees modeling time evolving images, can be represented in secondary memory by storing the linear FD codes of its leaves in a Linear Quadtree (in reality, in a B^+-tree). The OLQ structure is formed by overlapping consecutive Linear Quadtrees, that is by storing the common subtrees of the two trees only once [92]. Since in the same quadtree, a pair of black ancestor and descendant nodes cannot occur, two FD linear codes that coincide at all the directional digits cannot occur, either. This means that the directional part of the FD codes is sufficient for building Linear Quadtrees at all the levels. At the leaf-level, the depth of each black node should also be stored so that images are accurately represented and that overlapping can be correctly applied. The top part of Figure 6.17 depicts the Linear Quadtrees that correspond to two Region Quadtrees, and the bottom part depicts the resulting overlapping linear structure. Note that with the OLQ structure, there is no extra cost for accesses in a specific Linear Quadtree.

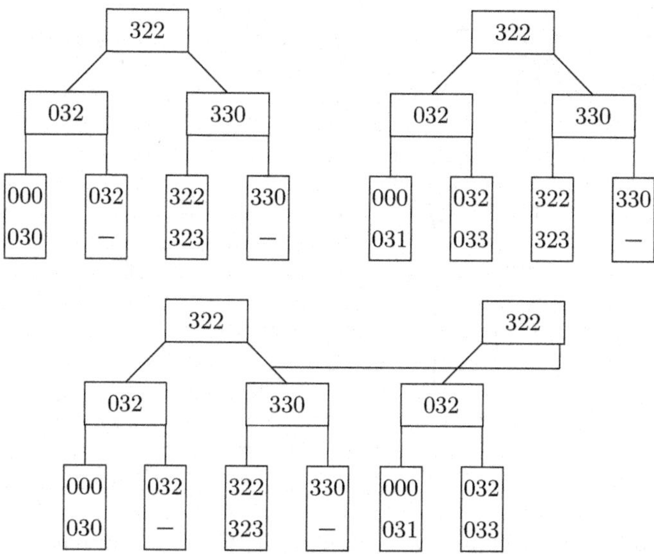

Fig. 6.17. Two B$^+$-trees storing linear quadtree codes and the corresponding OLQ structure

All nodes of the OLQ structure have an extra field, *StartTime*, that can be used to detect whether a node is being shared by other trees. We assign a value to *StartTime* during the creation of a node. There is no need for future modification of this field. In addition, leaf-nodes have one more field, *EndTime*, that is used to register the transaction time when a specific leaf changes and becomes historical.

In order to keep track of the image evolution (in other words, the evolution of quadcodes) and efficiently support spatio-temporal queries over the stored raster images, we embed additional "horizontal" pointers in the OLQ leaves. This way there will be no need to top-down traverse consecutive tree instances to search for a specific quadcode, thus avoiding excess page accesses. In particular, we embed two forward and two backward pointers in every OLQ leaf to support spatio-temporal queries. The F-pointer of a node points to the first of a group of leaves that belong to a successive tree and have been created from this node after a split/merge/update. The FC-pointers chain this group of leaves together. The B- and BC-pointers play analogous roles when traversing the structure backwards.

Figure 6.18 shows the chaining of the leaves of three successive Linear Quadtrees. The leaf on the left-top corner of the figure corresponds to the first time instant, t=1, and contains 3 quadcodes. Suppose that during time instant t=2, 8 new quadcodes are inserted. In such a case, we have a node split. During time instant t=3, a set of 5 quadcodes is deleted. Thus, two nodes of the tree corresponding to time instant t=2 are merged to produce a new node as depicted in the figure.

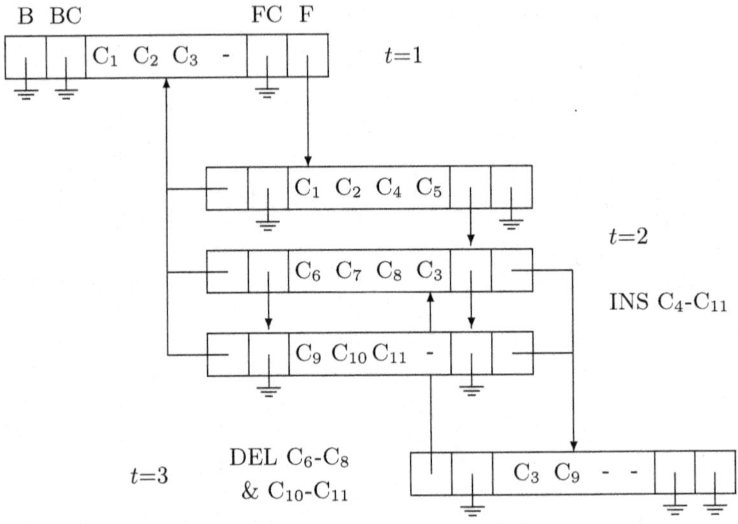

Fig. 6.18. Forward and backward chaining for the support of temporal queries

Spatio-temporal Query Processing

In this subsection, we present five different temporal window queries for evolving regional data that can be answered efficiently by using the OLQ structure. Given a window belonging in the area covered by our images and a time interval, the following spatio-temporal queries can be expressed:

The Strict Containment Window Query. Find the black regions that totally fall inside the window (including the ones that touch the window borders from inside), at each time point within the time interval.

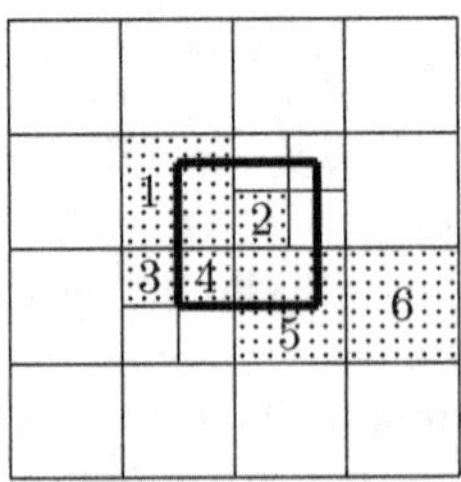

Fig. 6.19. The quadblocks of a binary raster image and a query window (thick lines)

Figure 6.19 depicts a raster image corresponding to a specific time point, partitioned in quadblocks, and a query window. The Strict Containment Window Query for this time point would return quadblocks 2 and 4.

The General Border Intersect Window Query. Find the black regions that completely fall inside the window or intersect a border of the window (including the ones that touch a border of the window from inside or outside), at each time point within the time interval.
The General Border Intersect Window Query for the time point corresponding to Figure 6.19 would return quadblocks 1, 2, 3, 4, and 5.

The Border Intersect Window Query. Find the black regions that intersect a border of the window (including the ones that touch a border of the window from inside or outside), at each time point within the time interval.
The Border Intersect Window Query for the time point corresponding to Figure 6.19 would return quadblocks 1, 3, 4, and 5.

The Cover Window Query. Find out whether or not the window is totally covered by black regions at each time point within the time interval.
The Cover Window Query returns YES/NO answers. For the time point corresponding to Figure 6.19, it would return No as an answer.

The Fuzzy Cover Window Query. The Cover Window Query algorithm can be extended so as to work for partially black windows, where the black percentage exceeds a specified threshold. That is, we could answer a query of one of the following two forms.

- Find out whether or not the percentage of the window area that is covered by black regions is larger than a given threshold, at each time point within the time interval.
- Find out the percentage of the window area that is covered by black regions, at each time point within the time interval.

The second kind of Fuzzy Cover Window Query for the time point corresponding to Figure 6.19 would return 80% as its answer. The first kind would return YES or NO depending on the comparison of 80% with the threshold given.

For such queries, [95] presented algorithms that take into account the horizontal pointers. Extensive experiments showed remarkable improvements of the response times of these sophisticated algorithms in comparison to those of the corresponding straightforward algorithms. All the presented algorithms can easily be transformed to work forward or backward: by starting from the beginning or the end of the time interval and by using the F- and FC-pointers or the B- and BC-pointers, respectively.

6.3.4 Multiversion Linear Quadtree

Description and Implementation of MVLQ

As proposed by Tzouramanis et al. [85], the technique for transforming Overlapping B^+-trees to a STAM can also be applied to other classical temporal access methods [84], especially to those that are modifications of the B-tree family. In this section, we present the *Multiversion Linear Quadtree* (MVLQ, for short) [94], which is based on hierarchical decomposition of space and adapts ideas from the Linear Region Quadtrees [28,79] and the Multiversion B-tree (MVBT) [7].

MVLQ associates time intervals with spatial objects in each node. Data records residing in leaves contain records of the form $\langle (C, L), T \rangle$, where (C, L) is the FD code of a black node of the Region Quadtree and T represents the time interval when this black node appears in the image sequence. Non-leaf nodes contain entries of the form $\langle C', T', Ptr \rangle$, where Ptr is a pointer to a descendent node, C' is the smallest C recorded in that descendent node and T' is the time interval that expresses the lifespan of the latter node. For reasons explained in [94], the FD implementation was chosen for the linear representation of the black nodes of a quadtree (the same choice as for the OLQ).

In each MVLQ node, we added a new field, *StartTime*, to hold the time instant when it was created. This field is used by the manipulation algorithms, which will be examined in the sequel. In addition, in each leaf we add a field *EndTime* that registers the transaction time when a specific leaf changes and becomes historical. The structure of the MVLQ is accompanied by two additional main memory sub-structures:

- The *root* table*: it is built on top of the MVLQ structure. MVLQ hosts a number of version trees and has a number of roots in such a way that each root stands for a time/version interval $T''=[T_i, T_j)$, where $i, j \in \{1, 2, ..., N\}$ and $i < j$. Each record in the root* table represents the root of a MVLQ and has the form $\langle T'', Ptr' \rangle$, where T'' is the lifespan of that root and Ptr' is a pointer to its physical disk address.
- The *Depth First-expression* (in short DF-expression, [37]) of the most recently inserted image: its usage is to keep track of all the black quadblocks of the most recently inserted image, and to be able to know at no I/O cost the black quadrants that are identical between this image and the one that will appear next. Thus, given a new image, we know beforehand which exactly are the FD code insertions, deletions, and updates. The DF-expression is a compacted array that represents an image based on the preorder traversal of its quadtree.

Manipulation Algorithms

As stated earlier, the basis for the new access method is the MVBT. However, its algorithms of insertion, deletion, and update processes are significantly different from the corresponding algorithms in the MVBT.

Insertion

If during a quadcode insertion at time point t_i, the target leaf is already full, a *node overflow* occurs. Depending on the *StartTime* of the leaf, the structural change can be triggered in two ways:

- If *StartTime* $= t_i$, then *a key split* occurs and the leaf splits. Assuming that b is the node capacity, after the key split the first $\lceil b/2 \rceil$ entries of the original node are kept in this node and the rest are moved to a new leaf.
- Otherwise, if *StartTime* $< t_i$, a copy of the original leaf must first be allocated, since it is not acceptable to change past states of the spatio-temporal structure. In this case, we remove all non-present (past) versions of quadcodes from the copy node. This operation is called *version split* [7], and the number of present versions of quadcodes after the version split must be in the range $[(1+e)d, (k-e)d]$, where k is a constant integer, $d = b/k$ and $e > 0$. If a version split leads to less than $(1+e)d$ quadcodes, then a merge is attempted with a sibling or a copy of that sibling containing only its present versions of quadcodes (the choice depends on the *StartTime* of the sibling). If a version split leads to more than $(k-e)d$ quadcodes in a node, then a key split is performed.

Deletion
Given a "real world" deletion of a quadcode at time point t_j, its implementation depends on the *StartTime* of the corresponding leaf:

- If *StartTime* $= t_j$, then the appropriate entry of the form $\langle C, L, T \rangle$ is removed from the leaf. After this *physical* deletion, the leaf is checked to see whether it holds enough entries. If the number of entries is above d, the deletion is completed. If the number is below, the *node underflow* is handled as in the classical B^+-tree, with the one difference that if a sibling exists (preferably the right one), then we have to check its *StartTime* before proceeding to a merge or a key redistribution.
- Otherwise, if *StartTime* $< t_j$ then the quadcode deletion is handled as a *logical* deletion, by updating the temporal information T of the appropriate entry from $T = [t_i, *)$ to $T = [t_i, t_j)$, where t_i is the insertion time of that quadcode. If an entry is logically deleted in a leaf with exactly d present quadcode versions, then a *version underflow* [7] occurs that causes a version split of the node, copying the present versions of its quadcodes into a new node. Evidently, the number of present versions of quadcodes after the version split is below $(1+e)d$, and a merge is attempted with a sibling or a copy of that sibling.

Update
Updating (i.e., changing the value of the level L of) an FD code leaf entry at time point t_j is implemented by:
(i) the logical deletion of the entry, and
(ii) the insertion of a new version of that entry; this new version of the entry has the same quadcode C, but a new level value L'.

Example
Consider the two consecutive images (with respect to their timestamps t_1=1 and t_2=2) on the left of Figure 6.20. The MVLQ structure after the insertion of the

first image is given in Figure 6.21(a). At the MVLQ leaves, the level L of each quadcode should also be stored, but for simplicity only the FD-locational codes appear. The structure consists of three nodes: a root R and two leaves A and B. The node capacity b equals 4 and the parameters k, d, and e equal 2, 2, and

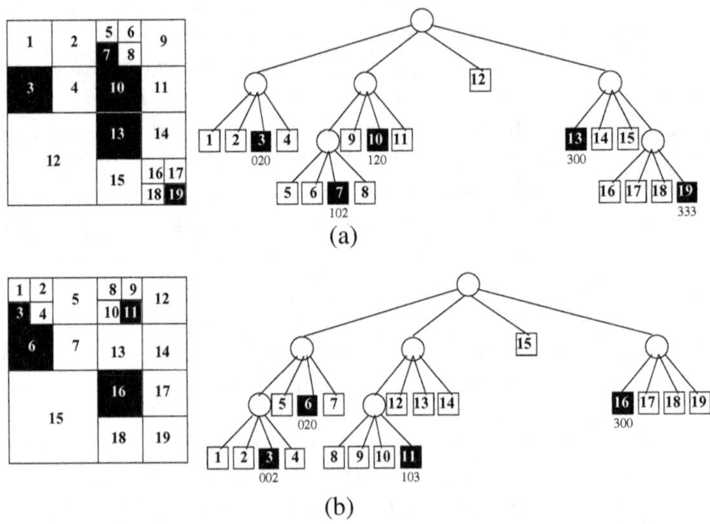

Fig. 6.20. Two similar binary raster images and their corresponding region quadtrees

0.5, respectively. The second version of the structure is constructed based on the first one, by inserting the FD code $\langle 002, 0\rangle$ (in the form $\langle C, L\rangle$), the deletion of $\langle 102, 0\rangle$, the insertion of $\langle 103, 0\rangle$, and the deletion of FD codes $\langle 120, 1\rangle$ and $\langle 333, 0\rangle$.

Figure 6.21(b) shows the intermediate result of the insertion of FD code $\langle 002, 0\rangle$, the deletion of $\langle 102, 0\rangle$, and the insertion of FD code $\langle 103, 0\rangle$. When we attempt to insert the quadcode 103 in the leaf A of Figure 6.21(b), the leaf overflows, and a new leaf C is created after a version split. All present versions of quadcodes of leaf A are copied into leaf C, and the parent R is updated for the structural change. Leaf C holds now more than $(k - e)d = 3$ entries, and a key split is performed producing a new leaf D. Again, the parent R is updated.

The final status of MVLQ after the insertion of the second image is illustrated in Figure 6.21(c). The quadcode 120 is deleted from leaf D of Figure 6.21(b) and a node underflow occurs (the number of entries is above d), which is resolved by merging this node with its right sibling B or a copy of it, containing only its present versions of quadcodes. After finding that the *StartTime* of leaf B is smaller than t_2, a version split on that leaf is performed, which is followed by a merge of the new (but temporary) leaf E and leaf D, in leaf D. The process terminates after the physical deletion of quadcode 333 from leaf D. The final number of entries in leaf D equals d. Both versions of MVLQ (Figure 6.21(a)

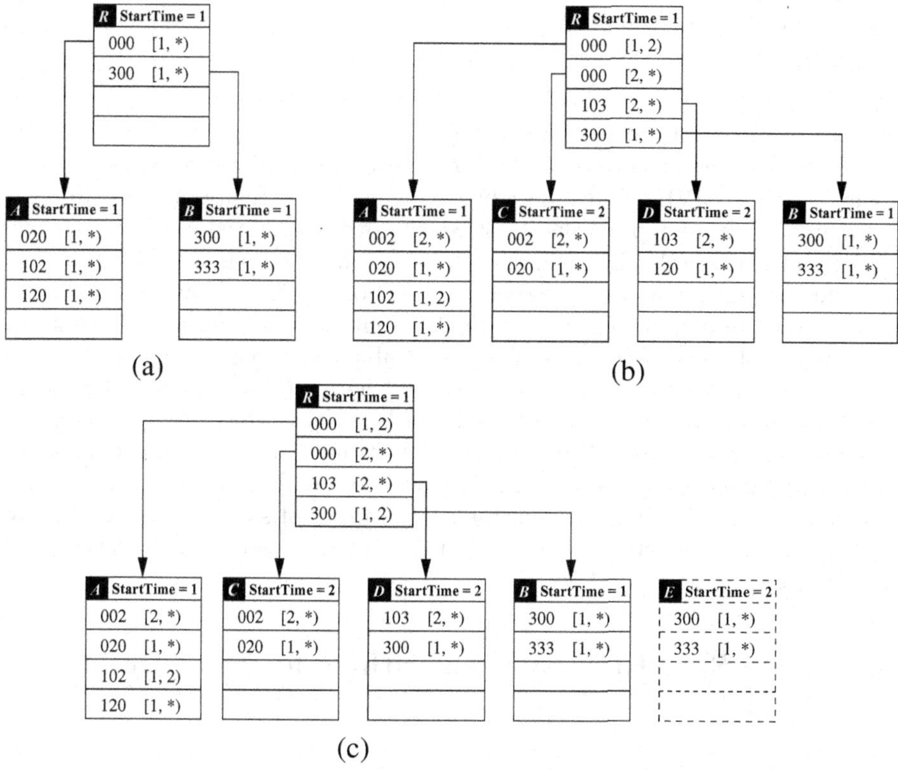

Fig. 6.21. (a) The MVLQ structure after the insertion of the first image (b) a preliminary result during the insertion of the second image, and (c) the final result after the insertion of the second image

and Figure 6.21(c)) have the same root R, although in general, more than one roots may exist.

Comments on Insertion, Deletion, and Update Algorithms

Generally, insertion of a new image occurs in two stages. The first stage is to sort the quadcodes of the new image and compare this sequence against the set of quadcodes of the last inserted image, using the binary table of its DF-expression. Thus, there is no I/O cost for black quadrants that are identical between the two successive images. During the next stage, we use the root* table to locate the root that corresponds to the most recently inserted image. Then, following ideas of the approach of [45], we build the new tree version by performing all the quadcode insertions, updates, and deletions in a batched manner, instead of performing them one at a time. (We did not follow this approach in the example of Figure 6.21 for simplicity reasons). It is obvious that after a batch operation with insertions, deletions, and updates at a specific time point, we may have conceptual node splittings and mergings. Thus, a specific leaf may split in more than two nodes, and, similarly, more than two sibling leaves may merge during FD code deletions.

Spatio-temporal Query Processing

The new indexing structure of MVLQ is based on transaction time and it is an extension of MVBT and Linear Quadtree for spatio-temporal data. In order to improve spatio-temporal query processing over the stored raster images, we added four horizontal pointers in every MVLQ leaf. Their names, roles, and functions are the ones that were presented in Section 6.3.3, and they are described in full detail in [95]. The structure supports all the well-known spatial queries for quadtree-based spatial databases (spatial joins, nearest neighbor queries, similarity and spatial selection queries, etc.) without taking into account the notion of time. It can also support efficiently all the typical temporal queries for transaction-time databases (most of which have been examined in [7,14]) without considering issues of space. However, the major feature of the MVLQ is that it can efficiently handle all the special types of spatio-temporal window queries for quadtree-based spatio-temporal databases, described in Section 6.3.3 and analyzed further in [94,95].

6.4 Data Structures and Algorithms for the Discrete Model

The discrete model for spatio-temporal data types [26] presented in Section 4.4 offers a precise basis for the implementation of data structures for a spatio-temporal database management system; it is in fact a high-level specification of such data structures. In this section, we briefly explain how these definitions translate into data structures and present a couple of algorithms on the data structures that implement operations specified in Section 4.4.

6.4.1 Data Structures

Two general issues need to be considered in the translation from data type specifications to physical data structures. First, some requirements arise because the data structures are to be used within a DBMS and must serve to represent attribute data types within some given data model. This means that values are placed under control of the DBMS into memory, which in turn implies that: (i) one should not use pointers, and (ii) representations should consist of a small number of memory blocks that can be moved efficiently between secondary and main memory. One way to fulfill these requirements is to implement each data type by a fixed number of records and arrays; arrays are used to represent the varying size components of a data type value and are allocated to the required size. All pointers are expressed as array indices.

The Secondo extensible DBMS (see Section 7.4), in which we are implementing this model, offers a specific concept for the implementation of attribute data types. Such a type has to be represented by a record (called the *root record*), which may have one or more components that are (references to) the so-called *database arrays*. Database arrays are basically arrays with any desired field size and number of fields; additionally, they are automatically either represented "in-line" in a tuple, or outside in a separate list of pages, depending on their size [19]. The root record is always represented within the tuple. In our subsequent design of data structures we will apply this concept. Each data type will be represented by a record and possibly some (database) arrays. In other DBMS environments, one can store the arrays using the facilities offered there for large-object management.

On the other hand, many of the data types presented in Section 4.4 are set-valued. Sets will be represented in arrays. We always define a unique order on the set domains and store elements in the array in that order. This way, we can enforce that two set values are equal if and only if their array representations are equal, which enables efficient comparisons.

Non-temporal Data Types

For the discrete base types and the time type, the implementation is straightforward: they are represented as a record consisting of the given programming language value[3] plus a Boolean flag indicating whether the value is defined. Type *point* is represented similarly by a record with two reals and a flag.

A *points* value is represented as an array containing records with two `real` fields, representing points. Points are in lexicographic order. The root record contains the number of points and the (database) array.

The data structures for *line* and *region* values are designed similar to corresponding structures reported in [29]. A *line* value is a set of line segments. This is represented as a list of *halfsegments*. The idea of halfsegments is to store every segment twice: once for the left end point and once for the right end point. These are called the *left* and *right halfsegment*, respectively, and the relevant point in the halfsegment is called the *dominating point*. The purpose is to support plane-sweep algorithms, which traverse a set of segments from left to right and have to perform an action (e.g., insertion into a sweep status structure) on encountering the left and another action on meeting the right end point of a segment. A total order is defined on halfsegments, which is the lexicographic order extended to treat halfsegments with the same dominating point (see [29] for a definition).

Hence, we represent the *line* value as an array containing a sequence of records, each of which represents a halfsegment (four reals plus a flag to indicate the dominating point); these are ordered as just mentioned. The root record manages the array plus some auxiliary information such as the number of segments, total length of segments, bounding box, etc.

A *region* value can be viewed as a set of line segments with some additional structure. This set of line segments is represented by an array of *halfsegments* containing the ordered sequence of halfsegment records, as for *line*. In addition, all halfsegments belonging to a cycle and to a face are linked together (via extra

[3] For *string* we assume an implementation as a fixed length array of characters.

fields such as *next-in-cycle* within halfsegment records). Two more arrays, *cycles* and *faces*, represent the structure. The array *cycles* contains records representing cycles by a pointer[4] to the first halfsegment of the cycle and a pointer to the next cycle of the face. The latter is used to link together all cycles belonging to one face. Array *faces* contains for each face a pointer into the *cycles* array to the first cycle of the face. A unique order is defined on cycles and faces, but is not described here.

The root record for <u>*region*</u> manages the three arrays and has additional information such as bounding box, number of faces, number of cycles, total area, perimeter, etc. Algorithms constructing region values generally compute the list of halfsegments and then call a *close* operation offered by the <u>*region*</u> data type, which determines the structure of faces and cycles and represents it by setting pointers. More details on the representation strategy can be found in [29], although some details are different here.

Intervals (s, e, lc, rc) are represented by corresponding records. A value of type <u>*range*</u>(α) is represented as an array of interval records ordered by value (all intervals are disjoint, hence there exists a total order). A value of type <u>*intime*</u>(α) is represented by a corresponding record.

Unit Types

We have to distinguish between units that can be represented in a fixed amount of space, called *fixed size units*, and those that cannot, *variable size units*. Fixed size units are <u>*const*</u>(<u>*int*</u>), <u>*const*</u>(<u>*string*</u>), <u>*const*</u>(<u>*bool*</u>), <u>*ureal*</u>, and <u>*upoint*</u>[5]. Variable size units are <u>*upoints*</u>, <u>*uline*</u>, and <u>*uregion*</u>.

Fixed size units can be represented simply in a record that has two component records to represent the time interval and the unit function, respectively. For example, for <u>*ureal*</u> the second record represents the quadruple (a, b, c, r).

For the representation of variable size units, we introduce *subarrays*. Conceptually, a subarray is just an array. Technically it consists of a reference to a (database) array together with two indices identifying a subrange within that array. The idea is that all units within a <u>*mapping*</u> (i.e., a sliced representation) share the same database arrays. Variable sized units are also all represented by a record whose first component is a time interval record. In the sequel we only describe the second component.

A <u>*upoints*</u> unit function is stored in a subarray containing a sequence of records representing *MPoint* quadruples, in lexicographic order on the quadruples. The <u>*upoints*</u> unit is represented in a record whose second component record contains a subarray reference and a three-dimensional bounding "box" (the number of points can be inferred from the subarray indices).

A <u>*uline*</u> unit function is stored similarly in a subarray containing a sequence of records representing *MSeg* pairs, which in turn are *MPoint* quadruples. Pairs are ordered lexicographically by their two component quadruples on which again

[4] From now on, by "pointer" we mean an integer index of a field of some array.

[5] We omit the other <u>*const*</u>(α) types, as they are not so relevant here.

lexicographic order applies. Again the uline unit is represented in a record whose second component consists of a subarray reference and a bounding cube.

A uregion unit function is basically a set of *MSeg* values (moving segments, trapeziums in 3D) with some additional constraints. We store these *MSeg* records in the same way and order in a subarray *msegments* as for uline. In addition, each record has two extra fields that allow for linking together all moving segments within a cycle and within a face. Furthermore, uregion has two additional subarrays *mcycles* and *mfaces* identifying cycles and faces, as in the region representation. The second component record of a uregion unit contains the three subarrays and a bounding cube for the unit.

For both uline and uregion one might add further summary information in the second component record, such as the (a, b, c, r) quadruples for the time-dependent length (for uline) or for perimeter and size (for uregion).

Sliced Representation

The data structure associated with the mapping type constructor organizes a collections of units (slices) as a whole. This data structure is parameterized by the unit data structures. We observe that all unit data structures are records whose first component represents a time interval, and whose second component may contain one or more subarrays.

The mapping data structure is illustrated in Figure 6.22. It is basically a (database) array *units* containing the unit records ordered by their time intervals. If the unit type uses k subarrays, then the mapping data structure has k additional database arrays. The database arrays mentioned in the unit subarray references will be the database arrays provided in the mapping data structure. The main array *units* as well as the k additional arrays are referenced from a single root record for the mapping data structure. Note that the structure has the general form required for attribute data types.

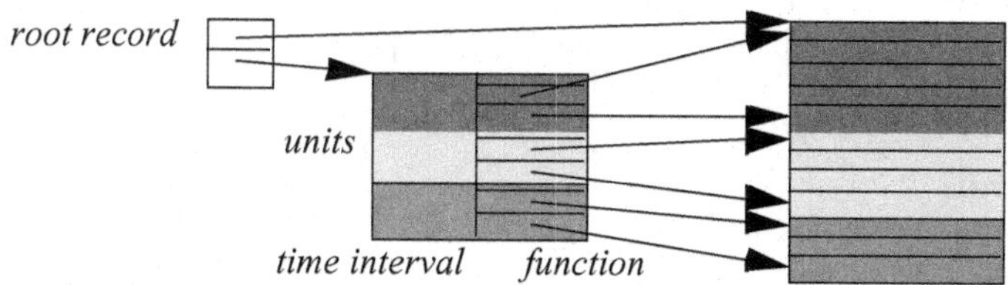

Fig. 6.22. A mapping data structure containing three units, for a unit type with one subarray, such as upoints

6.4.2 Two Example Algorithms

We proceed to briefly describe two algorithms thereby illustrating the use of the data model described in Section 4.4 and of the data structures just defined.

The first one implements the **atinstant** operation on a moving region, i.e., it determines the region value at a given time instant. The second one implements the **inside** operation on a moving point and a moving region, hence it returns a moving Boolean capturing when the point was inside the region.

Algorithm *atinstant*
The moving region is represented as a value of type <u>*mapping*</u> (<u>*uregion*</u>). The idea of the algorithm is to perform binary search on the array containing the region units to determine the unit u containing the argument time instant t. Then, a subalgorithm is called that evaluates each moving segment within the region unit at time t resulting in a line segment in two dimensions. These are composed to obtain the region value returned as a result.

> **algorithm** *atinstant* (mr, t)
> **input**: a moving region mr as a value of type *mapping*(*uregion*), and an *instant* t
> **output**: a *region* r representing mr at instant t
> **method**:
>   determine $u \in mr$ such that its time interval contains t;
>   **if** u exists **then return** *uregion_atinstant*(u, t) **else return** $\emptyset$ **endif**
> **end** *atinstant*.

> **algorithm** *uregion_atinstant*(u, t)
> **input**: a moving region unit ur (of type *uregion*) and an *instant* t
> **output**: a *region* r, the function value of ur at instant t
> **method**:
>   let $ur = (i, F)$; $r := \emptyset$;
>   **for each** mface $(c, H) \in F$ **do**
> $c' := \{\iota(s, t) | s \in c\}$;
> $H' := \emptyset$;
>   **for each** $h \in H$ **do**
> $h' := \{\iota(s, t) | s \in h\}$; $H' := H' \cup \{h'\}$
>   **endfor**;
> $r := r \cup \{(c', H')\}$
>   **endfor**;
>   **return** r
> **end** *uregion_atinstant*.

In the second algorithm the ι function defined in Section 4.4 is used to evaluate a moving segment at an instant of time to get a line segment.

The time complexity of this algorithm is basically $O(\log n + r)$, where n is the number of units in mr, and r is the size of the region returned (the number of segments). This is so because in the first step of *atinstant*, the unit can be found by binary search in $O(\log n)$ time, and because the traversal of the unit data structure takes linear time. However, to construct a proper region data structure as described in Section 6.4.1, one has to produce the list of halfsegments in lexicographic order, and hence needs to sort the r result segments. This results in a time complexity of $O(\log n + r \log r)$. Note that if the region value is just needed for output (e.g., for display on a graphics screen) then $O(\log n + r)$ is indeed sufficient.

The above algorithm works assumes instant t to be internal to the unit time interval. For simplicity, we have ignored the problem of possibly degenerated region values in the end points of the unit time interval. This necessitates a more complex cleanup after finding the line segments, as sketched at the end of Section 4.4. This problem can be avoided altogether if we spend a little more storage space, and represent a unit with a degenerated region at one end instead by two units, one with an open time interval, and the other with a correct region representation for the single instant at the end.

Analogous implementations of the *atinstant* operation can be obtained for all other moving data types. The first algorithm *atinstant* is in fact generic; one only needs to plug in other subalgorithms for the other data types.

Algorithm *inside*
Here the arguments are two lists (arrays) of units, one representing a moving point, the other a moving region. The idea is to traverse the two lists in parallel, computing the refinement partition of the time axis on the way (see Figure 6.23).

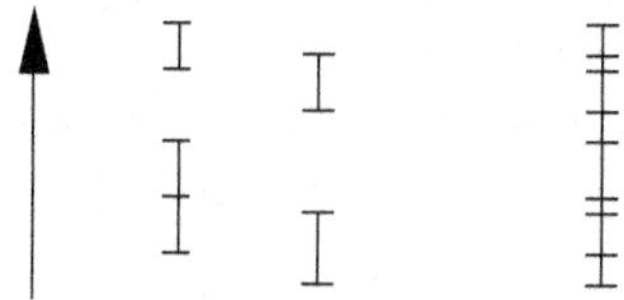

Fig. 6.23. Two sets of time intervals on the left, their refinement partition on the right

For each time interval i in the refinement partition, an *inside* algorithm is invoked on the point and region units valid during that time interval. A set of Boolean units results, which capture when the point was inside the region. Note that even a linearly moving point within a single <u>*upoint*</u> unit can enter and leave the region of the region unit several times.

algorithm *inside* (mp, mr)
input: a moving point mp (of type <u>*mapping*</u>(<u>*upoint*</u>)), and a moving region mr (of type <u>*mapping*</u>(<u>*uregion*</u>))
output: a moving Boolean mb, as a value of type <u>*mapping*</u>(<u>*const*</u>(<u>*bool*</u>)), representing when mp was inside mr
method:
let $mp = \{up_1, \ldots, up_n\}$ such that the list $\langle up_1, \ldots, up_n \rangle$ is ordered by time intervals;

let $mr = \{ur_1, \ldots, ur_m\}$ such that the list $\langle ur_1, \ldots, ur_m\rangle$ is ordered by time intervals;
$mb := \emptyset$;
scan the two lists $\langle up_1, \ldots, up_n\rangle$ and $\langle ur_1, \ldots, ur_m\rangle$ in parallel, determining in each step a new refinement time interval i and from each of the two lists either a unit up or ur, respectively, whose time interval contains i, or *undefined*, if there is no unit in the respective list overlapping i:
for each refinement interval i **do**
  if both up and ur exist **then**
    ub:=*upoint_uregion_inside*(up, ur);
    mb:=*concat*(mb, ub)
  endif
endfor;
return mb
end *inside*.

The operation *concat* on two sets of units is essentially the union, but merges adjacent intervals with the same unit value into a single unit. On the array or list representations, as given in the mapping data structure, this can be done in constant time (comparing the last unit of mb with the first unit of ub).

algorithm *upoint_uregion_inside*(up, ur)
input: a <u>upoint</u> unit up, and a <u>uregion</u> unit ur
output: a set of moving Boolean units, as a value of type <u>mapping</u>(<u>const</u>(<u>bool</u>)), representing when the point of up was inside the region of ur during their intersection time interval
method:
  let $up = (i', mpo)$ and $ur = (i'', F)$ and let $i = (s, e, lc, rc)$ be the intersection time interval of i' and i''; [6]
  if the 3d bounding boxes of mpo and F do not intersect **then return** $\emptyset$
  else
    determine all intersections between mpo and msegments occurring in (the cycles of faces of) F. Each intersection is represented as a pair $(t, action)$ where t is the time instant of the intersection, and $action \in \{enter, leave\}$;[7]
    sort intersections by time, resulting in a list $\langle (t_1, a_1), \ldots, (t_k, a_k)\rangle$ if there are k intersections. Note that actions in the list must be alternating, i.e., $a_i \neq a_{i+1}$;
    let t_0=s and t_{k+1}=e;
    if k=0 **then**

[6] For simplicity, the remainder of the algorithm assumes the intersection interval is closed. It is straightforward, but lengthy, to treat the other cases.

[7] The *action* can be determined if we store with each msegment (trapezium or triangle in 3D) a face normal vector indicating on which side is the interior of the region.

```
      if mpo at instant s is inside F at instant s then
        return {((s, e, true, true), true)}
      else return {((s, e, true, true), false)}
      endif
    else
      if a_1=leave then
        return {((t_i, t_{i+1}, true, true), true)|i ∈ {0, ..., k}, i is even}
          ∪{((t_i, t_{i+1}, false, false), false)|i ∈ {0, ..., k}, i is odd}
      else
        return {((t_i, t_{i+1}, true, true), true)|i ∈ {0, ..., k}, i is odd}
          ∪{((t_i, t_{i+1}, false, false), false)|i ∈ {0, ..., k}, i is even}
      endif
    endif
  endif
end upoint_uregion_inside.
```

Here, the moving point *mpo* is a line segment in 3D that may stab some of the moving segments of F, which are trapeziums in 3D. In the order of time, with each intersection the moving point alternates between entering and leaving the moving region represented in the region unit. Hence a list of Boolean units is produced that alternates between *true* and *false*. In case no intersections are found ($k = 0$), one needs to check whether at the start time of the time interval considered the point was inside the region. This can be implemented by a well-known technique in computational geometry, the "plumbline" algorithm, which counts how many segments in 2D are above the point in 2D.

The first algorithm *inside* requires time $O(n+m)$, where n, m are the numbers of units in the two arguments, except for the calls to algorithm *upoint_uregion_inside*. This second algorithm requires $O(s)$ time for finding all intersections, with s the number of msegments in F. Furthermore, $O(k \log k)$ time is needed to sort the k intersections, and to return the $k+1$ Boolean units. If no intersections are found, the check whether *mpo* is inside F at the start time s requires $O(s)$ time. The total time for all calls to *upoint_uregion_inside* is $O(S + K \log k')$, where S is the total number of msegments in all units, K is the total number of intersections between the moving point and faces of the moving region, and k' is the largest number of intersections occurring in a single pair of units. In practical cases, k' is likely to be a small constant, and $K \log k'$ will be dominated by S, hence the total running time will be $O(n + m + S)$. If the moving point and the moving region are sufficiently far apart, so that not even the bounding boxes intersect, then the running time is $O(n + m)$.

This algorithm illustrates nicely how algorithms for binary operations on moving objects can generally be reduced to simpler algorithms on pairs of units. Again, the first algorithm is generic; one only needs to plug in algorithms for specific operations on pairs of units.

6.5 Benchmarking and Data Generation

6.5.1 Benchmarking

As already presented, spatio-temporal data management concerns the design and implementation of access methods that aim at reducing query response time. The performance of an access method depends on the setting it is subjected to, which can be characterized by, e.g., the type of the dataset (points, rectangles, line segments), the distribution of the dataset, the available buffering strategy, the disk page size, and the queries.

A benchmark is composed of a dataset, an access method, and a set of queries. The output of a benchmark is a set of values describing the performance of the access method for the given dataset and queries. Often, the values describe separately the I/O time and CPU time needed to compute the queries. In order to run benchmarks and thus observe the behavior of access method under varying settings, is is advantageous to have available a flexible benchmarking environment that enables the experimentation with differing (i) datasets, (ii) access methods and (iii) query types.

In order to compare different spatial join strategies, the authors of [32] propose the exploitation of a spatial data generator that is capable of producing datasets of rectangles in two-dimensional space with different characteristics such as rectangle size, data distribution, and dataset size. Datasets are specified by means of so-called models that consist of a set of parameter settings. Models can be reused and modified to generate similar datasets. Often, the objective is to simulate real-life datasets by synthetic ones. The comparison of spatial join techniques is achieved by executing each algorithm on the same datasets and by collecting the results (query response time).

The generator proposed in [32] is limited to spatial datasets. However, a spatio-temporal dataset generator has been proposed in [89]. This generator is capable of producing datasets consisting of moving points that simulate, e.g., the movement of airplanes or ships.

In addition to a data generator, a benchmarking environment needs implementations of access methods and execution of queries. Based on preliminary work [33], a benchmarking environment for spatial query processing is proposed by Gurret et al. in [31]. The system is called BASIS (A Benchmarking Approach for Spatial Index Structures). The application of the system for spatial join processing strategies is studied in [73].

The main parts of the system are depicted in Figure 6.24 and are explained below.

- *Buffer Manager.* It is used to manage the buffers defined for each argument file. The user can control the buffer size.
- *File Manager.* It is used to manage the files stored in the system. A file can be either a dataset file or an access method file. Each BASIS file has a specific internal representation. External files containing datasets must be transformed into this representation before use.

- *Access Methods.* Many access methods can be implemented and integrated into the system. Currently, the system supports R-trees, R*-trees, B-trees, and Grid-files. However, new access methods can be implemented using the API provided.
- *Query Processor.* This component provides the tools needed for executing spatial queries. The query processor is based on iterators. Each complex query is decomposed into a set of more primitive queries, and each primitive query is assigned to an iterator. The complex query is composed by combining the iterators together as a tree. Currently, the system supports range, point, and spatial join queries. However, one can build new iterators in order to compare the access methods.
- *Datasets*: The BASIS system uses either real-life datasets (e.g., from TIGER or Sequoia 2000) or synthetic ones. The dataset generators that have been described previously can be used for this purpose. The only requirement is that these datasets must be transformed to the BASIS internal representation.

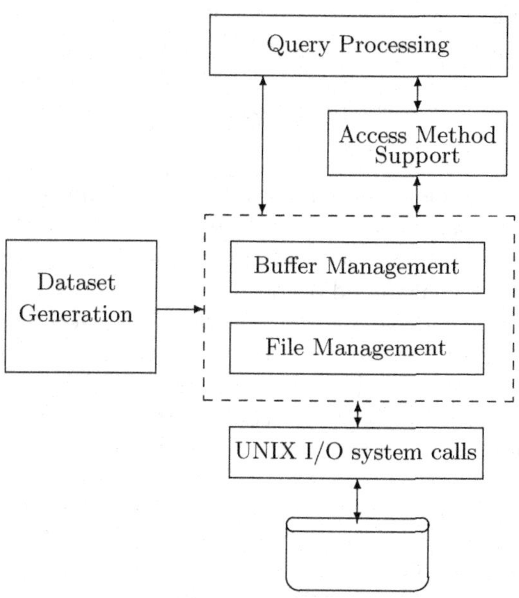

Fig. 6.24. The main components of the BASIS architecture

Although a lot of work has been performed for spatial benchmarking, spatio-temporal benchmarking must also be investigated thoroughly. STAMs can be implemented using BASIS, thus enabling the execution of spatio-temporal benchmarks.

6.5.2 Data Generation

In order for the user of a benchmarking environment to conduct an extensive series of experiments under a variety of conditions, one should be able to generate a variety of datasets. A fundamental issue in the generation of synthetic spatio-temporal datasets is the availability of a rich set of parameters that control the data generation.

The GSTD Rationale

Theodoridis et al. proposed the GSTD ("Generate Spatio-Temporal Data") algorithm for building sets of moving point or rectangular objects [89]. For each object o, GSTD generates tuples of the format (id, t, pl, pu, f), where id is the object identifier, t is the object timestamp, pl and pu are the lower-left and upper-right corners, respectively, of the object spacestamp (an MBR, assuming a two-dimensional scenario), and f is a flag denoting whether the spacestamp is (spatially) valid or not.

In the GSTD algorithm, three parameters are available.

- *duration* of an object instance; involving change of timestamps between consecutive instances,
- *shift* of an object; involving change of spatial location (in terms of center point shift), and
- *resizing* of an object; involving change of an object size (only applicable to non-point objects).

The GSTD methodology is as follows. Initially, all objects are given starting locations, such that their center points are distributed in the workspace with respect to a chosen distribution, and their extents are either set to zero (in case of point data) or calculated according to the desired density of the data. After the initialization phase, each new instance of an object is generated as a function of the current instance and the values of the three parameters, which are calculated according to a desired distribution.

In this scenario, it is possible that a coordinate may fall outside the workspace [90]; GSTD manipulates invalid instances according to one among three alternative approaches:

- the *"radar"* approach, where coordinates remain unchanged, although falling beyond the workspace,
- the *"adjustment"* approach, where coordinates are adjusted (according to linear interpolation) to fit the workspace, and
- the *"toroid"* approach, where the workspace is assumed to be toroidal, so that when an object leaves at one edge of the workspace, it enters back at the "opposite" edge.

In the first case, the output instance is appropriately flagged (f=0 in the generated tuple) to denote its invalidity, although the subsequent instance is still calculated with respect to it. In the other two cases, it is the modified instance that is stored in the resulting data file and used for the generation of the subsequent instance. Notice that in the "radar" approach, the number of objects present (i.e., valid) at each time may vary. The three alternative approaches are illustrated in Figure 6.25. For the sake of simplicity, only centers of spacestamps are illustrated; black (gray) locations represent valid (invalid) instances. In the example of Figure 6.25(a), the "radar" fails to detect s_3, hence s_3 is invalid, although the next location s_4 is based on that. Unlike for "radar", the other two approaches always calculate a valid instance s_3' to be stored in the data file which, in turn, is used by GSTD for the generation of s_4.

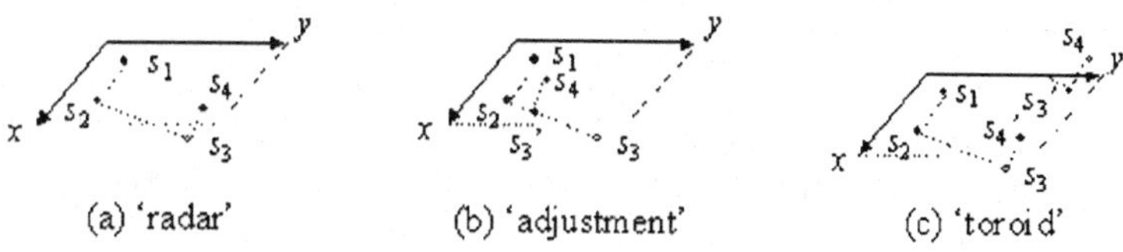

Fig. 6.25. GSTD manipulation of invalid instances

Through the careful use of the different distributions for the above parameters, GSTD may simulate a variety of interesting scenarios. For instance, using a random distribution for *duration* and *shift*, all objects move equally fast (or slow) and uniformly in the workspace. In contrast, using a skewed distribution for *duration*, a relatively large number of slow objects moving randomly results.

Examples of Generated Datasets

Figure 6.26 presents six different scenarios that illustrate the GSTD capability at simulating desired headings (scenarios 1 through 3) and speeds of objects (scenarios 4 through 6) [90]. Moreover, scenarios 1 and 2 follow the "toroid" and "radar" approach, respectively, while scenarios 3 through 6 follow the "adjustment" approach.

More specifically, scenarios 1 and 2 illustrate points with initial Gaussian spatial distribution moving towards the East and Northeast, respectively. In the former case, where the "toroid" approach was adopted, the points that leave at the right side re-enter on the left side of the workspace. Scenario 3 illustrates an initially skewed distribution of points and their movement towards the Northeast. Since the "adjustment" approach is used, the points concentrate around the upper-right corner. In Scenario 4, rectangles initially located around the middle of the workspace are moving and being resized randomly. The randomness of *shift* and *resizing* are obtained by applying a uniform distribution to these. Finally, scenarios 5 and 6 exploit the speed of objects as a function of the GSTD input parameters. By increasing (in absolute values) the minimum and maximum values of *shift*, users can generate "faster" objects while the same behavior

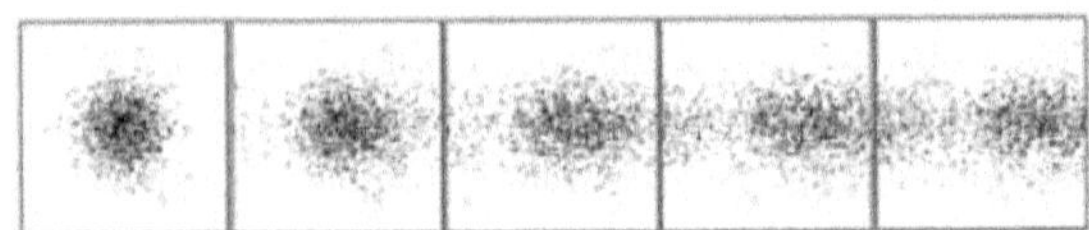

Scenario 1: Points moving from the center to the East ("toroid approach")

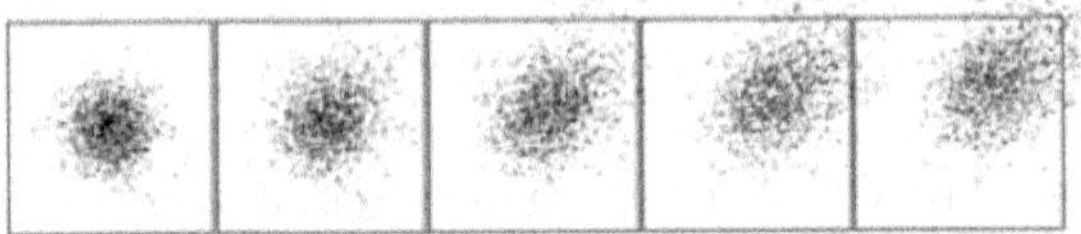

Scenario 2: Points moving from the center to the Northeast ("radar approach")

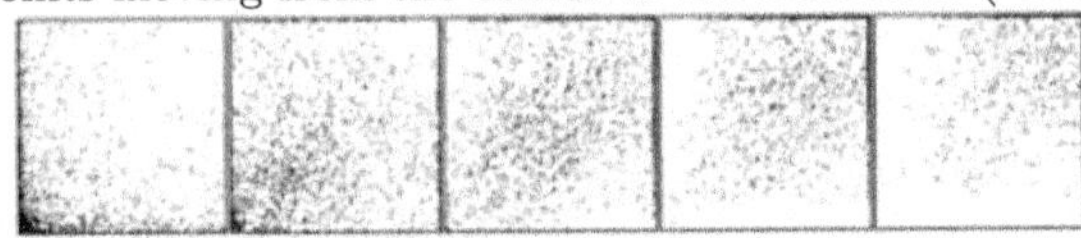

Scenario 3: Points moving from the Southwest to the Northeast

Scenario 4: Rectangles moving and resizing randomly

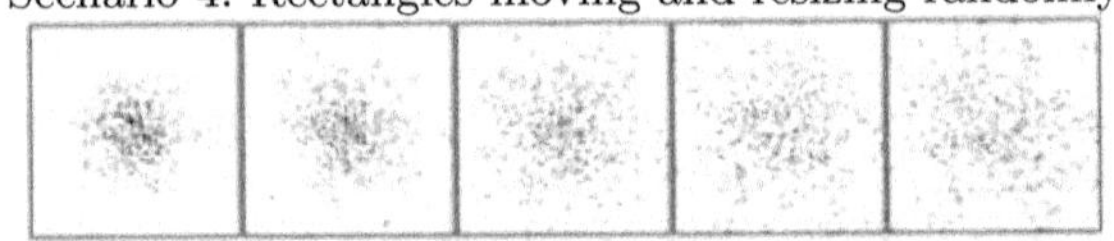

Scenario 5: Points moving randomly at low speed

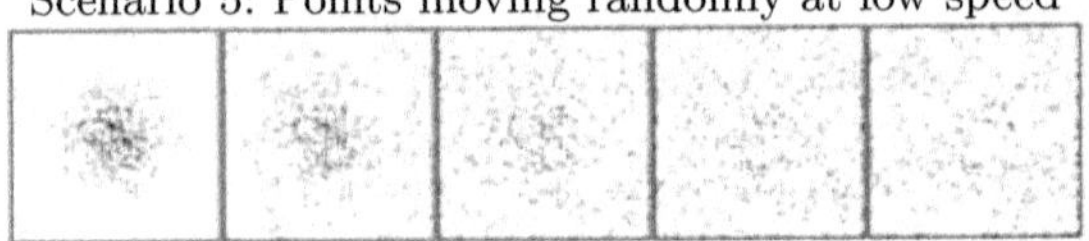

Scenario 6: Points moving randomly at high speed

Fig. 6.26. Example files generated by GSTD

could be achieved by decreasing *duration*. Similarly, the heading of objects can be controlled, as in scenarios 1 through 3.

6.6 Distribution and Optimization Issues

6.6.1 Distributed Indexing Techniques

A novel architectural choice for obtaining acceptable performance of spatio-temporal DBMSs subjected to huge volumes of data and high frequencies of updates is the so-called *network computing*: many powerful and inexpensive workstations connected through a fast communication network.

Several characteristics make this environment attractive. The most important one is that a set of sites has more computing power and resources than a single site, independently from the equipment of a site. Moreover, the network offers a transfer speed that is not comparable with those of magnetic or optical disks. Hence, in this framework it is possible and realistic to efficiently implement main memory applications using the main memory of distributed machines. This solution has performances that are not comparable with the traditional centralized ones.

In this paradigm of *Scalable Distributed Data Structures* (SDDSs) data objects are distributed among a variable number of servers and accessed by a set of clients. Both servers and clients are distributed among the nodes of the network. Clients and servers communicate by sending and receiving messages using *point-to-point* or *multicast* protocols[8]. Servers store objects uniquely identified by a key. Every server stores a single block (called *bucket*) of at most b data items, for a fixed number b.

A critical aspect of this solution is to accommodate the dynamic growth of a data file with scalable performance. The key to scalability is to be able to dynamically distribute data across multiple servers of a distributed system. This redistribution of objects should take place continuously as the numbers of objects and requests in the system grow.

An SDDS has to satisfy the following properties:

1. A file expands to new servers only if the used servers are loaded enough. This ensures an efficient use of resources.
2. There is no distinguishable server acting as a centralized controller. This avoids that a server becomes a bottleneck with the increase of the size of the file.
3. No operation involves the execution of an action on more than one client. This is required since clients are autonomous but not continuously available in general.

Efficiency in SDDSs is evaluated with respect to the communication network. This means that performance is measured in terms of the overall number of messages on the network.

Since there is no centrally located address structure that binds keys of all objects into one or more server locations, each client as well as each server is required to have a *local index*. This index is the client's or server's version of the address structure and represents its viewpoint on the latest information about object locations. Consequently, clients can make *addressing errors*, and mechanisms to cope with and to recover them have to be introduced. The goal of a distributed access method is to minimize the number of client and server address errors, as well as the number of local index correction messages between servers and clients.

Litwin et al. were the first to define an SDDS, by proposing a distributed version of linear hashing, namely LH* [47], supporting insert and exact search of one-dimensional objects. Other proposals have been advanced for SDDSs

[8] Multicast is a restricted version of broadcast, where only a subset of all machines on the network are collectively addressed.

supporting also range-queries on one-dimensional objects, namely RP* [48], DRT [43], RBST [13], BDST [20], and the distributed B^+-tree [15].

To be useful for the management of spatio-temporal data, though, an SDDS has to be able to deal with k-dimensional data. The first SDDS for managing k-dimensional points over a network where multicast is available was proposed by Nardelli in [53] and analyzed in [54,55]. The solution was based on a data structure, named *lazy k-d-tree*, for managing a collection of k-d-trees. The solution featured optimal algorithms for exact, partial, and range search. Distributed k-d-trees are also able to operate in a network where multicast is not available, like in other proposals [46].

In the SDDS model, the split of a server is the typical way to scale up when the number of objects grows. Whenever a server s is in *overflow*, meaning that (due to insertions) it manages a number of objects greater than its capacity b, half of its objects is transferred to a new server s'. The split of a server is a local operation, and clients and other servers are not kept, in general, up-to-date with the evolution of the structure. This means that it is possible for client requests to be sent to a wrong server s because the clients local index does not contain the latest object location information. This *address error* is managed by s by forwarding the received request to the server s' that is the *pertinent* server in the viewpoint of s. But s' can be a wrong server as well. The process then continues until the actual pertinent server s^* is found. This server manages the request and sends also local index updates back to the client and to the involved server (see Figure 6.27).

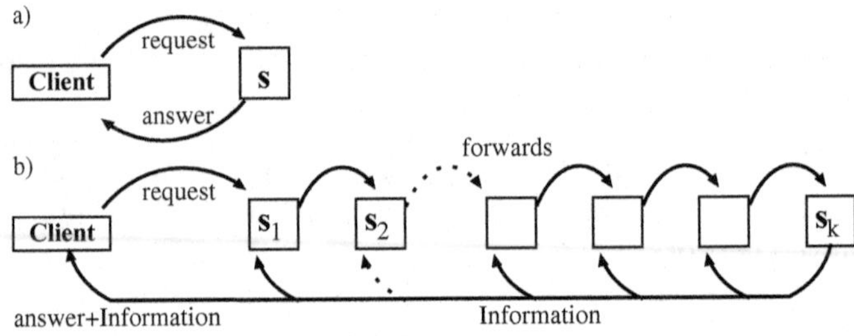

Fig. 6.27. Requests (a) without and (b) with address errors

Due to the impossibility of carrying over to the distributed environment both a balancing technique and a monotonic search process, the worst case number of messages for search in distributed search trees has a lower bound of $\Omega(\sqrt{n})$ [44]. But in [13] a slight relaxation of these requirements allowed to introduce the first distributed data structure with a poly-logarithmic search time for point and range queries while supporting both the insertion and deletion of elements. This was further improved in [20].

In a more recent proposal [21], amortized analysis of the performance of distributed searching for both one-dimensional and multi-dimensional data has been considered. The result is the definition of an SDDS, namely DRT*, representing a variant of DRT [43] for the one-dimensional case and a variant of the distributed k-d-tree [53] for the multi-dimensional case. In [21] it was proved that both for the one-dimensional and the k-dimensional case, inserts and exact searches have in DRT* an amortized cost of $O\left(\log_{(1+m/n)} n\right)$ messages, where m is the number of requests and n is the total number of servers of the structure. This was obtained by showing that the way local indices change during the evolution of the DRT* structure is similar to the structural changes happening in the set union problem [49], and that request management and splits are strictly related to operations used in the set union problem. Moreover, since in an SDDS m and n are related, inserts and exact searches in DRT* have an amortized almost constant costs, namely $O\left(\log_{(1+A)} n\right)$ messages, while a range query has an amortized cost of $O\left(\log_{(1+A)} n + \lceil k/b \rceil\right)$ messages, where k is the number of items returned by the search, b is the capacity of each server, and $A = b/2$. Considering that in real application environments, A is a large value, of the order of thousands, we can assume to have a constant cost for exact searches in practical cases. Only worst case analysis was previously considered and the result of an almost constant cost for the amortized analysis of the general k-dimensional case appears to be very promising in the light of the well known difficulties in proving optimal worst case bounds for k-dimensions.

Since one of the approaches to obtain efficient indexing techniques for spatio-temporal data is to consider time as a 'spatial' dimension, DRT* offers a very efficient and promising technique for efficient spatio-temporal indexing.

6.6.2 Query Optimization

Query optimization, the task of selecting a suitable strategy for executing an operation, e.g., a query, is an essential task for any DBMS. This is even more true for a spatio-temporal DBMS, due to the huge volumes of data involved. Initial steps towards spatio-temporal query optimization have been taken, but much research remains to be done.

An important part of query optimization is to have available an accurate *statistical model* enabling concise descriptions of datasets with few parameters. This is useful in spatio-temporal DBMSs for analyzing STAM characteristics (e.g., how many nodes there are in a MOF-tree), which is an essential ingredient to estimate cost and selectivity of spatio-temporal queries. This task is performance critical: for instance, query optimizers use query result size estimates to select query execution strategies. In the following, we briefly review some of the results achieved in recent years with respect to query optimization for spatial data; many of these results need to be extended to spatio-temporal data.

Concerning selectivity estimation queries, an analytical formula to compute the *selectivity of a window query* as a function of the underlying data morphology and distribution has been given in [38,74]. When formula parameters are unknown, one typically makes *uniformity* and *independence* assumptions.

Unfortunately, these assumptions do not hold for real datasets and generally lead to pessimistic results [17]. For one-dimensional data, some non-uniform distributions have been applied with success, but difficulties remain for multi-dimensional data. In fact, some of the proposed non-uniform models (e.g., the Gaussian distribution [60] or clustering ad-hoc methods [64,12]), show their limitations for data having a different nature from the data they were designed for.

The recent introduction of the concept of fractal dimension has allowed to better describe the statistical properties of data, thus enhancing selectivity estimation in several contexts. For point-data, using the fractal dimension, it is possible to accurately estimate the selectivity of (self) spatial joins [5] and nearest-neighbor queries [70].

Next, novel results for *region data* have been proposed in [65], where a realistic statistical model was proposed: more precisely, they showed that the *complementary cumulative distribution function*[9] (CCDF) of the region areas follow a power law, and this observation is used to compute the selectivity of window queries. When the enclosed spatial objects are *lines* (e.g., roads, rivers, and utilities), it has been observed that the CCDF of the lengths, obeys to an exponential law [66], and once again this result is useful in predicting query performances [59].

Finally, a recent paper by Acharya et al. [3] presents a novel technique based on the notion of *spatial skew* of rectangular data. Using this technique, the authors partition the input rectangles into subsets and approximate each partition, thus obtaining an accurate selectivity estimation over a broad range of spatial queries.

Concerning the analysis of SAMs, we briefly review results that relate to R-trees [34]. An early result was related to the optimal packing for R-trees construction [38], based on data distribution. More recently, the fractal dimension of a set of point has been used to estimate the performance of R-trees for range queries [27]. In [88], a model for the prediction of I/O cost of spatial queries is given, using the concept of *density* of data. In [67], the node distribution of an R-tree storing region data has been studied: the authors showed that the area distribution of the regions is recursively propagated up to the root. Based on this observation, the authors were able to accurately estimate the search effort for range queries and to predict the selectivity of a self spatial join posed on the dataset [68].

Finally, we mention another application of fractal theory to spatial data: the estimation of the number of quadtree blocks needed to store a spatial dataset consisting of a single region, once that the fractal dimension of the periphery of the region is known [25].

---

[9] Remember that the cumulative distribution function of $f(x) : \Re \rightarrow \Re$ is defined as $F(x) = \int_{-\infty}^{x} f(t)dt$, while the complementary cumulative distribution function is defined as $\overline{F}(x) = \int_{x}^{+\infty} f(t)dt$.

6.7 Related Work

We divide the work related to access methods and query processing in spatio-temporal databases into two categories: the ones related to general STAMs and those related to indexing moving points.

One of the first results in STAMs is reported in [102]. Specifically, the authors propose MR-trees. These structures are very similar to the HR-trees of [61], which were presented earlier in the chapter. The authors also an R-tree-based structure, termed RT-trees, which is quite different from the presented structures up to now.

RT-trees index objects in two-dimensional space and view time as complementary information that is incorporated as time intervals inside a two-dimensional R-tree structure. More specifically, each RT-tree node contains entries of the form (S, T, P), where S is the spatial information (i.e., the covering MBR), T is the temporal information (i.e., the covering interval), and P is a pointer to either a subtree or the detailed description of an object. Let $T = [t_i, t_j)$, where $i \leq j, lett_j$ be the current timestamp, and let t_{j+1} be the successor of t_j. If an object does not change its spatial location from t_j to t_{j+1}, then the spatial information S remains the same, and the temporal information T is updated to T', by increasing the interval upper bound, i.e., $T'{=}[t_i, t_{j+1})$. When an object changes its spatial location, a new entry with temporal information $T{=}[t_{j+1}, t_{j+1})$ is created and inserted.

The insertion strategy makes RT-trees efficient data that is mostly static. If the number of updates is large, many entries are created and the RT-tree grows considerably. It should also be observed that the RT-tree node construction depends primarily on the spatial information S, while T only plays a secondary role. Hence the RT-tree is not able to support efficiently temporal queries (e.g., "find all objects that exist in the database within a given time interval").

Several structures similar to MOF- and MOF$^+$-trees have also appeared in the literature. Cheiney at al. [18] have proposed a Region Quadtree-based structure, called *Fully Inverted Quadtrees* (FI-quadtrees, for short) to index an image database. FI-quadtrees are suitable for answering queries on image content (exact and fuzzy search). Vassilakopoulos et al. [97] have proposed *Dynamic Inverted Quadtrees* (DI-quadtrees, for short) which improved FI-quadtrees, since it requires far less disk space, image pattern is performed more efficiently, an it is dynamic since there is no demand for obligatory reorganization. Along the same line, lately in [87] Tourir has proposed *Multi-layer Quadtress*, a new access method based on PM1-quadtrees [79] to represent thematic layers with line segments. All these structures are designed for use in spatial applications, however, they could be used for spatio-temporal applications equally well. In particular, the later one could be used to index trajectories.

The recent work by Zimbrao et al. [104,103] is similar to RT-trees, as they propose another R-tree variant, the structure of *Temporal R-tree* (TR-tree, for short), which uses features from the MVBT structure (e.g. version split and block copy mechanism) [7]. Although TR-trees, like the MVLQ structure, are based on the MVBT structure, they differ since they are designed for vector data rather than raster data. Manipulation algorithms are given in [104] and performance evaluation results against 2+3 R-trees, HR-trees, and RT-trees can be found in [103]. In the latter paper, it is reported that for mixed sets of queries (i.e., time instant and time interval queries), TR-trees outperform their opponents.

In [86], a preliminary effort is described that aims to establish a STAM based on k-d-trees. The proposed structure is called *Multi-dimensional Persistent tree* (MP-tree) and implements the idea of partial persistence as do all the previously examined structures. This work is limited in that the structure is not an external balanced multi-way tree, but rather an main-memory resident unbalanced ternary tree. However, their approach can be accommodated in other structures beyond main-memory k-d-trees.

In [82] the *Adaptive tree structure* (AT structure) is put forward. This hybrid method consists of a pair of a spatial (i.e., k-d-trees) and a temporal (i.e. ternary trees) structure. Then, according to the demand, the method selects the data structure that is expected to perform most efficiently. In [56], the AT structure is used to index moving points. These works assume main-memory environments.

Related work on indexing the current and future positions has concentrated mostly on points moving in one-dimensional space. The authors in [91] use PMR-quadtrees [79] for indexing the future linear trajectories of one-dimensional moving point objects as line segments in (x, t)-space. The segments span the time interval that starts at the current time and extends *horizon* time units into the future. A tree expires after U time units, and a new tree must be made available for querying. This approach introduces data replication in the index; a line segment is usually stored in several nodes.

Kollios et al. [39] employ the dual data transformation where a line $x = x(t_{ref}) + v(t - t_{ref})$ is transformed to the point $(x(t_{ref}), v)$, enabling the use of regular spatial indices. It is argued that indices based on k-d-trees are well suited for this problem because these best accommodate the shapes of the (transformed) queries on the data. Kollios et al. suggest, but do not investigate in detail, how this approach can be extended to two and higher dimensions. They also propose two other methods that achieve better query performance at the cost of data replication. These methods do not seem to apply to more than one dimension.

The authors in [6] propose to use the notion of *kinetic* main-memory data structures for mobile objects. The idea is to schedule future events that update a data structure so that necessary invariants hold. Agarwal et al. [1] apply these ideas to external range trees [4]. Their approach may possibly be applicable to R-trees or time-parameterized R-trees where events would fix MBRs, although it is unclear how to contend with future queries that arrive in non-chronological order. Agarwal et al. address non-chronological queries using partial persistence techniques and also show how to combine kinetic range trees with partition trees to achieve a trade-off between the number of kinetic events and query performance.

Within the direction of indexing the past positions of moving objects, most approaches deal with spatial data changing discretely over time and do not take continuous changes into account.

6.8 Conclusions

The present chapter has described a significant number of STAMs and related query processing techniques. The chapter also examines other issues related to the physical database level, namely benchmarking, data generation, distributed indexing techniques, and query optimization. These contributions have appeared in the literature mostly during the last five years. The proliferation and diversity of access methods seen here stem from the very different requirements of the many practical applications involving spatio-temporal data.

Perhaps the nature of the data supported is the most fundamental characteristic that may be used to categorize the proposed access methods. For instance, spatial data can be of vector or raster type. To support these different kinds of data, a structure based on R-trees or on quadtrees, respectively, should be selected as most appropriate.

It appears that the access methods now available are able to contend with most static spatio-temporal data, which is spatial data that remains unchanged for a time interval. Much more work is necessary to support applications that involve the past and present positions of continuously moving objects.

It is evident that formal mathematical analysis of the many new access methods is not an easy task. More work, e.g., on parametric complexity analysis, is needed in the area of analytical studies. Because of the shortcoming of mathematical analyses, performance comparisons based on empirical experiments with real or synthetic data also play a significant role in the design and evaluation of access methods and query processing techniques. Better infrastructure for empirical studies and more comprehensive comparisons of ranges of access methods are desirable.

For all the access methods proposed for spatio-temporal data, several manipulation algorithms have been reported. Mostly, these algorithms concern standard operations, i.e., insertion, deletion, bulk loading, and fairly simple types of queries, such as window queries based on time intervals, space intervals, or both. There has only been presented little work on more sophisticated algorithms, e.g., for spatio-temporal nearest neighbor queries and spatio-temporal join. Finally, query optimization largely remains a *terra incognita* in relation to spatio-temporal data. Very few papers have appeared that address this topic, which should be investigated in the future, e.g., with the objective of obtaining cost models, heuristics, and algebraic transformation rules.

References

1. L. Arge, P. Argawal, and J. Erickson. Indexing moving points. In *Proceedings 19th ACM PODS Symposium (PODS 2000)*, pages 175–186, 2000.
2. D.J. Abel. A B^+-tree structure for large quadtrees. *Computer Vision, Graphics and Image Processing*, 27(1):19–31, 1984.
3. S. Acharya, V. Poosala, and S. Ramaswamy. Selectivity estimation in spatial databases. In *Proceedings ACM SIGMOD Conference on Management of Data*, pages 13–24, 1999.

4. L. Arge, V. Samoladas, and J.S. Vitter. On two-dimensional indexability and optimal range search indexing. In *Proceedings 19th ACM PODS Symposium (PODS 2000)*, pages 346–357, 1999.
5. A. Belussi and C. Faloutsos. Estimating the selectivity of spatial queries using the correlation fractal dimension. In *Proceedings 21st Conference on Very Large Data Bases (VLDB'95)*, pages 299–310, 1995.
6. J. Basch, L. Guibas, and J. Hershberger. Data structures for mobile data. In *Proceedings 8th ACM-SIAM Symposium on Discrete Algorithms (SODA'97)*, pages 747–756, 1997.
7. B. Becker, S. Gschwind, T. Ohler, B. Seeger, and P. Widmayer. An asymptotically optimal multiversion B-tree. *The VLDB Journal*, 5(4):264–275, 1996.
8. F.W. Burton, M.W Huntbach, and J. Kollias. Multiple generation text files using overlapping tree structures. *The Computer Journal*, 28(4):414–416, 1985.
9. R. Bliujūtė, C.S. Jensen, S. Šaltenis, and G. Slivinskas. R-tree based indexing of now-relative bitemporal data. In *Proceedings 24th Conference on Very Large Data Bases (VLDB'98)*, pages 345–356, 1998.
10. R. Bliujūtė, C.S. Jensen, S. Šaltenis, and G. Slivinskas. Light-weight indexing of bitemporal data. In *Proceedings 9th Conference on Statistical and Scientific Database Management Systems (SSDBM 2000)*, pages 125–138, 2000.
11. F.W Burton, J.G Kollias, and D.G. Matsakis. Implementation of overlapping B-trees for time and space efficient representation of collection of similar files. *The Computer Journal*, 33(3):279–280, 1990.
12. N. Beckmann, H.-P. Kriegel, R. Schneider, and B. Seeger. The R*-tree: An efficient and robust method for points and rectangles. In *Proceedings ACM SIGMOD Conference on Management of Data*, pages 322–331, 1990.
13. F. Barillari, E. Nardelli, and M. Pepe. Fully dynamic search trees can be balanced in $O(\log^2 n)$ time. Technical Report 146, Università di L'Aquila, 1997. Accepted in *Journal of Parallel and Distributed Computing.*
14. J. Bercken and B. Seeger. Query processing techniques for multiversion access methods. In *Proceedings 22nd Conference on Very Large Data Bases (VLDB'96)*, pages 168–179, 1996.
15. Y. Breitbart and R. Vingralek. Addressing and balancing issues in distributed B^+-trees. In *Proceedings 1st Workshop on Distributed Data and Structures (WDAS'98)*, 1998.
16. J. Clifford, C.E. Dyreson, T. Isakowitz, C.S. Jensen, and R.T. Snodgrass. On the semantics of "now". *ACM Transactions on Database Systems*, 22(2):171–214, 1997.
17. S. Christodoulakis. Implications of certain assumptions in database performance evaluation. *ACM Transactions on Database Systems*, 9(2):163–186, 1984.
18. J.P. Cheiney and A. Tourir. Fi-quadtree - A new data structure for content-oriented retrieval and fuzzy search. In *Proceedings 2nd Symposium on Spatial Databases (SSD'91)*, pages 23–32, 1991.
19. S. Dieker and R.H. Güting. Efficient handling of tuples with embedded large objects. Technical Report Informatik-236, FernUniversität Hagen, 1998. Also in *Data and Knowledge Engineering*, 32:247-268, 2000.
20. A. Di Pasquale and E. Nardelli. Balanced and distributed search trees. In *Proceedings 2nd Workshop on Distributed Data and Structures (WDAS'99)*, pages 73–90, 1999.
21. A. Di Pasquale and E. Nardelli. Distributed searching of k-dimensional data with almost constant cost. In *Proceedings 4th East European Conference on Advances in Databases and Information Systems (ADBIS 2000)*, volume 1884 Lecture Notes in Computer Science, pages 239–250, 2000.

22. J.R. Driscoll, N. Sarnak, D.D. Sleator, and R.E. Tarjan. Making data structures persistent. *Journal of Computer and System Sciences*, 38:86–124, 1989.
23. C.R. Dyer. The space efficiency of quadtrees. *Computer Graphics and Image Processing*, 19(4):335–348, 1982.
24. M. Erwig, R.H. Güting, M. Schneider, and M. Vazirgiannis. Spatio-temporal data types: An approach to modelling and querying moving objects in databases. *GeoInformatica*, 3(3):269–296, 1999.
25. C. Faloutsos and V. Gaede. Analysis of n-dimensional quadtrees using the Hausdorff fractal dimension. In *Proceedings 22nd Conference on Very Large Data Bases (VLDB'96)*, pages 40–50, 1996.
26. L. Forlizzi, R.H. Güting, E. Nardelli, and M. Schneider. A data model and data structures for moving objects databases. In *Proceedings ACM SIGMOD Conference on Management of Data*, pages 319–330, 2000.
27. C. Faloutsos and I. Kamel. Beyond uniformity and independence: Analysis of R-trees using the concept of fractal dimension. In *Proceedings 13th ACM PODS Symposium (PODS'94)*, pages 4–13, 1994.
28. I. Gargantini. An effective way to represent quadtrees. *Communications of the ACM*, 25(12):905–910, 1982.
29. R.H. Güting, T. de Ridder, and M. Schneider. Implementation of the ROSE algebra: Efficient algorithms for realm-based spatial data types. In *Proceedings 4th Symposium on Spatial Databases (SSD'95)*, pages 216–239, 1995.
30. V. Gaede and O. Günther. Multidimensional access methods. *ACM Computer Surveys*, 30(2):170–231, 1998.
31. C. Gurret, Y. Manolopoulos, A. Papadopoulos, and P. Rigaux. BASIS: A benchmarking approach for spatial index structures. In *Proceedings Workshop on Spatiotemporal Database Management (STDBM'99)*, pages 152–170, 1999.
32. O. Günther, V. Oria, P. Picouet, J.-M. Saglio, and M. Scholl. Benchmarking spatial joins a la carte. In *Proceedings 7th Conference on Statistical and Scientific Database Management Systems (SSDBM'98)*, pages 32–41, 1998.
33. C. Gurret and P. Rigaux. An integrated platform for the evaluation of spatial query processing strategies. In *Proceedings 9th Conference on Database and Expert Systems Applications (DEXA'98)*, pages 757–766, 1998.
34. A. Guttman. R-trees: A dynamic index structure for spatial searching. In *Proceedings ACM SIGMOD Conference on Management of Data*, pages 47–57, 1984.
35. O. Günther and E. Wong. A dual approach to detect polyhedral intersections in arbitrary dimensions. *BIT*, 31(1):3–14, 1991.
36. C.S. Jensen and R. Snodgrass. Semantics of time-varying information. *Information Systems*, 21(4):311–352, 1996.
37. E. Kawaguchi and T. Endo. On a method of binary picture representation and its application to data compression. *IEEE Transactions on Pattern Analysis and Machine Intelligence*, 2(1):27–35, 1980.
38. I. Kamel and C. Faloutsos. On packing R-trees. In *Proceedings 2nd Conference on Information and Knowledge Management (CIKM'93)*, pages 490–499, 1993.
39. G. Kollios, D. Gunopoulos, and V.J. Tsotras. On indexing mobile objects. In *Proceedings 18th ACM PODS Symposium (PODS'99)*, pages 261–272, 1999.
40. D.E. Knuth. *The Art of Computer Programming*, volume 3: Sorting and Searching. Addison-Wesley, 1973.
41. A. Kumar, V.J. Tsotras, and C. Faloutsos. Access methods for bi-temporal databases. In *Proceedings Workshop on Temporal Databases*, pages 235–254, 1995.
42. A. Kumar, V.J. Tsotras, and C. Faloutsos. Designing access methods for bitemporal databases. *IEEE Transactions on Knowledge and Data Engineering*, 10(1):1–20, 1998.

43. B. Kröll and P. Widmayer. Distributing a search tree among a growing number of processor. In *Proceedings ACM SIGMOD Conference on Management of Data*, pages 265–276, 1994.
44. B. Kröll and P. Widmayer. Balanced distributed search trees do not exists. In *Proceedings 4th Int. Workshop on Algorithms and Data Structures (WADS'95)*, volume 995 Lecture Notes in Computer Science, pages 50–61, 1995.
45. S.D. Lang and J.R. Driscoll. Improving the differential file technique via batch operations for tree structured file organizations. In *Proceedings 2nd IEEE Conference on Data Engineering (ICDE'86)*, pages 524–532, 1986.
46. W. Litwin and M.A. Neimat. k-RP$^*_s$ - A high performance multi-attribute scalable data structure. In *Proceedings 4th Conference on Parallel and Distributed Information System (PDIS'96)*, pages 120–131, 1996.
47. W. Litwin, M.-A. Neimat, and D.A. Schneider. LH* - linear hashing for distributed files. In *Proceedings ACM SIGMOD Conference on Management of Data*, pages 327–336, 1993.
48. W. Litwin, M.-A. Neimat, and D.A. Schneider. RP*: A family of order preserving scalable distributed data structures. In *Proceedings 20th Conference on Very Large Data Bases (VLDB'94)*, pages 342–353, 1994.
49. J. Van Leeuwen and R.E. Tarjan. Worst-case analysis of set union algorithms. *Journal of the ACM*, 31:245–281, 1984.
50. Y. Manolopoulos and G. Kapetanakis. Overlapping B$^+$-trees for temporal data. In *Proceedings 5th Jerusalem Conference on Information Technology (JCIT'90)*, pages 491–498, 1990.
51. Y. Manolopoulos, E. Nardelli, A. Papadopoulos, and G. Proietti. MOF-tree: A spatial access method to manipulate multiple overlapping features. *Information Systems*, 22(8):465–481, 1997.
52. Y. Manolopoulos, Y. Theodoridis, and V. Tsotras. *Advanced Database Indexing*. Kluwer Academic Publishers, 1999.
53. E. Nardelli. Distributed k-d trees. In *Proceedings 16th Conference of Chilean Computer Science Society (SCCC'96)*, pages 142–154, 1996.
54. E. Nardelli, F. Barillari, and M. Pepe. Design issues in distributed searching of multi-dimensional data. In *Proceedings 3rd International Symposium on Programming and Systems (ISPS'97)*, 1997.
55. E. Nardelli, F. Barillari, and M. Pepe. Distributed searching of multi-dimensional data: A performance evaluation study. *Journal of Parallel and Distributed Computing*, 49(1):111–134, 1998.
56. S. Nishida, H. Nozawa, and N. Saiwaki. Proposal of spatio-temporal indexing methods for moving objects. In *Proceedings Entity-Relationship Workshop (ER'98)*, pages 484–495, 1998.
57. E. Nardelli and G. Proietti. Managing overlapping features in spatial database applications. In *International Computer Symposium (ICS'94)*, pages 1297–1302, 1994.
58. E. Nardelli and G. Proietti. Efficient secondary memory processing of window queries on spatial data. *Information Sciences*, 84:67–83, 1995.
59. E. Nardelli and G. Proietti. Size estimation of the intersection join between two line segment datasets. In *Proceedings 4rd East-European Conference on Advances in Databases and Information Systems (ADBIS 2000)*, pages 229–238, 2000.
60. R. Nelson and H. Samet. A population analysis of quadtrees with variable node size. Technical Report CAR-TR-241, University of Maryland, Computer Science Department, 1986.
61. M.A. Nascimento and J.R.O. Silva. Towards historical R-trees. In *Proceedings 13th ACM Symposium on Applied Computing (ACM-SAC'98)*, 1998.

62. M.A. Nascimento, J.R.O. Silva, and Y. Theodoridis. Access structures for moving points. Technical Report TR-33, TimeCenter, 1998.
63. M.A. Nascimento, J.R.O Silva, and Y. Theodoridis. Evaluation for access structures for discretely moving points. In *Proceedings Workshop on Spatio-Temporal Database Management (STDBM'99)*, pages 171–188, 1999.
64. J. Orenstein. Spatial query processing in an object-oriented database system. In *Proceedings ACM SIGMOD Conference on Management of Data*, pages 326–336, 1986.
65. G. Proietti and C. Faloutsos. Accurate modeling of region data. Technical Report 98-137, Carnegie-Mellon University, 1998. Also in *IEEE Transactions on Knowledge and Data Engineering*, 13(6):874-883, November/December 2001.
66. G. Proietti and C. Faloutsos. Selectivity estimation of windows queries for line segment datasets. In *Proceedings 7th Conference on Information and Knowledge Management (CIKM'98)*, pages 340–347, 1998.
67. G. Proietti and C. Faloutsos. I/O complexity for range queries on region data stored using an R-tree. In *Proceedings 15th IEEE Conference on Data Engineering (ICDE'99)*, pages 628–635, 1999.
68. G. Proietti and C. Faloutsos. Analysis of range queries and self spatial join queries on real region datasets stored using an R-tree. *IEEE Transactions on Knowledge and Data Engineering*, 12(5):751-762, September/October 2000.
69. D. Pfoser, C.S. Jensen, and Y. Theodoridis. Novel approaches in query processing for moving objects. In *Proceedings 26th Conference on Very Large Data Bases (VLDB 2000)*, pages 395–406, 2000.
70. A. Papadopoulos and Y. Manolopoulos. Performance of nearest neighbor queries in R-trees. In *Proceedings 6th International Conference on Database Theory (ICDT'97)*, pages 394–408, 1997.
71. G. Proietti. The MOF$^+$-tree: A space efficient representation of images containing multiple overlapping features. *Journal of Computing and Information*, 2:42–56, 1996.
72. G. Proietti. An optimal algorithm for decomposing a window into its maximal blocks. *Acta Informatica*, 36(4):257–266, 1999.
73. A. Papadopoulos, P. Rigaux, and M. Scholl. A performance evaluation of spatial processing strategies. In *Proceedings 6th Symposium on Spatial Databases (SSD'99)*, pages 286–307, 1999.
74. B. Pagel, H. Six, H. Toben, and P. Widmayer. Towards an analysis of range query performance. In *Proceedings 12th ACM PODS Symposium (PODS'93)*, pages 214–221, 1993.
75. S. Ravada and J. Sharma. Oracle8i spatial: Experiences with extensible databases. In *Proceedings 6th Symposium on Spatial Databases (SSD'99)*, pages 355–359, 1999.
76. R.T. Snodgrass and T. Ahn. A taxonomy of time in databases. In *Proceedings ACM SIGMOD Conference on Management of Data*, pages 236–246, 1985.
77. Y. Sagiv. Concurrent operations on B*-trees with overtaking. *Journal of Computer and System Sciences*, 3(2):275–296, 1986.
78. H. Samet. *Applications of Spatial Data Structures*. Addison-Wesley, 1990.
79. H. Samet. *The Design and Analysis of Spatial Data Structures*. Addison-Wesley, 1990.
80. S. Šaltenis and C.S. Jensen. R-tree based indexing of general spatio-temporal data. Technical Report TR-45 and Chorochronos CH-99-18, TimeCenter, 1999.
81. S. Šaltenis, C.S. Jensen, S. Leutenegger, and M. Lopez. Indexing the positions of continuously moving objects. In *Proceedings ACM SIGMOD Conference on Management of Data*, pages 331–342, 2000.

82. N. Saiwaki, A. Naka, and S. Nishida. Spatio-temporal data management for highly interactive environment. In *Proceedings 6th IEEE Workshop on Robot and Human Communication (ROMAN'97)*, pages 571–576333, 1997.
83. R.T. Snodgrass. The temporal query language TQuel. *ACM Transactions on Database Systems*, 12(2):247–298, 1987.
84. B. Salzberg and V. Tsotras. A comparison of access methods for time evolving data. *ACM Computing Surveys*, 31(2):158–212, 1999.
85. T. Tzouramanis, Y. Manolopoulos, and N. Lorentzos. Overlapping B^+-trees: An implementation of a temporal access method. *Data and Knowledge Engineering*, 29(3):381–404, 1999.
86. T. Teraoka, M. Maruyama, Y. Nakamura, and S. Nishida. The MP-tree: A data structure for spatio-temporal data. In *Proceedings 14th IEEE Annual Phoenix Conference on Computers and Communications*, pages 326–333, 1995.
87. A. Tourir. A multi-layer quadtree: A spatial data structure for multi-layer processing. *Geoinformatica*, 2001.
88. Y. Theodoridis and T. Sellis. A model for the prediction of R-tree performance. In *Proceedings 15th ACM PODS Symposium (PODS'96)*, pages 161–171, 1996.
89. Y. Theodoridis, J.R.O. Silva, and M.A. Nascimento. On the generation of spatiotemporal datasets. In *Proceedings 6th Symposium on Spatial Databases (SSD'99)*, pages 147–164, 1999.
90. Y. Theodoridis, T. Sellis, A. Papadopoulos, and Y. Manolopoulos. Specifications for efficient indexing in spatiotemporal databases. In *Proceedings 7th Conference on Statistical and Scientific Database Management Systems (SSDBM'98)*, pages 123–132, 1998.
91. J. Tayeb, O. Ulusoy, and O. Wolfson. A quadtree based dynamic attribute indexing method. *The Computer Journal*, 41(3):185–200, 1998.
92. T. Tzouramanis, M. Vassilakopoulos, and Y. Manolopoulos. Overlapping linear quadtrees: A spatio-temporal access method. In *Proceedings 6th ACM Symposium on Advances in Geographic Information Systems (ACM-GIS'98)*, pages 1–7, 1998.
93. T. Tzouramanis, M. Vassilakopoulos, and Y. Manolopoulos. Processing of spatio-temporal queries in image databases. In *Proceedings 3rd East-European Conference on Advances in Databases and Information Systems (ADBIS'99)*, pages 85–97, 1999.
94. T. Tzouramanis, M. Vassilakopoulos, and Y. Manolopoulos. Multiversion linear quadtrees for spatio-temporal data. In *Proceedings 4rd East-European Conference on Advances in Databases and Information Systems (ADBIS 2000)*, pages 279–292, 2000.
95. T. Tzouramanis, M. Vassilakopoulos, and Y. Manolopoulos. Overlapping linear quadtrees and window query processing in spatio-temporal databases. *The Computer Journal*, 43(4):325–344, 2000.
96. Y. Theodoridis, M. Vazirgiannis, and T. Sellis. Spatio-temporal indexing for large multimedia applications. In *Proceedings 3rd IEEE Conference on Multimedia Computing and Systems (ICMCS'96)*, pages 441–448, 1996.
97. M. Vassilakopoulos and Y. Manolopoulos. Dynamic inverted quadtrees - A structure for pictorial databases. *Information Systems*, 20(6):483–500, 1995.
98. M. Vassilakopoulos, Y. Manolopoulos, and K. Economou. Overlapping for the representation of similar images. *Image and Vision Computing*, 11(5):257–262, 1993.
99. M. Vassilakopoulos, Y. Manolopoulos, and B. Kröll. Efficiency analysis of overlapped quadtrees. *Nordic Journal of Computing*, 2:70–84, 1995.
100. M. Vazirgiannis, Y. Theodoridis, and T. Sellis. Spatio-temporal composition and indexing large multimedia applications. *Multimedia Systems*, 6(4):284–298, 1998.

101. O. Wolfson, B. Xu, S. Chamberlain, and L. Jiang. Moving objects databases: Issues and solutions. In *Proceedings 10th Conference on Scientific and Statistical Database Management*, pages 111–122, 1998.
102. X. Xu, J. Han, and W. Lu. RT-tree - An improved R-tree index structure for spatiotemporal databases. In *Proceedings 4th Symposium on Spatial Data Handling (SDH'90)*, pages 1040–1049, 1990.
103. G. Zimbrao, J. Moreira de Souza, R. Chaomey Wo, and V. Teixeira de Almeida. Efficient processing of spatiotemporal queries in temporal geographical information systems. In *Proceedings 4th Multiconference on Systemics, Cybernetics and Informatics, 6th Conference on Information Systems, Analysis and Synthesis (SCI/ISAS 2000),* Vol.8, Part.II, pages 46–51, 2000.
104. G. Zimbrao, J. Moreira de Souza, and V. Teixeira de Almeida. The temporal R-tree. Technical Report ES-429/99, Federal University of Rio de Janeiro, Computer Science Department, 1999.

7 Architectures and Implementations of Spatio-temporal Database Management Systems

Martin Breunig[1], Can Türker[2], Michael H. Böhlen[5], Stefan Dieker[4], Ralf Hartmut Güting[4], Christian S. Jensen[5], Lukas Relly[3], Philippe Rigaux[6], Hans-Jörg Schek[2], and Michel Scholl[6]

[1] Uni Vechta, Germany
[2] ETH Zürich, Switzerland
[3] UBS AG Zürich, Switzerland
[4] FernUniversität Hagen, Germany
[5] Aalborg University, Denmark
[6] CNAM, Paris, France

7.1 Introduction

This chapter is devoted to architectural and implementation aspects of spatio-temporal database management systems. It starts with a general introduction into architectures and commercial approaches to extending databases by spatio-temporal features. Thereafter, the prototype systems CONCERT, SECONDO, DEDALE, TIGER, and GEOTOOLKIT are presented. As we will see, the focus of these systems is on different concepts and implementation aspects of spatial, temporal, and spatio-temporal databases, e.g. generic indexing, design of spatial, temporal, and spatio-temporal data types and operations, constraint modeling, temporal database management, and 3D/4D database support. A comparison of the prototype systems and a brief résumé conclude the chapter.

7.2 Architectural Aspects

To support spatio-temporal applications, the adequate design of a system architecture for a spatio-temporal database management system (STDBMS) is crucial. Spatio-temporal applications have many special requirements. They deal with complex objects, for example objects with complex boundaries such as clouds and moving points through the 3D space, large objects such as remote sensing data, or large time series data. These complex objects are manipulated in even more complex ways. Analysis and evaluation programs draw conclusions combining many different data sources.

To build an STDBMS, the traditional DBMS architecture and functionality have to be extended. Managing spatio-temporal data requires providing spatio-temporal data types and operations, extensions to the query and data manipulation language, and index support for spatio-temporal data. Such issues arise not only in a spatio-temporal context but also when building spatial only or temporal only systems. Over the recent years we witnessed three base variants of extending system architectures (see Figure 7.1):

T. Sellis et al. (Eds.): Spatio-temporal Databases, LNCS 2520, pp. 263–318, 2003.
© Springer-Verlag Berlin Heidelberg 2003

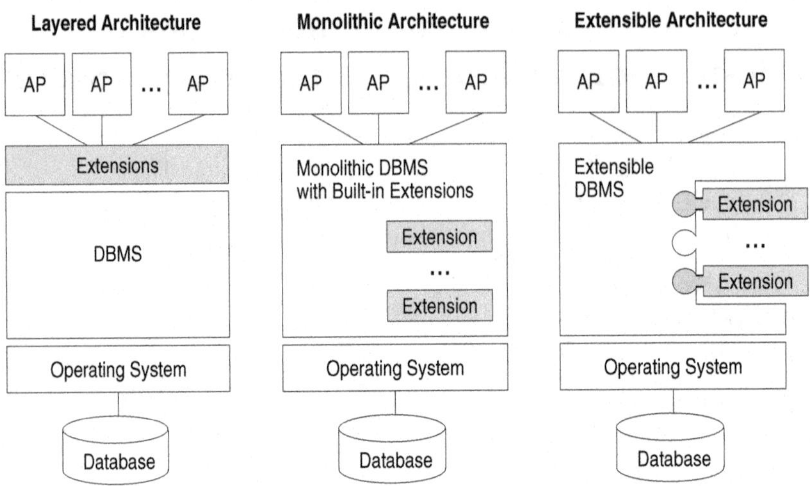

Fig. 7.1. Comparison of system architectures

1. The *layered* approach uses an off-the-shelf database system and extends it by implementing the missing functionality on top of the database system as application programs.
2. In the *monolithic* approach, the database manufacturer integrates all the necessary application-specific extensions into the database system.
3. The *extensible* approach provides a database system which allows to plug user-defined extensions into the database system.

The following subsections present these variants in more detail.

7.2.1 The Layered Architecture

A traditional way of designing an information system for advanced data types and operations is to use an off-the-shelf DBMS and to implement a layer on top providing data types and services for the specific application domain requirements. The DBMS with such a generic component is then used by different applications having similar data type and operation requirements. Such enhanced DBMSs exploit the standard data types and data model — often the relational model — as a basis. They define new data types and possibly a new layer including data definition and query language, query processing and optimization, indexing, and transaction management specialized for the application domain. The new data types are often mapped to low-level data storage containers usually referred as binary large objects of the underlying DBMS. Applications are written against the extended interface. Data definition, queries, and update operations are transformed from the application language to the underlying DBMS interface.

Many of the layered architecture systems have originally been designed as stand-alone applications directly using the operating system as their underlying storage system. In order to exploit generic database services such as transaction management, their low-level storage system has been replaced by an off-the-shelf database system. Since most of these systems supported only a few standard data types, the layered architecture systems have to use the operating system's file system to store the application-specific data types while using the DBMS to store the standard data types. This architecture is also called the *dual* architecture (see Figure 7.2). Todays advanced DBMSs support binary large objects, and thus get rid of the need to store a part of the data directly in the operating system bypassing the DBMS.

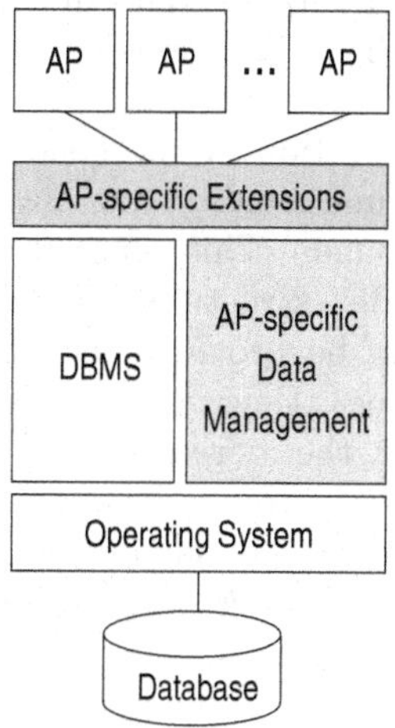

Fig. 7.2. The dual system architecture

The layered approach has the advantage of using standard off-the-shelf components reusing generic data management code. There is a clear separation of responsibilities: application-specific development can be performed and supported independently of the DBMS development. Improvements in one component are directly available in the whole system with almost no additional effort. On the other hand, the flexibility is limited. Development not foreseen in the standard component has to be implemented bypassing the DBMS. The more effort is put into such an application-specific data management extension, the more difficult it gets to change the system and to take advantage of DBMS improvements.

Transaction management is only provided for the standard data types handled by the DBMS. Transactional guarantees for advanced data types as well as application-specific transaction management have to be provided by the application-specific extension. Query processing and optimization has to be performed on two independent levels. Standard query processing and optimization can be handled by the DBMS while spatio-temporal query processing has to take place in the extension. Because system-internal statistics and optimization information are only available inside the DBMS, global query processing and optimization of combined queries is hard to implement. Indexing of standard data types takes place inside the DBMS while indexing spatio-temporal data has to be dealt with in the extension. Therefore, combined index processing cannot be used.

7.2.2 The Monolithic Architecture

As in the layered architecture, many systems using a monolithic architecture have originally been designed as stand-alone applications. In contrast to the layered architecture, the monolithic architecture extends an application system with DBMS functionality instead of porting it to a standard DBMS. In this way, specialized DBMSs with query functionality, transaction management, and multi-user capabilities are created. The data management aspects traditionally associated with DBMS and the application-specific functionality are integrated into one component.

Because of the tight integration of the general data management aspects and the application-specific functionality, monolithic systems can be optimized for the specific application domain. This generally results in good performance. Standard and specialized index structures can be combined for best results. Transaction management can be provided in a uniform way for standard as well as for advanced data types. However, implementing an integrated system becomes increasingly difficult, the more aspects of an ever-growing application domain have to be taken into account. It might be possible — even though it has proven difficult — to build a monolithic system for a spatial only or a temporal only database. Combining spatial and temporal data management adds another dimension of complexity such that is very difficult to provide a satisfactory solution using the monolithic approach.

7.2.3 The Extensible Architecture

The layered as well as the monolithic architecture do not support an easy adaptation of the DBMS to new requirements of advanced applications. The user, however, should be able to "taylor" the DBMS flexibly according to his specific requirements. Extensible database systems provide a generic system capable of being extended internally by application-specific modules. New data types and functionality required for specialized applications is integrated as close as possible into the DBMS. Traditional DBMS functionality like indexing, query optimization, and transaction management is supported for user-defined data types and functions in a seamless fashion. In this way, an extensible architecture takes the advantage of the monolithic architecture while avoiding its deficiencies. It thus provides the basis for an easy integration of advanced spatio-temporal data types, operations, and access methods which can be used by the DBMS analogously to its standard data types and access methods.

The first extensible system prototypes have been developed to support especially non-standard DBMS applications like geographical or engineering information systems. Research on extensible systems has been carried out in several projects, e.g. Ingres [57], DASDBS [69,68], STARBURST [40], POSTGRES [74,75], OMS [10], Gral [35], Volcano [29,30], and PREDATOR [70]. These projects addressed, among other, data model extensions, storage and indexing of complex objects as well as transaction management and query optimization in the presence of complex objects.

Another way to provide extensible database systems are database toolkits. Toolkits do not prescribe any data model, but rather identify and implement a common subset of functionality which all database systems for whatever data model must provide, e.g., transaction management, concurrency control, recovery, and query optimization. Key projects of that category are GENESIS [5], EXODUS [18], and SHORE [19].

While toolkits are very generic and can be used for the implementation of many different data models, they leave too much expenditure at the implementor to be accepted as a suitable means for fast system implementation. Extensible systems, on the other hand, pre-implement as much functionality as possible at the expense of flexibility, since they usually prescribe a specific data model, e.g., an object-relational model.

The SQL99 standard [48] specifies new data types and type constructors in order to better support advanced applications. As we will see in the next section, commercially leading database vendors already support the development of generic extensions — even though not in its full beauty. These extensible relational systems are referred as *object-relational* systems.

7.2.4 Commercial Approaches to Spatial-temporal Extensions

In the following, we briefly sketch the commercial approaches of Informix, Oracle, and IBM DB2, which exploit extensible, object-relational database technology to provide spatio-temporal extensions. Also, we briefly sketch commercial approaches in the field of geographic information systems.

Informix Datablades. The Informix Dynamic Server can by extended by *datablades* [47]. Datablades are modules containing data type extensions and routines specifying the behavior of the data and extending the query language. New data types can be implemented using built-in data types, collection types like `SET`, `LIST`, or `MULTISET`, or large unstructured data types like `BLOB`. User-defined functions encapsulated in the datablades determine the behavior and functionality of new data types.

A new access method must create an *operator class* containing both a set of strategy functions and a set of support functions. The strategy functions are used for decisions of the optimizer to build execution plans. The support functions are needed to build, maintain, and access the index. Access methods are used by the database server every time an operator in the filter of a query matches one of the strategy functions and results in lower execution costs than other access methods.

The datablade development requires expendable implementation work and a deep understanding of the internals of the system. Therefore, the creation of new access methods and their integration into the optimizer is mostly done by companies offering commercial products like the text indexing and retrieval datablade of Excalibur [43].

The Informix *Geodetic* datablade [44] extends the server with spatio-temporal data types like `GeoPoint`, `GeoLineseg`, `GeoPolygon`, `GeoEllipse`, `GeoCircle`, and `GeoBox`, and associated operators like `Intersect`, `Beyond`, `Inside`, `Outside`, and `Within`. An R-tree [39] is provided to index spatio-temporal data. A specialized operation class, called `GeoObject_ops`, is available, which associates several operators to the R-tree access method. An R-tree index for a spatio-temporal attribute is defined as follows:

```
CREATE INDEX index_name
ON table_name(column_name GeoObject_ops)
USING RTREE;
```

Informix also provides the *Spatial* datablade [45] as well as the *TimeSeries* datablade [46] to extend the server with data types and functions referring to spatial data only and to temporal data only, respectively.

Oracle Cartridges. The Oracle8 server can be extended by *cartridges* [60]. These modules add new data types and functionality to the database server. As in Informix with datablades, data types and functions can easily be implemented within a cartridge but the integration of indexes into the query optimizer needs deep knowledge of the database server. In addition, access method extensions and their integration into the query optimizer are limited to a small amount of simple query constructions.

Based on the built-in data types, array type, reference type, and large object types, new data types can be implemented within a cartridge. The complexity of data types is limited and not all data types can be combined orthogonally. An index structure is defined within the cartridge. In contrast to Informix, index maintenance – insertion, deletion, and update – must be implemented within the cartridge; no genericity is provided here. The usage of new index structures can be plugged into the optimizer by user-defined functions returning information about execution costs, statistics information, and selectivity of the index for a given predicate. This allows the cost-based optimizer to use new indexes and to build alternative execution plans.

To deal with spatial data, the *Oracle Spatial* cartridge [58] is available. This cartridge extends the database server with data types representing two-dimensional points, lines, or polygons. Based on these primitive data types, composite geometric data types like point cluster and polygons with holes are provided. A linear quad-tree [26] is provided for spatial indexing. A spatial index on a column of type `MDSYS.SDO_GEOMETRY` is defined as follows:

```
CREATE INDEX index_name
ON table_name(column_name)
INDEXTYPE IS MDSYS.SPATIAL_INDEX;
```

This index is then exploited when executing geometry-specific operations like `RELATE`, `SDO_INTERSECTION`, `SDO_Difference`, and `SDO_UNION`.

The *Oracle Time Series* cartridge [59] is provided for temporal data. This cartridge extends the server with calendar and time series data types.

IBM DB2 Extenders. IBM's modules which extend the DB2 server with abstract data types are called *extenders*. Extenders can be implemented using built-in data types, reference types, and large object types.

Spatial data is supported by IBM's *DB2 Spatial* extender. This module extends the DB2 server with spatial data types like `Point`, `Linestring`, and `Polygon` and a rich repertoire of operations like `Contains`, `Intersects`, `Overlap`, `Within`, `Disjoint`, `Touch`, and `Cross`. A grid file [55] is used as spatial index. Such a spatial index is defined as follows:

```
CREATE INDEX index_name
ON table_name(column_name)
USING spatial_index;
```

The database server is then aware of the existence of this index and the query optimizer takes the advantage of the index when accessing and manipulating spatial data. In addition to specifying the costs of a user-defined function, the user can indicate whether or not the function can be taken as predicate in the where clause.

A DB2 *Time Series* extender, which provides similar functionality as its Informix and Oracle counterparts, is offered by a Swedish company [25].

Geographical Information Systems. Geographical Information Systems (GIS) have also been designed according to the different architecture variants discussed in the previous subsections. ARC/INFO [2], for example, is a representative of the dual architecture, whereas ARC/INFO's spatial database engine and SMALLWORLD GIS [72] have a monolithic system architecture. However, hitherto there are no commercial GIS that allow an extension of their database kernel.

As GIS have originally been developed to support the construction of digital maps, they have made much more progress in handling spatial data than temporal data. The key issue with GIS during the last years has been the combined handling of spatial and non-spatial data [54]. Database support for dynamic maps, however, is still a research topic. To the best of our knowledge, there is no efficient support for the handling and processing of temporal data in today's commercial GIS. These systems have not been designed for temporal data support and currently are not easily extensible for temporal data management. A prototype module for the management of temporal data on the basis of a commercial GIS has been proposed by [50].

7.3 The Concert Prototype System

7.3.1 Introduction

CONCERT [9,65,64] is a database kernel system based on an extensible architecture. However, CONCERT propagates a new paradigm called "exporting database functionality" as a departure from traditional thinking in the field of databases. Traditionally, all data is loaded into and owned by the database in a format determined by the database, whereas according to the new paradigm data is viewed as abstract objects with only a few central object methods known to the database system. The data may even reside outside the database in external repositories or archives. Nevertheless, database functionality, such as query processing and indexing, is provided. The new approach is extremely useful for spatio-temporal environments, as there is no single most useful data format to represent spatio-temporal data. Rather, many different spatio-temporal data formats coexist. The new paradigm presented in CONCERT allows them to be viewed in a uniform manner.

While traditional indexing is based on data types — an integer data type can be indexed by a B-tree, a polygon data type can be indexed by an R-tree — CONCERT's indexing is based on the conceptual behavior of the data to be indexed: data that is associated with an ordering can be indexed by a B-tree. Data implementing spatial properties (such as overlaps and covers predicates) can be indexed by an R-tree. CONCERT uses four classes of behavior, `SCALAR`, `RECORD`, `LIST`, and `SPATIAL`, covering all relevant concepts that are the basis of today's indexing techniques.

With these classes conceptual behavior of data is described in contrast to type dependent properties like in GIST [41]. On top of the CONCERT classes, arbitrary indexing structures can be build which are valid for a variety of data types. New indexing structures are implemented for concepts and use concept typical operations. In this way, indexing techniques of the CONCERT approach are independent of the concrete implementation of the data types — in particular it is not limited to tree structures alone. This is the major difference between GIST and the indexing framework provided by CONCERT.

7.3.2 Architecture

Following the extensible system architecture, CONCERT consists of a kernel system for low-level data management and query processing as well as an object manager providing advanced query functionality. The kernel is responsible for the management of individual objects within collections, resource and storage management, and low-level transaction management. The object manager combines different collections adding join capability to the kernel collections. Figure 7.3 shows an overview of the CONCERT architecture.

The CONCERT kernel system consists of two components, the Storage Object Manager (SOM) and the Abstract Object Manager (AOM). The SOM provides DBMS base services such as segment management, buffer management,

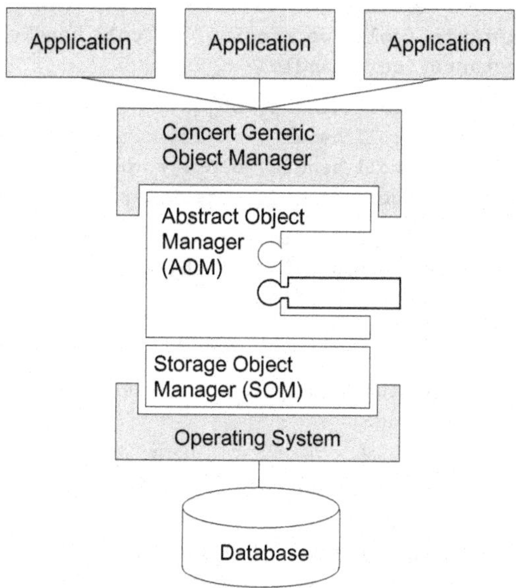

Fig. 7.3. CONCERT system architecture

and management of objects on database pages. It is tightly integrated into the underlying operating system exploiting multi-threading and multi-processing. It uses a hardware-supported buffer management exploiting the operating system's virtual memory management by implementing a memory-mapped buffer [8]. Details of the CONCERT SOM can be found in [63].

The AOM provides the core functionality for extending the kernel with application-specific code. It implements a fundamental framework for the management of collections. It uses the individual objects handled internally by the SOM or, through its interoperability capability, by an external storage system, combining them into collections. The AOM collection interface defines operations to insert, delete, and update individual objects. Retrieval of objects is performed through direct object access or through collection scans. Scans are initiated using an optional predicate filtering the qualifying objects. Figure 7.4 shows the (simplified) collection interface definition.

Indexes are treated similar to base collections. From the AOM's perspective, they are collections of objects containing the index entry (the attribute to be indexed on) and the object key of the base collection. Accessing data using an index is performed by an index scan identifying all qualifying object keys that are subsequently used to retrieve the objects themselves. Depending on the kind of the index and the query, the objects retrieved might have to be filtered further performing a false drops elimination.

On object insertion, the object is inserted into the base collection first. The insertion operation of the base collection returns an object key that can be used for the index entry. Because the index uses the same interface as the base collection, hierarchies of indexes can be built easily. This allows, for example, an

```
createInstance (coll_description) :- coll_handle
deleteInstance (coll_handle)

insertObject   (coll_handle, object) :- object_key
deleteObject   (coll_handle, object_key)
updateObject   (coll_handle, old_key, object) :- new_key
getObject      (coll_handle, object_key) :- object

scanStart      (coll_handle, predicate) :- scan_handle
scanGet        (scan_handle) :- object
scanNext       (scan_handle)
scanClose      (scan_handle)
```

Fig. 7.4. (Simplified) AOM collection interface

inverted file index to be indexed itself by a B-tree index. The conceptual equality of base collections of objects and the index collections of their keys enables the components to be combined in many different ways providing a powerful tool that goes far beyond simple indexing and includes things such as constraint validation and trigger mechanisms.

A special aspect of the CONCERT approach is the fact that indexing is performed for abstract data types. Obviously it is not possible to index completely unknown objects. Some knowledge of the user-defined objects has to be available to the storage system. A small set of concepts is identified to be sufficient to allow most physical design decisions for spatio-temporal DBMS. The next subsection introduces an R-tree like generic index framework with minimal restrictions to the flexibility of external objects.

7.3.3 Spatio-temporal Extensions

To explain CONCERT's database extensibility idea, let us look at a standard B-tree index [21], as it is implemented in most database systems. A B-tree stores keys of objects ordered by their values. The ordering operation is chosen depending on the data type of the key value: for data type `NUMBER`, standard ordering over numbers is used whereas `STRING` data types are lexicographically ordered. Another ordering operation is used for `DATE` data types. Much more data types are possible but the central aspect when data is indexed by a B-tree is that there is an attribute which can be ordered [73]. On the other hand, data types which cannot be ordered cannot be indexed by a B-tree.

The generalization of this observation is the basis of the *abstract object storage type* concept in CONCERT: The knowledge about the type of the data is not needed to manage it but its conceptual behavior and the *concept typical operations* associated with these concepts must be known. These operations have to be provided by the user to allow CONCERT to interpret a given abstract object storage type. Four concepts can be identified:

- SCALAR: A data type belongs to the SCALAR concept if there is an ordering operation. All data types behaving like a SCALAR can be indexed, for instance, by a B-tree.
- LIST: A data type belongs to the LIST concept if it consists of a set of components over which an iterator is defined. A data type behaving like a LIST might be a list of keywords, forming the inverted index of a text document. The concept typical operations are FIRST and NEXT.
- RECORD: A data type belongs to the RECORD concept if it is an ordered concatenation of components which themselves are arbitrary concepts. The concept typical operation is the SUB_OBJECT operation returning a component of the RECORD. A RECORD implements object partitioning.
- SPATIAL: A data type belongs to the SPATIAL concept if it is extended in an arbitrary data space and can be indexed with respect to its spatial properties. This is the most important concept in the context of spatio-temporal data and will be explained in detail in the next subsection. Note that this concept is not limited to geometric or time space alone. The concept typical operations OVERLAPS, SPLIT, COMPOSE, and APPROX are explained in the following.

The user has to implement extensions by plugging new data types into the existing framework. Spatio-temporal indexing structures can be implemented and plugged into CONCERT independent of the data type indexed by just using the SPATIAL concept typical operations.

In traditional systems, the class of space covering objects is predefined by the index method provided and thus "hard-wired" in the system, e.g., a space covering rectangle in the R-tree. Therefore, the classes of application objects and query objects also have to be predefined by the DBMS requiring the user to convert his data into the DBMS format. This clearly is not desirable in the context of extensible systems. CONCERT addresses this problem. Rather than predefining spatial index objects and thereby forcing the data structure of user objects and user queries, it allows the user to implement not only the application and query objects but also corresponding index objects. The spatial index algorithm is written in a generic way only exploiting the spatial space-subspace relationships through method invocation.

These spatial[1] relationships define the concept typical operations of the SPATIAL concept as follows:

- The operation OVERLAPS checks for spatial overlap of two objects or data spaces. It is used by index tree navigation to prune the search space. It also helps finding the appropriate leaf node on index insertion. Therefore it is not defined as a simple predicate operation, but rather as an operation returning

[1] Note that the spatial concept is not restricted to objects of the geometric space. The spatial conceptcan also be used to model object of the temporal space as well as other data spaces whose objects exhibit similar behavior as geometric objects, e.g., supports inclusion and overlapping predicates as well as split and compose operations.

an integer value indicating the degree of overlap. Negative values are used as a measure for indicating the distance between two objects[2]. Finally, the OVERLAPS operation is used to approximate spatial and temporal overlaps, intersects, and covers predicates.

- The SPLIT operation divides a spatial object into several spatial objects. This operation can be used to a priori divide large user objects into smaller ones before inserting them or while reorganizing the search tree. It is also used to split index nodes when they become too large. The question "when nodes are split" and "which concrete spatial object is returned" is determined by the implementation of the SPLIT operation of the associated data type following the SPATIAL concept. In this way, a split operation can return a simple set of rectangles resulting from splitting a "large" rectangle (like in R+-trees) but also a set of complex polygons or spheres.
  Here the terms small and large are used in two contexts. Large objects can be objects occupying a lot of storage space, as well as objects covering a large data space. In both cases, it might be beneficial to divide a large object into several smaller ones. In the first case, memory allocation and buffering is easier. In the second case, the data space in index nodes is smaller allowing a better pruning in the index structure. The behavior depends on the implementation of the SPLIT operation.
- The COMPOSE operation recombines spatial objects that have previously been split to an "overall" spatial object. This operation is used for the reconstruction of large data objects which have been split on insertion, as well as for index node description objects when index nodes are merged. Note that we make no assumption about the implementation of the operations for a given data type. We only consider the conceptual behavior of the data. The COMPOSE operation is the inverse of the SPLIT operation, which means that if O is an object of concept SPATIAL then O=COMPOSE(SPLIT(O)) holds.
- Finally, the APPROX operation approximates a spatial object or a set of spatial objects with a new spatial object. This new spatial object is a representative of the data space covered by the objects to be approximated. The typical and most common approximation is the n-dimensional bounding rectangle. However, the APPROX operation is not restricted to the bounding rectangle. Arbitrary operations can be used as long as they satisfy the condition of dominating predicates [77]. In our context, this means that, for example, if the approximations of two objects do not overlap, the two original object must not overlap. Figure 7.5 summarizes these four concept typical operations of the SPATIAL concept.

The user implements these four operations for an application, query and index objects and registers them to the CONCERT kernel system.

---

[2] Note that there is no restriction about the implementation of these operations. Programmers might decide to only distinguish two values — overlapping and non-overlapping — for the OVERLAPS operation. The concepts described here will work in the same way. However, the optimizations described are not possible.

CONCEPT SPATIAL		
Operation	**Parameter**	**Result**
OVERLAPS	spatial_object1, spatial_object2	SCALAR
SPLIT	spatial_object	{ spatial_object }
COMPOSE	{ spatial_object }	spatial_object
APPROX	{ spatial_object }	spatial_object

Fig. 7.5. Typical operations of the SPATIAL concept

7.3.4 Implementation Details

CONCERT can deal with each data type belonging to one of the concepts for which the necessary management and concept typical operations are implemented, independently of the concrete implementation of the data types. Indexing and query processing in the DBMS kernel is performed based on these operations only. For more information about the concepts, the operations, and the CONCERT system in general see [63,66,65,9]. In the following, we describe how spatio-temporal indexes can be plugged into CONCERT by using a framework for generic spatio-temporal data indexing.

Although the well-known R-tree [39] may not be optimal for indexing spatio-temporal objects (particularly for temporal indexes; see [67] for more information), its simple and well-understood structure is useful to explain the extensibility aspects. The R-tree is implemented and generalized within the CONCERT framework. Abstracting from the R-tree approach results in an index structure which is generic in the sense of data because it is based only on the "behavior" of the concept it belongs to. It is also generic in the sense of algorithms because of the use of generic heuristic functions which determine the tree-internal processes (varying the heuristics can change the R-tree like behavior into an R+-tree or another derivation). Finally, it is also generic in the sense of nodes covering search spaces because of the use of spatial objects to approximate the space of subtrees and no fixed spatial shapes.

Whereas the R-tree is restricted to store rectangular data objects, the CONCERT approach allows any objects conforming to the SPATIAL concept to be stored in the *generic spatio-temporal tree.* The CONCERT low-level storage capabilities provides an efficient multi-page storage system. Therefore, the generic spatio-temporal tree does not need to have a fixed node size. Nodes can be enlarged dynamically to virtually unlimited[3] size using a multi-page secondary storage allocation scheme [8]. The R-tree nodes have minimal bounding rectangles associated with them, whereas the generic spatio-temporal tree uses abstract spatial objects instead (e.g. spheres, convex polygons, or just rectangles). These spatial objects are usually computed using the APPROX operation. The only assumption made here, which is implicit given by the SPATIAL concept, is the existence of an APPROX operation which can be evaluated on node objects.

[3] There is a hard limit of 4 GByte per node. However, in order to be efficient, inner nodes should not become larger than about 1 MByte.

7.3.5 Case Studies

In the following we describe how a generic spatio-temporal tree can be used in CONCERT with its typical operations of object-lookup, insertion with splitting of a node and deletion including the reorganization of the tree.

Object Lookup. The generic spatio-temporal tree uses a SPATIAL object to describe the data space of the query window, or more generally, the arbitrary query space. With this knowledge the R-Tree algorithms as given in [39] can be extended to a generic algorithm using the SPATIAL concept and its concept typical operation OVERLAPS, as shown in Figure 7.6. Such algorithms can be incorporated easily into the CONCERT system.

```
find (query, node)
begin
  for all e ∈ {node.entries}
    if leafnode(node)
      if OVERLAPS (query, e.object)
        report e.object
      endif
    else
      if OVERLAPS (query, e.region)
        find (query, e)
      endif
    endif
  endfor
end
```

Fig. 7.6. Lookup in a generic spatial index

The concept typical operation for spatial index lookup is the OVERLAPS operation. Note that the CONCERT spatial index is much more flexible than any given tree variant. If the abstract objects stored in the nodes are minimum bounding rectangles, and query objects are rectangles as well, the algorithm depicted in Figure 7.6 behaves exactly as the R-tree lookup. Since CONCERT makes almost no assumption about the objects in the tree, the algorithm works the same way also for arbitrary n-dimensional spatial objects, as long as the subtrees form hierarchies of data spaces. Certain applications might prefer to use overlapping convex polygons to partition the data space or a sphere, if the Euclidean distance is important like in nearest neighbor queries for point data.

If the objects contain, beside the spatial, a temporal dimension, the algorithm can directly be used for spatio-temporal objects. Note that it is the responsibility of the user implementing the OVERLAPS operation to distinguish between the spatial and the temporal dimension.

Splitting a Node. One of the important issues often discussed for R-trees in the context of spatio-temporal applications is the problem of dead space in the rectangles. The larger the rectangles are with respect to the data space covered by the objects contained in the node, the less efficient the R-tree becomes. Our generic tree provides an easy and flexible solution to this problem.

Index nodes as well as data objects stored in the index are spatially extended objects implementing the operations of the `SPATIAL` concept. Therefore, large objects can be split into several (smaller) ones by using the concept typical operation `SPLIT`. This operation can be called many times until the resulting objects have a good size from an application point of view. It is not the database system and its index structure that determines the split granularity or the split boundaries. If, from an application point of view, there is no point in splitting an object further, the `SPLIT` operation just returns the unchanged object.

Note that the `SPLIT` operation is much more powerful than just splitting a rectangle into many others. The `SPLIT` operation needs an object of an arbitrary data type following the `SPATIAL` concept and returns a set of objects. The only requirement is that the resulting objects follow the `SPATIAL` concept. Such objects might be, for instance, rectangles as in R-trees or spheres as in M-trees. The exact behavior of a `SPLIT` operation is determined by its implementation.

As discussed earlier, CONCERT has virtually no size restriction for its index nodes. Using the `OVERLAPS` operation, the spatial index code can therefore handle arbitrary large objects — it just might not be very efficient, if the `SPLIT` operation is not actually splitting the objects. In any case, splitting is done by exploiting application semantics rather than following a node-space constraint.

Splitting is possible in different situations. One important situation is the a priori splitting of objects at insertion or update time. Such approaches are included in R-tree derived trees. By using the concept typical `SPLIT` operation these well-studied split procedures are generalized. The concrete implementation of the operation can determine different application dependent heuristics adapted to the requirements.

Insertion of Objects. In contrast to [41], we develop our algorithm based on the concept typical operations and the tree typical operations adapt to the concept driven approach whereas GiST follows a tree structure driven approach, generalizing operations in the context of tree management. Figure 7.7 shows the generic insertion procedure for our index.

Although similar to its outline, the generic spatial index has some important differences to the R-tree. Whereas in the R-tree nodes are of fixed size and, hence, a node has to be split according to application requirements, the generic index is more flexible. The operation ***Consider Split*** can implement a flexible splitting heuristics considering not only the size of the node but also the spatial extension of the objects and the amount of dead space in the node. In this way, the concrete choice of a heuristic determines the behavior of the tree such that it behaves like an R-tree, an R+-tree, or any other indexing tree for spatial data.

The splitting itself is also more flexible. It can be performed not only by distributing the entries among the nodes (using an arbitrary splitting strategy

```
insert (object, node)
begin
  if leafnode(node)
    Consider Split {Heuristics for splitting or enlarging leaf node}
    Insert Object to leaf node
  else
    Choose Subtree {Heuristics for choice of best subtree}
    insert (object, subtree)
    Consider Subtree Split {Heuristics for splitting or enlarging inner node}
  endif
  Adjust Node
end
```

Fig. 7.7. Insertion into a generic spatial index

such as one of the strategies discussed in [39]) but also by splitting large objects using the concept typical operation `SPLIT` reducing the dead space further. In this way, the implementation of `SPLIT` controls a part of the behavior of our indexing framework.

Even the insertion of an object in a subtree is more flexible than in a concrete tree implementation. The insertion is handled by recursively passing the object down the tree. At each non-leaf node, an appropriate subtree has to be chosen for the recursion. This is done by the operation *Choose Subtree*. In the standard R-tree algorithm, the subtree is chosen based on the least enlargement of the bounding rectangle necessary. In the generic spatial index, the concept typical operation `OVERLAPS` is used. In order to optimize the choice for a subtree, `OVERLAPS` is not defined as a simple predicate but rather returns an integer value indicating the amount of overlap. In addition to the amount of overlap, the current size of the subtree and the amount of free space in the subtree can also be considered.

The additional flexibility over fixed multi-dimensional indexing trees gained with the operations *Choose Subtree*, *Consider Split* and *Consider Subtree Split* together with the mechanism avoiding dead space using the concept typical operation `SPLIT` makes the CONCERT approach a useful framework of an R-tree like index for spatio-temporal applications.

Reorganizing the Tree. After insertion, nodes have to be adjusted using the `AdjustNode` operation as shown in Figure 7.8. Insertion into the generic spatial tree can bring the tree out of balance. This can be avoided by reorganizing the nodes by moving some of its entries to sibling nodes. However, such a reorganization can be very expensive. Therefore, a complete reorganization is performed at given points in time when the tree gets too much out of balance.
Remembering that each node has a covering subspace object associated with it which follows the `SPATIAL` concept (e.g. a bounding rectangle in R-trees), this object has to be adjusted on insertion using operation `APPROX` if reorganization is necessary. Whether a node has to be adjusted is determined by the *Consider*

```
adjust node (node)
begin
  Consider Reorganization {Heuristics for reorganization with sibling nodes}
  Adjust Covering Subspace {using the APPROX operation}
  if node has been split
    Create Covering Subspace of new node
    Propagate Split to parent node
  endif
end
```

Fig. 7.8. Adjust an index node

Reorganization operation which implements the desired heuristic. If the node has been split, the APPROX operation has to be calculated for both nodes and the split has to be propagated to the parent node. If the root node is split, the tree grows by one level.

Deletion of Objects. For object deletion, two strategies can be followed. It is always possible to recompute the spatial extent of each node using the APPROX operation. This keeps the space covered by each node minimal, but it requires substantial overhead each time an object is deleted.

Alternatively, the spatial extent of the nodes is left unchanged. Deletion is more efficient since no deletion or adjustment of inner nodes is necessary. If a node becomes too small, it is merged with one of its siblings. This keeps the overhead of reorganizing the tree low, but at the same time decreases the efficiency of the index due to possible dead space in the covering objects of a node.

The CONCERT prototype has been tested, among others, with an application from photogrammetry targeting raster image management [65]. The focus has been on physical database design and query processing for raster image management.

7.4 The Secondo Prototype System

7.4.1 Introduction

SECONDO is a data-model independent environment for the implementation of non-standard database systems. Its goal is to offer the advantages of both the extensible system and the toolkits approach (see Section 7.2.3) while avoiding their disadvantages.

The core component of SECONDO is a generic "database system frame" that can be filled with implementations of a wide range of data models, including, for example, relational, object-oriented, graph-oriented, sequence-oriented, or spatio-temporal database models.[4] The key strategy to achieve this is the separa-

[4] Some of the goals of SECONDO and Concert are similar. A comparison of the two systems is given at the end of Section 7.4.3.

tion of the data model independent components and mechanisms in a DBMS (the *system frame*) from the data-model dependent parts. Nevertheless, the frame and the "contents" have to work together closely. With respect to the different levels of query languages in a DBMS, we have to describe the system frame:

- the *descriptive algebra*, defining a data model and query language,
- the *executable algebra*, specifying a collection of data structures and operations capable of representing the data model and implementing the query language, and
- the *rules* to enable a query optimizer to map descriptive algebra terms to executable algebra terms, also called *query plans* or *evaluation plans*.

A general formalism serving all these purposes has been developed earlier, called *second-order signature (SOS)* [36]. It is reviewed in Section 7.4.2.

On top of the descriptive algebra level there may be some syntactically sugared language, e.g., in an SQL-like style. We assume that the top-level language and the descriptive algebra are entirely equivalent in expressive power; only the former may be more user-friendly whereas the latter is structured according to the SOS formalism. A compiler transforming the top-level language to descriptive algebra can be written relatively easily using compiler generation tools, since it just has to perform a one-to-one mapping to the corresponding data definitions and operations.

At the system level, definitions and implementations of type constructors and operators of the executable algebra are arranged into *algebra modules*, interacting with the system frame through a small number of well-defined support functions for manipulation of types and objects as well as operator invocation. Those algebra support functions dealing with type expressions will be created automatically from the corresponding SOS specification.

7.4.2 Second-Order Signature

Since the SECONDO system implements the framework of second-order signature (SOS) [36], it is necessary to recall the essential concepts here. The basic idea of SOS is to use two coupled signatures to describe a data model as well as an algebra over that data model. To distinguish the two levels of signature, we call the first *type signature* and the second *value signature*. A signature in general has *sorts* and *operators* and defines a set of *terms*.

Specifying a Descriptive Algebra. The type signature has so-called *kinds* as sorts and *type constructors* as operators. The terms of the type signature are called *types*. In the sequel, we show example specifications for the relational model and a relational execution system. Although the purpose of SECONDO is not to reimplement relational systems, it makes no sense to explain an unknown formalism using examples from an unknown data model. The structural part of the relational model can be described by the following signature:

kinds IDENT, DATA, TUPLE, REL
type constructors

	$\rightarrow$	DATA	$\underline{int}$, $\underline{real}$, $\underline{string}$, $\underline{bool}$
(IDENT $\times$ DATA)$^+$	$\rightarrow$	TUPLE	$\underline{tuple}$
TUPLE	$\rightarrow$	REL	$\underline{rel}$

Here $\underline{int}$, $\underline{real}$, $\underline{string}$, and $\underline{bool}$ are type constructors without arguments, or *constant* type constructors, of a result kind called DATA. A kind stands for the set of types (terms) for which it is the result kind. For DATA this set is finite, namely DATA = $\{\underline{int}, \underline{real}, \underline{string}, \underline{bool}\}$. In contrast, there are infinitely many types of kind TUPLE or REL. For example,

$\underline{tuple}$([(name, $\underline{string}$), (age, $\underline{int}$)])
$\underline{rel}$($\underline{tuple}$([(name, $\underline{string}$), (age, $\underline{int}$)]))

are types of kind TUPLE and REL, respectively. The definition of the $\underline{tuple}$ type constructor uses a few simple extensions of the basic concept of signature that are present in the SOS framework. For example, if $s_1, \ldots, s_n$ are sorts, then $(s_1 \times \cdots \times s_n)$ is also a sort (*product* sort), and if s is a sort, then s^+ is a sort (*list* sort). The term $(t_1, \ldots, t_n)$ belongs to a product sort $(s_1 \times \cdots \times s_n)$ if and only if each t_i is a term of sort s_i; the term $[t_1, \ldots, t_m]$, for $m \geq 1$, is a term of sort s^+ if and only if each t_i is a term of sort s. The kind IDENT is predefined (and treated in a special way in the implementation in SECONDO). Its type constructors are drawn from some infinite domain of "identifiers". Hence they can be used as attribute names here.

The notion of a relation *schema* has been replaced by a relation type, and "relation" is not considered to be a single type, but a type constructor. Hence operations like selection or join are viewed as polymorphic operations. Note that the choice of kinds and type constructors is completely left to the designer of a data model. In contrast to the CONCERTapproach, we are not offering a toolbox with a fixed set of constructors such as *tuple*, *list*, *set*, etc., but instead a framework where new constructors can be defined. Hence the SECONDO system frame as such knows nothing about $\underline{rel}$ or $\underline{tuple}$ constructors, in contrast to CONCERTwhere, for instance, the record concept requires to provide component access.[5] In summary, the terms of the type signature of an SOS specification define a *type system*, which within the descriptive algebra is equivalent to a DBMS data model.

Now the value signature is used to define operations on the types generated by the type signature. Whereas a signature normally has only a small, finite number of sorts, the second level of signature in SOS generally has to deal with infinitely many sorts. Because of this, we write *signature specifications*. The basic tool is *quantification over kinds*. We define a few example operations for the relational model above:

[5] The only predefined kinds and type constructors in SECONDO are IDENT with its "type constructors" and the type constructor $\underline{stream}$ which plays a special role in query evaluation.

operators
$\forall$ *data* in DATA.
$data \times data \rightarrow \underline{bool}$ $=, \neq, <, \leq, >, \geq$

Here, *data* is a type variable ranging over the types in kind DATA. Hence, it can be bound to any of these types which is then substituted in the second line of the specification. So, we obtain comparison operators on two integers, two reals, etc. Relational selection is specified as follows: **pipes**

$\forall$ *rel*: $\underline{rel}(tuple)$ in REL.
$rel \times (tuple \rightarrow \underline{bool}) \rightarrow \underline{rel}$ **select**

Here, $\underline{rel}(tuple)$ is a *pattern* in the quantification which is used to bind the two type variables *rel* and *tuple* simultaneously. Hence, the first argument to **select** is a relation of some type *rel*, and the second argument is a function from its tuple type to $\underline{bool}$, that is, a predicate on this tuple type. The result has the same type as the first argument. The second argument of **select** is based on the following extension of the concept of signature defined in SOS: If, for $n \geq 0$, $s_1, \ldots, s_n$ and s are sorts, then $(s_1 \times \cdots \times s_n \rightarrow s)$ is a sort (*function* sort). Furthermore,

$$\mathbf{fun}(x_1 : s_1, \ldots, x_n : s_n)\ t$$

is a term of sort $(s_1 \times \cdots \times s_n \rightarrow s)$ if and only if t is a term of sort s with free variables $x_1, \ldots, x_n$ of sorts $s_1, \ldots, s_n$, respectively.

An operator **attr** allows us to access attribute values in tuples:

$\forall$ *tuple*: $\underline{tuple}(list)$ in TUPLE, *attrname* in IDENT, *member*(*attrname*, *attrtype*, *list*).
$tuple \times attrname \rightarrow attrtype$ **attr**

Here, *member* is a *type predicate* that checks whether a pair (x, y) with $x =$ *attrname* occurs in the *list* making up the tuple type definition. If so, it binds *attrtype* to y. Hence, **attr** is an operation that for a given tuple and attribute name returns a value of the data type associated with that attribute name. Type predicates are implemented "outside" the formalism in a programming language. Precisely the same formalism (although we have not yet seen all of it) can be used to define an *executable algebra*, thereby specifying an execution system for some data model. In this case, type constructors represent data structures and operators represent query processing algorithms implemented in the system.

Commands. In the SOS framework, a *database* is a pair (T, O), where T is a finite set of *named types* and O is a finite set of *named objects*. A named type is a pair, consisting of an identifier and a type of the current (descriptive or executable) algebra. A named object is a pair, consisting of an identifier and a value of some type of the current algebra. SOS defines six basic commands to manipulate a database, regardless of the data model:

type <identifier> = <type expression>
delete type <identifier>

create <identifier> : <type expression>
update <identifier> := <value expression>
delete <identifier>
query <value expression>

A command can be given at the level of descriptive or executable algebra. In the first case, it is subject to optimization before execution; in the second, it is executed directly. In these commands, a *type expression* is a type of the current type signature, possibly containing names of (previously defined) types in the database. A *value expression* is a term of the current value signature, which may also contain constants and names of objects in the database. The **type** command adds a new named type, **delete type** removes an existing type. The **create** command creates a new object of the given type; its value is yet undefined. The **update** command assigns a value resulting from the value expression which must be of the type of the object. The **delete** command removes an object from the database. The **query** command returns a value resulting from the value expression to the user interface or application. Here are some example commands at the executable algebra level:

type city = *tuple*([(name, *string*), (pop, *int*), (country, *string*)])
type city_rel = *srel*(city)
create cities: city_rel
update cities := {enter values into the cities relation, omitted here}
query cities **feed filter** [**fun** (c: city) **attr**(c, pop) > 1000000] **consume**

More details about the SOS framework can be found in [36]. In [6], a descriptive algebra has been defined for GraphDB [37], an object-oriented data model that integrates a treatment of graphs, which shows that the framework is powerful enough to describe complex, advanced data models.

7.4.3 Architecture

Second-order signature is the formal basis for specifying data models and query languages. In this section, we present the SECONDO system frame, providing a clean extensible architecture, implementing all data-model independent functionality for managing SOS type constructors and operators, and supporting persistent object representations. Extending the frame with algebra modules results in a full-fledged DBMS. In addition to the basic commands, SECONDO provides several other commands, e.g., for transaction management, system configuration, administration of multiple databases, and file input and output.

Overview. Figure 7.9 shows a coarse architecture overview of the SECONDO system. We discuss it level-wise from bottom to top. White boxes are part of the fixed *system frame*, which is independent of the currently implemented data model. Gray-shaded boxes represent the extensible part of the SECONDO system. Their contents differ with specific database implementations.

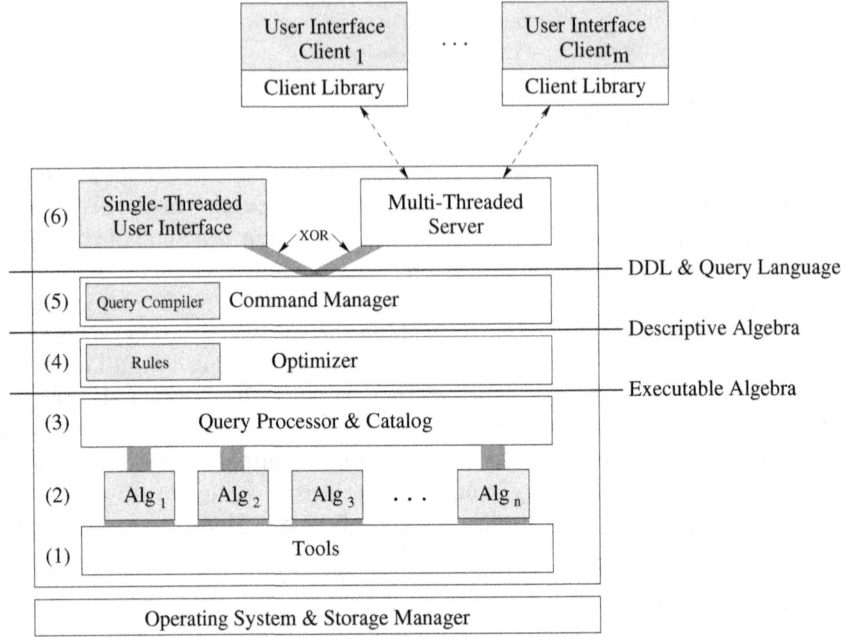

Fig. 7.9. The SECONDO architecture

The system is built on top of the Solaris operating system. Since we want to offer a full-fledged DBMS using a storage manager for dealing with persistent data is essential. In fact, we use the storage manager component of SHORE [19].

At level 1 of the SECONDO architecture, we find a variety of tools, for instance:

- *Nested lists*, a library of functions for easy handling of nested lists, the generic format to pass values as well as type descriptions.
- *SecondoSMI*, a simplified storage manager interface to the SHORE functions used most often. It can be used together with original SHORE function calls whenever the simplified functionality is not sufficient.
- *Catalog tools* for easy creation of system catalogs and algebra-specific catalogs.
- The *Tuple Manager* is an efficient implementation for handling tuples with embedded large objects.
- The *SOS Parser* transforms an SOS term to the generic nested list format used in the system.
- The *SOS specification compiler* creates the source code for the `TypeCheck` and `TransformType` algebra support functions from a valid SOS specification.

Level 2 is the algebra module level. To some extent, an algebra module of SECONDO is similar to ADTs of PREDATOR [71] or Informix datablades [47]. Using the tools of level 1, a SECONDO algebra module defines and implements type constructors and operators of an executable query algebra. SECONDO allows for implementations in C++, Modula-2, and C. To be able to use a module's types and operators in queries, the module must be registered with the system frame, thereby enabling modules in upper levels to call specific support functions provided by the module. In Figure 7.9, modules 1, 2, and n are *active* since they are connected to the frame, while module 3 is *inactive*. C++ modules are activated by linking them to the system frame. In addition, the activation of C and Modula-2 modules requires insertion of some standardized lines into the body of a predefined startup function.

Level 3 contains the query processor, the system catalog of types and objects (remember that a database is just a set of named types and named objects), and the mechanism for module registration. During query execution, the query processor controls which support functions of active algebra modules are executed at which point of time. Input to the query processor is a query plan, i.e. a term of the executable algebra defined by active algebra modules. A detailed description of query processing techniques in SECONDO, including the powerful stream concept, can be found in [22] and, on a more technical level, in [38].

The query optimizer depicted at level 4 transforms a descriptive query into an efficient evaluation plan for the query processor by means of transformation rules. For each algebra module, the database implementor provides a corresponding set of rules as well as algebra support functions supplying information on estimated query execution costs.

The command manager at level 5 provides a procedural interface to the functionality of the lower levels. Depending on the command level, the query (or other command) is passed either to the query compiler, provided by the database implementor, to the optimizer or to the query processor.

At level 6 we find the front end of a SECONDO installation, providing the user interface. In general, there are two mutually exclusive alternatives: either the user interface is linked with the frame and active algebra modules to a self-contained program, or the SECONDO process is made a server process serving requests of an arbitrary number of client processes which implement the user interfaces. In the first case, SECONDO is a single-user, single-process system, while in the latter case SECONDO is a multi-user capable client-server system, exploiting the multi-threaded environment offered by SHORE. To support the implementation of user clients, SECONDO provides comfortable client libraries for C++ and Java.

Algebra Modules. An algebra module has to implement a set of support functions for all of its type constructors and operators. Figure 7.10 lists all support functions for type constructors implemented in algebra modules. In addition to these functions, also the type constructor name is passed to the system frame. During query processing, the `In` function is used to convert an external value (in nested list form), given as part of an input file or in an interactive query command, to its corresponding internal value. The internal value typically is a catalog index essentially referencing either a main memory pointer or a persistent identifier. The `Out` function is the reverse of the `In` function, producing external values for user interfaces or output files.

`In/Out`	Conversion from external to internal value representation and vice versa.
`Create/Delete`	Allocate/deallocate memory for internal value representation.
`TypeCheck`	Validation of type constructor applications in type expressions.
`InModel/OutModel`	Conversion from nested list to internal model representation and vice versa.
`ValueToModel`	Computes a model for a given value.

Fig. 7.10. Support functions for type constructors

The `Create/Delete` pair of support functions is used to allocate memory for a default value of the given type or to delete an object of that type, respectively. Default value creation is performed by the query processor to reserve space for single stream elements, thereby avoiding multiple allocation and deallocation of memory for stream elements of a common stream, which are never accessed simultaneously. The `Delete` function is called later on to deallocate this default memory space as well as all intermediate results generated while processing a query.

The `TypeCheck` function is called whenever a new database object is created to check whether its type conforms to the signature of the underlying data model.

Furthermore, for each type constructor a *model* may be registered which is a data structure containing summary information about a value of the type. The model is a place to keep statistical information such as expected number of tuples, histograms about attribute value distribution, etc. For maintaining models there are three support functions `InModel`, `OutModel`, and `ValueToModel`, as listed in Figure 7.10.

Figure 7.11 presents all support functions for operators. For each operator, its name and the number of `Evaluate` functions are also passed to the system frame. During query execution, the query processor calls the `TransformType` function for type checking. `Select` is needed for the resolution of overloaded operators. The `Evaluate` function(s) do(es) the "real work" by computing an operator's result value from its argument values.

`TransformType`	Computes the operator's result type from given argument types.
`Select`	Selects the correct evaluation function in case of overloading by means of the actual argument types.
`Evaluate`	Computes the result value from input values. In case of overloading, several evaluation functions exist.
`MapModel`	Computes the result model from argument models.
`MapCost`	Computes the estimated cost of an operator application from argument models and costs.

Fig. 7.11. Support functions for operators

Concerning cost estimation, there are two support functions called `MapModel` and `MapCost`. For an operator application, `MapModel` takes the models of the operator's arguments and returns a model for the result. `MapCost` also takes the models of the operator's arguments as well as the estimated costs for computing the arguments and, based on these, estimates and returns the cost for the operator application. By calling these support functions, the optimizer can estimate properties of intermediate results and the cost of query plans or subplans. Furthermore, there is a startup routine for each algebra module which is used to associate type constructors with the kinds containing them, to perform initializations of global arrays, etc.

The registration mechanism for support functions differs from the implementation language. Registration is most comfortable in C++: For each type constructor, an instance of the predefined class `TypeConstructor` must be defined, passing operator support functions as constructor arguments. The same happens with operators and a predefined class `Operator`. For a complete algebra, an instance of a class derived from the predefined class `Algebra` is defined. The constructor of this class is the startup routine of the modules.

Algebra modules need not only to cooperate with the system frame, but also with other algebra modules. Modules implement certain signatures. At the type level, *kinds* are the functions between different signatures. For instance, each type in kind DATA will be a valid attribute type for the *tuple* type constructor. Thus, a type constructor *polygon* is made a type constructor for attribute types by simply adding *polygon* to the type constructors for DATA.

At the implementation level, the interface between system frame and algebra modules does not impose any specific inter-module interaction conventions on the algebra implementation, but rather the algebra implementor is free to define the protocol for interaction with type constructors and operators of his algebra. For C++ implementations there is a general strategy, based upon the inheritance and virtual method mechanisms provided by C++, which allows one to define generic interfaces between modules in a uniform manner as follows.

The basic observation is that the relationship between kinds and type constructors corresponds to the relationship between base classes and derived classes. For each kind K, the algebra module *alg* requiring an interface to values of types in K defines an abstract base class *k_base*. For the implementation of operators in *alg*, typically some support functions for dealing with values of kind K will be necessary. Just these support functions are defined as abstract virtual methods of *k_base*. Whenever a class *tc* in any algebra module is defined to implement a type constructor in kind K, *tc* must be derived from *k_base*: `class tc : public k_base`. For instance, the base class `Data` corresponding to kind DATA contains a virtual method `Compare` which has to be defined within all attribute data type implementations, thereby enabling the generic implementation of the `sort` operator of the relational algebra module.

SECONDO *versus Concert.* A common goal of SECONDO and Concert is to support an easy and efficient implementation of new application-specific data types.

In particular, both systems provide facilities for managing collections of related types, namely *kinds* in SECONDO and concepts in Concert.

However, the facilities offered differ as follows. SECONDO enables a database implementor to define arbitrary new type constructors, operators, and, through the notion of *kinds*, sets of type constructors for data types with common properties. The system frame does not implement any particular data model, but rather provides a generic query processor and powerful tools to implement type constructors and operators.

Concert *concepts* are similar to SECONDO kinds in two major aspects. First, concepts group sets of data types with common properties, and second, a concept sets up a programming interface, which has to be implemented by those new data types that shall fit the concept. That way, general processing techniques are implementable for groups of data types rather than specific types only. In Concert, this is basically exploited for indexing.

As opposed to SECONDO kinds, Concert concepts are hardwired with the system. A database implementor is neither able to add new concepts nor to provide further implementations of concept-specific index structures and operators. On the other hand, Concert concepts already cover a broad range of requirements of new data models, and hardwired implementations are often slightly more efficient than those registered through a general interface, because the adaptation to details of other modules in the system is easier.

Regarding SECONDO, introducing a new kind naturally requires some effort to implement the type constructors and operators applicable to these kinds. However, with a growing number of algebra module implementations the probability grows that a database implementor finds kinds with their operators and type constructors already implemented that fit pretty well the requirements of new data types.

In summary, Concert offers a higher degree of implemented functionality than SECONDO. On the other hand, SECONDO is more flexible with respect to new data models since it does not hardwire a fixed number of concepts, but is open for arbitrary "concepts". In fact, SECONDO could be used to implement the Concert approach by defining all concepts as kinds and index structures as type constructors applicable to these kinds.

7.4.4 Implementing Spatio-temporal Algebra Modules

Because of its generality, particularly with respect to type constructor definitions, SECONDO is very well suited to the implementation of spatio-temporal data models. In fact, the signatures given in Sections 4.3 and 4.4 (see Chapter 4) defining the abstract and the discrete data model, are genuine second-order signatures. Thus, the `TypeCheck`, `TransformType`, and `Select` support functions can be implemented in a straightforward manner. Later on, they will even be generated automatically by the SOS specification compiler.

Apart from the type level, a database implementor has to provide *operator implementations* as well as efficient *memory representations* of data type instances. In doing so, he is supported by SECONDO's implementation tools. Implementors of spatial data types often encounter the problem that instances of a given type might vary in size heavily. For example, a *polygon* value may consist of only three vertices as well as many thousands of them. SECONDO's *tuple manager* (see Section 7.4.3) can be used to handle tuples with embedded attribute values of varying size in an efficient and comfortable way. For that purpose it offers a new large object abstraction called *FLOB* (Faked Large OBject). FLOBs are implemented on top of the large object abstraction of the underlying storage manager. Depending on a FLOB value's size and access probability, it is either stored as a large object or embedded within the tuple [23].

Allowing for type constructors arbitrarily nesting other types, however, requires implementation techniques beyond those of basic FLOBs. Consider a data type that arises from the application of a type constructor to some other type, e.g., *list*(*polygon*). The representation of such a data type needs to organize a set of representations of values of the argument types. If the argument types employ FLOBs themselves, we immediately arrive at a tree of storage blocks which may be small or large, i.e., a *FLOB tree.* The generalized problem is then to determine an efficient storage layout or *clustering* of the tree, i.e., a partitioning into components stored as large objects. Thus, we extended the basic FLOB functionality, offering *nestable FLOBs*, so supporting a direct and elegant implementation of type constructors [24].

Due to the aspired generality of our clustering tool, we cannot exploit specific schema information to decide on a good clustering. Instead, we consider a FLOB's *size* and *access probability*, as well as the general *FLOB access speed* parameters of the underlying storage manager and operating system. In [24], we deduced a *rank function* $R(C, f)$ that for a given cluster C and a FLOB f not yet assigned to any cluster returns a measure indicating whether it is more efficient to insert f into C or to store it in another large object.

Our tool uses the function R in an algorithm that finds a clustering of a FLOB tree in time linear in the number of FLOBs as follows. The root FLOB starts a new cluster. Then, each son of the root either is included into the root's cluster or starts a new cluster, according to the return value of R. This is done recursively for the entire tree. Even though this algorithm cannot guarantee to return an optimal solution, in all tests performed in an experimental comparison with other algorithms it returned high-quality results.

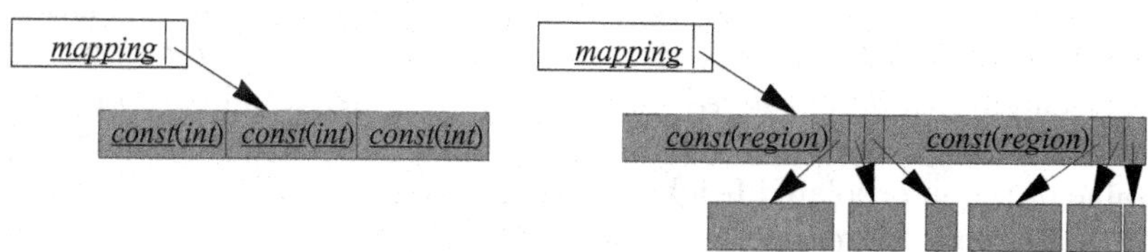

Fig. 7.12. Storage layouts for mapping values

As an application example, consider the type constructor <u>*mapping*</u> presented in Section 4.4. The implementation of this type constructor basically provides a persistent *dynamic array* of *unit* values. As the size of the array differs with varying instances of <u>*mapping*</u>, it is implemented as a FLOB. Now the level of nesting depends on the argument types of the mapping, as shown in Figure 7.12. In the case of <u>*const*</u>(<u>*int*</u>) values, there is only one level of nesting. However, in the case of <u>*const*</u>(<u>*region*</u>) values, which are in turn implemented using FLOBs for the different polygons defining a region, another nesting level arises. For database implementors, all shaded boxes in Figure 7.12 are distinct large objects. The much more efficient physical clustering set up by the SECONDO tuple manager is completely hidden to them, so that their focus of attention is left on data type functionality issues.

In summary, SECONDO with its concept of extensibility by algebra modules, fits very well the designs of spatio-temporal algebras presented in Chapter 4 and is an excellent environment for implementing them. Beyond extensibility, it offers powerful tools for managing large objects occurring in such designs, and even supports type constructors organizing collections of such large objects.

7.5 The Dedale Prototype System

7.5.1 Introduction

The DEDALE prototype [32,31] follows the *extensible approach*. It is one of the first implementations of a database system based on the linear constraint model (see Chapter 5). It is intended to demonstrate the relevance of this model to handle geometric applications in areas such as spatial or spatio-temporal applications. One of the interesting features of DEDALE is its ability to represent and manipulate multi-dimensional point-sets as linear constraint relations. [32,31] gave examples of spatio-temporal applications of DEDALE. However, the major contribution of DEDALE to spatio-temporal databases lies on its recent extension to handle *interpolated* spatial data [33]. This model captures the class of geometric objects embedded in a d-dimensional space such that one of the attributes can be defined as a function of a subset of the other attributes. Moreover, this function can be obtained as a *linear* interpolation based upon some finite set of sample values.

DEDALEcovers, among other applications, *moving objects.* Indeed, a trajectory can be represented by a sample of points with time and position. If we make the reasonable assumption that the speed is constant on each segment, the full trajectory can then be recovered from these points using linear interpolation. As another example of interpolated data, we can mention *elevation data* which can be represented by a *Triangulated Irregular Network*, i.e., a finite set of points P along with their elevation. An interpolation based on a triangulation of P gives the value of the interpolated height at any location.

The data model proposes, in the spirit of the constraint data model, to see interpolated pointsets as infinite relations and to express queries on these pointsets with the standard SQL language. It also includes an accurate finite representation for interpolated data and an algorithm to evaluate queries at a low cost. We currently proceed with the implementation of the model in DEDALE, as explained in the sequel.

We begin with a brief description of the model (a detailed presentation of the constraint model can be found in Chapter 5). All the remainder of the presentation is devoted to the physical aspects of spatio-temporal data management in DEDALE. We first describe the architecture of the prototype and then give detailed explanations on data storage, indexing and query processing. A simple example summarizes the presentation.

7.5.2 Interpolation in the Constraint Model: Representation of Moving Objects

The constraint model can be used efficiently to model both interpolated and non-interpolated data, and allows to query interpolated and non-interpolated data uniformly without the need of specific constructs.

Let us illustrate this with two examples. In case of elevation data, assume that a partition of the plane into triangles T_i is given together with the height of each of the triangles' summits. The interpolated height h of a point p in the plane is defined by first finding i such that T_i contains p. The h value is linearly interpolated from the three heights of the summits of T_i. This latter function depends only upon i and can be defined as a linear function $f_i(x, y)$, valid only for points in T_i. There is a very natural and simple symbolic representation for the three dimensional relation TIN in the linear constraint model:

$$TIN(x, y, h) = \bigvee_i \; t_i(x, y) \wedge h = f_i(x, y)$$

where $t_i(x, y)$ is the semi-linear symbolic representation of the triangle T_i (as a conjunction of three inequalities).

In case of moving objects assume that the position of an object is known at finitely many time points. This defines finitely many time intervals T_i. If the speed of an object is assumed to be constant during each time interval, its position at any time t has coordinates $x = v_i t + x_i$ and $y = w_i t + y_i$ where i is the index of the interval T_i which contains t, v_i, w_i the speed on the axis of the object during that interval, and (x_i, y_i) are chosen appropriately so that the position of the object is correct at the beginning and end of the interval. Thus, the object trajectory $TRAJ(x, y, t)$ can be represented in the linear constraint model as follows:

$$TRAJ(x, y, h) = \bigvee_i \; t_i(t) \wedge x = f_i(t) \wedge y = g_i(t)$$

where $t_i(t)$ are the constraints defining the time interval T_i and f_i, g_i are the linear equations mentioned above. It is easy to see that interpolated relations have similar properties in both cases:

- A subset of attributes which is used as a basis to compute the interpolated values: (x, y) for terrain modeling, t for the trajectories. We denote it as the *key* in the sequel.
- A disjunction of conjuncts, each consisting of equality or inequality constraints restricted to the attributes of the *key*, t_i, and constraints on the interpolated value defined as linear functions on the *key*, f_i and g_i.

This defines a normal form to represent interpolated relations. An important aspect of the model is that interpolated relations can be seen, from the user's point of view, as classical relations, and queried by means of standard query languages. Moreover, it can be shown that evaluating queries upon interpolated databases can be done by manipulating only the key of interpolated relations, while interpolated functions remain identical. A practical consequence is that the cost of query evaluation does not depend on the dimension of the embedding space but on the dimension of the key.

In summary, the model allows to express SQL queries on infinite relations, finitely represented with linear constraints. The formulae which represent objects, such as moving objects, have the specific form outlined above, and this permits to use efficient geometric algorithms for query evaluation.

7.5.3 Architecture

Figure 7.13 depicts the architecture of the DEDALE system. A data server is in charge of data storage and query processing while Java clients propose a graphical interface to express queries and visualize geometric data.

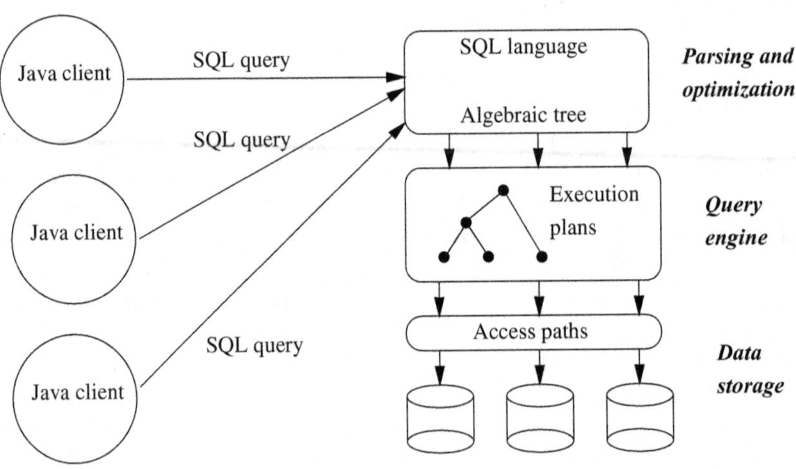

Fig. 7.13. The DEDALEprototype architecture

Our implementation is built on the standard technology for data storage, database indexing, and for the parsing and optimization of the SQL query language.

Since, however, the storage provides a finite representation of conceptually infinitely many points, the query evaluation process features specific algorithms which simulate, upon the finite representation, the semantics of the regular SQL language over infinite sets.

For this part of query processing, we rely on a constraint manipulation engine as already implemented in the first prototype of DEDALE[31]. Note that a vector-based representation could be used as well, but the constraint representation offers a nice framework for the symbolic manipulation of data.

The data storage and query engine levels are built on the BASIS System [34], an extensible C++ implementation of the classical storage management functionalities, as well as an open query processor based on an iteration query execution model, to be described below.

7.5.4 Implementation Details

The implementation mainly aims at showing that the proposed data model for interpolated data can be integrated in a relational DBMS with minimal effort.

Buffer Management and Data Storage. The storage manager provides standard I/O and caching services to access a *database.* A database is a set of binary files which store either datasets (i.e., sequential collection of records) or Spatial Access Methods (SAM). A SAM or index refers to records in an indexed data file through *record identifiers.* Any binary file is divided into *pages* with size chosen at database creation.

The buffer manager handles one or several *buffer pools.* Data files and indexes are assigned to one, global buffer pool. Some operators (e.g. sort and hash) may require a specific buffer pool to the buffer manager. The buffer pool is a constant-size cache with LRU or FIFO replacement policy (LRU by default).

Datasets are stored in sequential files with constant-size records. Since DEDALE proposes a nested data model with only one level of nesting (geometric data are sorted as nested relations), indirection is used to store geometric data. The *main file* stores records with alphanumeric attributes. Spatial attributes are referred to via a `GeomRef` pointer which consists of (i) the bounding box of the spatial object and (ii) the physical address in the *geometric file* where this object can be found.

Although the processing of queries uses a constraint representation for geometric data, at the *data storage* level, we use a vector-based representation. For instance, TINs and trajectories are respectively stored as sequences of triangles and segments in the 3D space. The vector representation is compact and suitable for many basic tasks such as indexing and graphical display. A constant time conversion allows to get a constraint representation from the vector-based representation.

Indexing is based on R*-trees [7]. Since our pointsets are embedded in the 3D space, we could have used a 3D R-tree. However recall that the distinction between the key attributes and the interpolated attributes is semantically meaningful: a query may separately address the multidimensional domains represented by the key or interpolated variables, respectively. We believe that most queries need only the support of an index on a subset of the variables. We therefore chose to use separate indexing for key attributes and interpolated attributes. For instance, a TIN is indexed on (x, y) by a 2D R*-tree (i.e, we index the bounding rectangle of triangles), and on the h value with a 1D R*-tree (by indexing the intervals). This approach is not the best one in the presence of mobile objects, since rectangles are bad approximations of segments.

The Query Engine. The query engine is based on the pipelined query execution model described in [28]. In this model, query execution is a demand-driven process with iterator functions. Each operator in the query engine is an iterator, and iterators are grouped to form execution plans represented by trees as usual. This allows for a pipelined execution of multiple operations: a tuple goes from one tree node to another, from the disk up toward the root of the query tree, and is never stored in intermediate structures.

Another advantage of the iterator approach is the easy QEP creation by assembling iterators existing in BASIS and the easy extension of BASIS by a trivial integration of a new iterator in its library. All execution plans are *left-deep trees* [28]. In such trees the right operand of a join is always an index, as well as the left operand for the left-most node. The leaves represent data files or indexes, internal nodes represent algebraic operations and edges represent dataflows between operations. Examples of algebraic operations include data access (`FileScan` or `RowAccess`), spatial selections, spatial joins, etc. This simple scheme permits to use the simple indexed nested-loop join algorithm. This strategy is not always optimal, in particular with spatial joins [52,61], but our first goal was model validation rather than query processing optimization.

Query Processing on Interpolated Data. First *disk operators* retrieve the vector representation from files or indexes: "records" are vector-based lists of points. Typical disk operators perform a file or an index scan. They output data into linear constraints, thanks to a trivial conversion. The second category consists of *symbolic operators* which implement the constraint-solving algorithms upon the constraint-based representation. At this level a "record" is a conjunction of linear constraints, representing either a block, or the intermediate structure constructed from several blocks during a join operation.

This design introduces a new level between the physical and abstract representations called *symbolic level* (see Figure 7.14) and based on the linear constraint representation. Note that we could build the system with algorithms working on vector data. However the constraint approach has two advantages. First it provides a uniform representation of interpolated geometric data: both keys and interpolation functions are uniformly represented as linear constraints over a set of variables. Second, such a representation provides a nice support for the specific algorithms, required in our data model, and based on variable substitution.

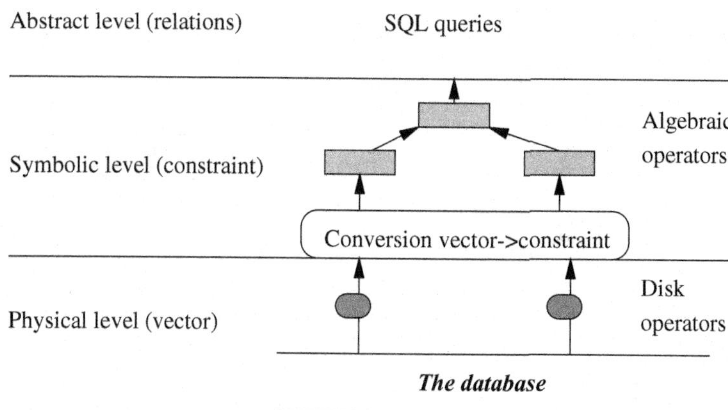

Fig. 7.14. The representation levels

7.5.5 Example of Query Evaluation

In the following, we illustrate the process of query evaluation with the query "*Select in the trajectory of a moving object the points on a TIN with elevation above 10000 meters*". The schema of the database consists of a relation TIN which stores elevation data describing the height above sea, and a relation TRAJ which describes the trajectory of the moving point.

The two point-sets TIN and TRAJ are seen at the query level as classical, although infinite, relations. Hence the query can be expressed with SQL as:

```
SELECT t1.x, t1.y, h
FROM   TIN t1, TRAJ t2
WHERE  t1.x = t2.x AND t1.y = t2.y AND t1.h >= 10000;
```

The data storage consists respectively of a list of points describing (in the 3D space) a trajectory, and a list of triangles describing (also in the 3D space) the TIN. A spatial join between TIN and TRAJ is at the core of the query evaluation. The join is evaluated by an indexed-nested loop join as shown in Figure 7.15. The *trajectory* relation is scanned sequentially, and the bounding box of each segment is used as an argument to carry out a window query on the R-tree indexing the TIN relation. These operations are implemented by *disk* operators. Data is read in a vector format and converted by the join node into a constraint representation. One obtains, at each step, a pair of symbolic tuples of the form shown in left side of Figure 7.16.

The conjunction of constraints from each pair gives a new symbolic tuple with variables x, y, t (see right side of Figure 7.16) as follows. Since x and y are linear functions of t, and h is a linear function of (x, y), a simple substitution of variables is possible. x and y are replaced by the proper function of t in all the inequality constraints. As a result, one obtains x, y, and h as functions of t (which is the *key* of the result), and a somewhat complex conjunction of constraints over the single variable t.

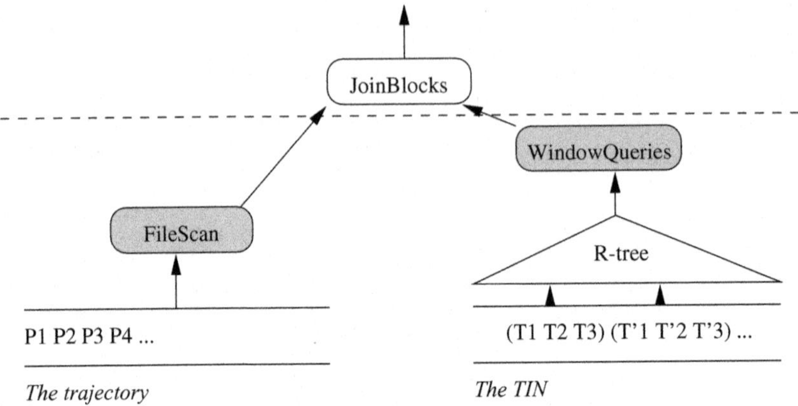

Fig. 7.15. A sample query evaluation

From $TRAJ$	From TIN
$0 \leq t \leq 2$	$-4x + y + 1 \leq 0$
$x = 2t + 1$	$y - 3 \leq 0$
$y = t + 2$	$3x + y - 18 \leq 0$
	$h = x + y + 3$

The New Symbolic Tuple
$0 \leq t \leq 2$
$-4(2t + 1) + (t + 2) + 1 \leq 0$
$(t + 2) - 3 \leq 0$
$3(2t + 1) + (t + 2) - 18 \leq 0$
$x = 2t + 1$
$y = t + 2$
$h = (2t + 1) + (t + 2) + 3$

Fig. 7.16. Joining two symbolic tuples

The process ends up by *normalizing* the result, i.e., applying a constraint solving algorithm which checks whether the representation is consistent, and delivers a normalized representation (see [31] for a detailed presentation of the algorithms). This allows to solve the system of equations on t while keeping the other constraints unchanged. The result is depicted in Figure 7.17. For this query, the result is obtained in linear time in the size of this system. The constraint solving algorithm delivers a time interval and linear functions for x, y, and h.

The final result
$0 \leq t \leq 1$
$x = 2t + 1$
$y = t + 2$
$h = 3t + 6$

Fig. 7.17. Final result

In conclusion, the data storage and query engine are currently available, as well as a simplified version of the algorithms presented in [33], restricted to the manipulation of TINs and mobile objects. The parser and the optimizer are under implementation. Our experimental setting consists of several TINs, ranging from 2000 to 20000 triangles. We plan to use the GSTD generator [76] when it is available to create a consistent sample of mobile objects.

7.6 The Tiger Prototype System

7.6.1 Introduction

TIGER is a temporal database system prototype, which can be extended for spatio-temporal data access [15]. This layered system adopts the dual system architecture (cf. Section 7.2.1). Its salient features include enhanced temporal support through statement modifiers, the intelligent reuse of existing database technology, and the seamless extension of database systems with external modules. Specifically, extensibility is supported by enabling users to plug external modules into the layer. Such modules take spatio-temporal relations as arguments and perform advanced operations on these, the objective being to obtain better overall query performance.

The system is available online via `http://www.cs.auc.dk/~tiger`. Because the web interface uses socket communication, users protected by a firewall might not be able to use it. The source code is therefore also available for downloading and local installation.

7.6.2 Architecture

TIGER follows the dual system architecture. It is implemented as a layer to the Oracle DBMS and systematically enhances that system with temporal functionality. Whenever feasible, the data processing is delegated to the DBMS. However, if functionality and efficiency concerns make the processing by the database system inappropriate, external modules are used instead. External modules seamlessly extend the functionality of the database system. Existing modules perform advanced functions such as coalescing, temporal aggregation, and temporal difference.

The general architecture, depicted in Figure 7.18, identifies three main parts. At the bottom, the DBMS is used as an enhanced storage manager. It is responsi-

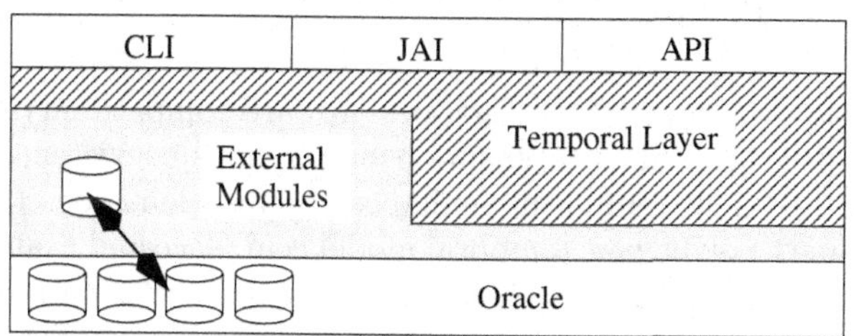

Fig. 7.18. The general layered system architecture of TIGER

ble for all standard database activities. The middle part consists of the temporal layer and the external modules. Together, they provide the core of the temporal functionality. At the top, several interfaces are provided. This includes a command line interface (CLI), a Java applet interface (JAI), and an application program interface (API).

7.6.3 Spatio-temporal Extensions

The TIGER system focuses on temporal database extensions. However it can also be extended for spatio-temporal data access, as can be seen with the query language STSQL [15].

This section briefly sketches the basics of ATSQL—the temporal query language hitherto supported by TIGER. The crucial concept of ATSQL is *statement modifiers*. These control the basic semantics of statements. Four classes of statements are distinguished: upward compatibility, temporal upward compatibility, sequentiality, and non-sequentiality. Each class is described below. For clarity, new syntactic constructs are underlined.

The meaning of a statement modifier naturally divides into four orthogonal parts, namely the specification of the statement class, the time-domain specification, the time-range specification, and the specification of coalescing. We focus on the classes and on coalescing; the reader is referred to [14] for a detailed coverage of domain and range specifications.

Upward Compatibility. It is fundamental that all code without modification will work unchanged with the new spatio-temporal system. A data model is upward compatible with another data model iff all the data structures and legal query expressions of the old model are contained in the new model and iff all queries expressible in the old model evaluate to the same results in the new model. The following statements illustrate upward compatibility:

```
CREATE TABLE p (A INTEGER);
INSERT INTO p VALUES (7);
INSERT INTO p VALUES (8);
COMMIT;
```

These statements are simple legacy SQL–92 statements that must be supported by any reasonable temporal extension of SQL–92. The semantics is the one dictated by SQL–92 [53].

Temporal Upward Compatibility. Temporal upward compatibility, which is easily extended to space [15], aims to ensure a harmonious coexistence of legacy application code and new, temporally-enhanced application code. To illustrate the problem, assume that the new temporal model is in place and that an application needs temporal support, for which reason a snapshot relation must be changed to become a temporal relation. Clearly, it is undesirable (or even impossible) to change the legacy application code that accesses the snapshot relation that has become temporal. Temporal upward compatibility ensures that this is unnecessary. Essentially, tables can be rendered temporal without changing application code. To illustrate, the statements from the previous section are assumed:

```
ALTER TABLE p ADD VT;
INSERT INTO p VALUES (6);
DELETE FROM p WHERE A = 8;
COMMIT;
SELECT * FROM p;
```

The first statement extends p to capture valid time by making it a valid-time table, which contains a timestamping attribute. The insert statement adds 6. Temporal upward compatibility ensures that the past is not changed and that 6 will also be there as time passes by. Similarly, 8 is deleted without changing the past. The select statement returns current, but not past (and future), knowledge. Note that it may not return the valid time.

Sequentiality. Sequentiality protects the investments in programmer training while also providing advanced temporal functionality. Two properties are crucial: snapshot reducibility and interval preservation.

Briefly, snapshot reducibility implies that for all non-temporal queries q, a temporal query q' exists, such that at each snapshot, the result of the temporal query reduces to the result of the original query. For snapshot reducibility to be useful, the relationship between the non-temporal and the temporal query has to be restricted. We require $q' = S_1\, q\, S_2$ where S_1 and S_2 are constant (statement independent) strings. The strings S_1 and S_2 are termed *statement modifiers* because they change the semantics of the entire enclosed statement. ATSQL prepends statements with, e.g., the statement modifier SEQ VT:

```
SEQ VT SELECT * FROM p;

CREATE TABLE q (B INTEGER SEQ VT PRIMARY KEY) AS VT;

SET VT PERIOD "1974-1975" INSERT INTO q VALUES (6);
SET VT PERIOD "1976-1978" INSERT INTO q VALUES (6);
SET VT PERIOD "1977-1979" INSERT INTO q VALUES (6);

SEQ VT SELECT * FROM q;
```

The first and last query return all tuples together with their valid time. This corresponds to returning the content of a table at each state. The second statement defines a table q and requires column B to be a sequenced primary key, i.e., B must be a primary key at each state (but not necessarily across states). This constraint implies a conflict between the second and third insert statements: allowing both would violate the primary key constraint for the years 1977 and 1978.

Beyond snapshot reducibility, sequentiality is also preserves the intervals of the argument relations as much as possible in the results. Consider Figure 7.19.

Electricity Bill

Val	VT
150	1993/01–1993/03
150	1993/04–1993/06
70	1993/07–1993/09

Electricity Bill

Val	VT
150	1993/01–1993/06
70	1993/07–1993/09

Fig. 7.19. Snapshot-equivalent relations

Assuming that *Val* denotes the amount to be paid for electricity during the specified period, the two relations are quite different. Simply merging or splitting the intervals would be incompatible with the intended semantics. Sequentiality preserves intervals as much as possible, i.e., within the bounds of snapshot reducibility [13].

Finally, sequentiality also includes queries of the following type:

```
SEQ VT SELECT * FROM p, q
  WHERE p.X = q.X AND DURATION(VTIME(p),YEAR) > 5;
```

The query is quite natural and easy to understand. It constrains the temporal join to p-tuples with a valid time longer than 5 years. The temporal condition cannot be evaluated on individual snapshots because the timestamp is lost when taking a snapshot of a temporal database, and it thus illustrates how sequentiality extends snapshot reducibility to allow statement modifiers to be applied to all statements [14].

Non-sequentiality. As discussed above, sequenced statements are attractive because they provide built-in temporal semantics based on the view of a database as a sequence of states. However, some queries cannot be expressed as sequenced queries. Therefore, a temporal query language should also allow *non-sequenced* queries with no built-in temporal semantics being enforced. ATSQL uses the modifier `NSEQ VT` to signal non-sequenced semantics, i.e., standard semantics with full explicit control over timestamps:

```
NSEQ VT SELECT * FROM p, q
 WHERE VTIME(p) PRECEDES VTIME(q) AND A = B;
```

The query joins p and q. The join is not performed at each snapshot. Instead, it is required that the valid time of p precedes the valid time of q. A non-temporal relation results.

Coalescing. Coalescing merges tuples into a single tuple if they have overlapping or adjacent timestamps and identical corresponding attribute values [16]. Coalescing is allowed at the levels where the modifiers are also allowed. In addition, as a syntactic shorthand, a coalescing operation is permitted directly after a relation name in the from clause. In this case, the coalesced instance of the relation is considered.

```
SEQ VT SELECT * FROM q;
(SEQ VT SELECT * FROM q)(VT);
SEQ VT SELECT * FROM q(VT);
```

7.6.4 Tiger's Implementation

TIGER is an heterogeneous system that employs different programming paradigms. Specifically, the code is a mixture of C++ (81 KB), C (14 KB), Java (88 KB), and Prolog (160 KB). The implementation of TIGER is based on the architecture of ChronoLog [11], a temporal deductive database system, which uses a layer to translate temporal FOPL to SQL. Another related system is TimeDB [12]. TIGER is the first system that employs external modules, is accessible online and can be used for distance learning.

The temporal layer consists of thirteen modules as illustrated in Figure 7.20. An arrow indicates the direction of an import (e.g., module `meta` imports services from module `unparser`). The layer translates temporal statements to sequences of legacy SQL statements. It adheres to standard compiler implementation techniques [20]. We briefly discuss the prime functionalities associated with the main modules.

Module `interpret` acts as a dispatcher. It calls the scanner and parser to construct a parse tree. A conservative approach is pursued to not break legacy code. Thus, if a statement cannot be identified as being temporal and if the statement does not access temporal data structures, it is passed on to the DBMS. Modules `rewrite` and `deps` normalize and check statements. This includes the rewriting of subqueries, the verification of relation schemas, the lookup of missing table qualifiers, and the detection of dependencies implied by views and integrity constraints. Module `trans` translates temporal queries to sequences of non-temporal ones. For involved temporal statements, this includes the generation of calls to external modules. Data manipulation statements, views, and integrity constraints are handled in the modules `dml`, `views`, and `constraint`, respectively. Module `meta` provides a general-purpose interface to the DBMS and a special-purpose interface to the layer's additional temporal metadata, which is stored in the DBMS.

The purpose of the external modules exists to aid in the computation of queries that the underlying DBMS computes only very inefficiently. For example, although it is possible to perform coalescing, temporal difference, and temporal aggregation in the DBMS, this is exceedingly inefficient [16] and should be left to external modules. External modules fetch the required data from the DBMS,

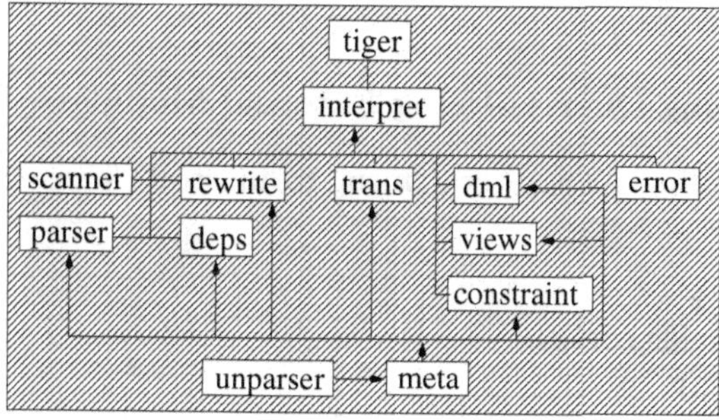

Fig. 7.20. The temporal layer

process it, and store the result back for further processing. To enable the use of external modules, the temporal layer isolates the subtasks that should be delegated to the modules, and an interface between the temporal layer and external modules is provided that consists of a set of procedures of the following form:

```
EM_coal_vt(char* sql, char* table)
EM_coal_tt(char* sql, char* table)
EM_coal_bi_tt(char* sql, char* table)
EM_coal_bi_vt(char* sql, char* table)
EM_diff_vt(char* sql, char* table)
```

All procedures take as input an SQL statement that defines the argument data and return the name of a temporary table that stores the computed result. Additional input parameters can be passed as needed.

The different components of the external module block are shown in Figure 7.21. The DB class handles connections to the DBMS. This service is used by

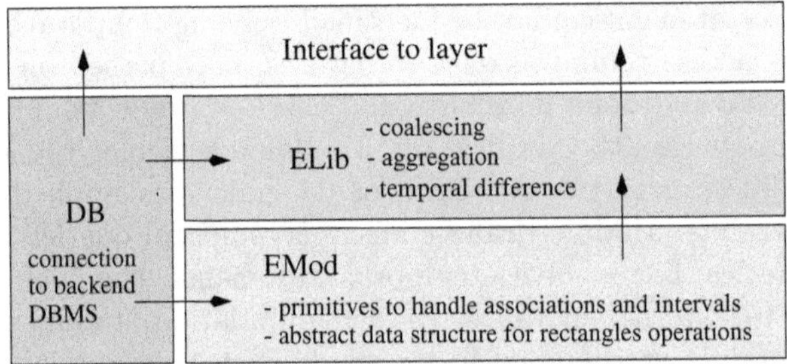

Fig. 7.21. The external modules

all other modules. The EMod class implements an abstract matrix structure that is used to implement advanced bitemporal algorithms. Essentially, the matrix keeps track of a set of rectangles. Together with an association structure, the matrix is used to merge rectangles (coalescing) or to split them (difference, aggregation). A coloring scheme is used to ensure interval preservation. The ELib class uses the services of the EMod and DB classes to provide external algorithms for coalescing, difference, and aggregation.

7.6.5 Processing Queries Using External Modules—Case Study

We use the following ATSQL statement to illustrate processing that includes external modules.

```
SEQ VT SELECT r.a
  FROM (SEQ VT SELECT p.a
          FROM p, q
          WHERE p.a = q.a )(VT) AS r
  WHERE VTIME(r) OVERLAPS PERIOD "10-20";
```

An external module is to perform the valid-time coalescing that occurs in the middle of the statement. Thus, the statement has to be split up. We also want to maximally use the underlying DBMS (e.g., for processing joins), thus using only the layer when the DBMS cannot be used with reasonable efficiency.

The innermost part is a sequenced valid-time join, which computes the intersection of overlapping valid times. The innermost part is translated into an SQL statement Q1 (next), which performs the necessary intersection of the valid times (Oracle's SQL with functions GREATEST and LEAST is used).

```
Q1 = SELECT GREATEST(P.s,Q.s) s, LEAST(P.e,Q.e) e, P.A A
     FROM P, Q
     WHERE GREATEST(P.s,Q.s)<=LEAST(P.e,Q.e) AND P.A=Q.A;
```

Next, the external module takes over, and Q1 is passed as the first argument to EM_coal_vt(char* sql, char* table) (cf. above), which implements the coalescing operation. An auxiliary table is first created:

```
CREATE TABLE T (s DATETIME, e DATETIME, a INTEGER);
```

Then the tuples identified by Q1 are retrieved. They are ordered so that coalescing can be performed on the fly [16].

```
SELECT A, s, e
FROM (Q1)
ORDER BY 1, 2, 3;
```

The coalesced tuples are computed and stored in the temporary table, the name of which is finally returned by the external module. The DBMS then takes over and finishes the computation, using the data stored in the temporary table:

```
SELECT R.s s, R.e e, R.A A
FROM (SELECT T.s, T.e, T.A FROM T) R
WHERE r.e > 10 AND 20 > r.s;
```

7.7 The GeoToolKit Prototype System

7.7.1 Introduction

A subsequent analysis of geoscientific and particularly geological domains showed that they use common data sources (3D data, cross sections, digital elevation models etc.) to a large extent, utilizing them for special objectives [3]. They also share a lot of functionality primarily concerning spatial data management and geometric operations. The idea to reuse already implemented components (sources) in the development of new applications was obvious. This was the starting point for designing a "geo" tool kit rather than implementing a series of specialized systems from scratch.

Extensibility in GeoToolKit refers to the extension of the 3D geometric data model classes and the plugging in of new spatial access methods. Contrary to the CONCERT prototype, the temporal access has not yet been integrated into the multi-dimensional access methods. However, GeoToolKit has been extended by temporal classes to support the animation of geological processes.

7.7.2 Architecture

GeoToolKit [3,4] follows the layered system architecture. It has been developed on top of the OODBMS ObjectStore [56]. The system architecture of GeoToolKit is shown in Figure 7.22. It is divided into two main parts: a C++-library and interactive tools. The class library consists of ObjectStore-based classes for spatial data maintenance. The graphical classes provide the visualization of 2D maps and 2D/3D areas which are based on motif. To communicate with external geoscientific tools, first a protocol on top of UNIX-sockets has been realized. The case study on geological basin evolution which is shown below uses this low-level communication. However, an advanced CORBA-based communication architecture has been developed, too.

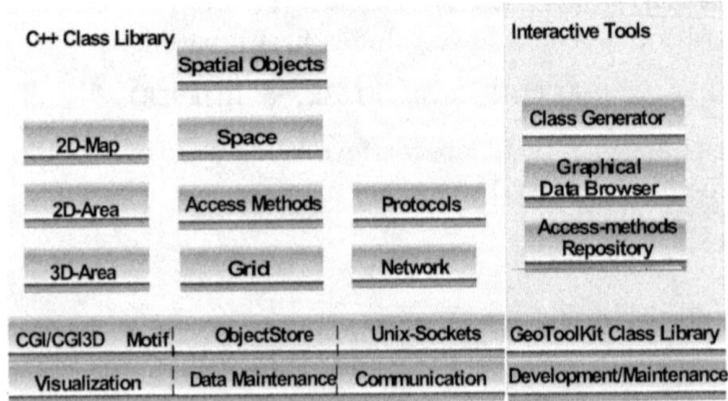

Fig. 7.22. GeoToolKit system architecture

GeoToolKit's architecture has been extended via wrapper technology and a CORBA-based infrastructure to enable remote data and operation access from other 3D geo-information components. The approach has been tested with GOCAD [51,27] a geological 3D modeling and visualization tool. The prototype implementation uses Orbix [49] for transparent network access between the heterogeneous computing platforms. Covered by a wrapper, GeoToolKit can be used as part of a distributed geo-database. The wrapper enables the easy access to the DBMS from other remote CORBA clients. Using CORBA, clients may concurrently access the database on condition that they keep CORBA compatibility. The approach was evaluated and proved by the development of an object database adapter (ODA) prototype and a specialized spatial ODA (SODA). For more detail see [17].

GeoToolKit's interactive tools consists of modules for comfortable loading of databases and navigation through data sources. The class generator provides the generation of user-defined classes without C++-knowledge.

7.7.3 Spatio-temporal Extensions

Spatial Extensions. GeoToolKit[6] primarily deals with two basic notions: a `SpatialObject` and a collection of spatial objects referred to as a `Space`.

On the abstract level a spatial object is defined as a point set in the three-dimensional Euclidean space. Diverse geometric operations can be applied to a spatial object. However, they cannot be implemented unless a spatial object has a concrete representation. There is a direct analogy with the object-oriented modeling capabilities. An abstract spatial object class exclusively specifies the interface inherited by all concrete spatial objects. A concrete object is modeled as a specialization of the abstract spatial object class. It provides an appropriate representation for the object as well as the implementation for the functions.

The geometric functionality involves geometric predicates returning `true` or `false` (e.g. contains), geometric functions (e.g., distance) and geometric operations (e.g. intersection). The geometric operations are algebraically closed. The result of a geometric operation is a spatial object which can be stored in the database or used as an argument in other geometric operations. Naturally every spatial object class includes a set of service facilities required for the correct maintenance of objects (clone, dynamic down cast). The GeoToolKit class hierarchy includes the following classes:

- 0D-3D spatial simplexes: `Point`, `Segment`, `Triangle`, `Tetrahedron`;
- 1D-3D spatial complexes: `Curve`, `Surface`, `Solid`;
- Compound objects: `Group`;
- Analytical objects: `Line`, `Plane`.

Usually complexes are approximated (digitized) and represented as homogeneous collections of simplexes. A `Curve` (1D complex) is approximated through a polyline, a `Surface` and a `Solid` - as a triangle and tetrahedron network, respectively. However, it is not intended to restrict users only with the representations supplied with GeoToolKit. Complex spatial objects are designed in such a way that they do not predefine a physical layout of objects. They contain a reference to a dependent data structure referred to as a representation. The two layer architecture allows for the object to have multiple representations (e.g., one representation for the compact storage, another more redundant one for efficient computations). An object can change its representation without changing the object identity. This feature is of extreme importance in the database context since an object can be referred to from multiple sources. Following certain design patterns a user is able to integrate his own special-purpose representation within GeoToolKit's standard classes.

Spatial objects of different types can be gathered into an heterogeneous collection, called `Group`, which is further treated as a single object. A group is a construction for the representation of the results of geometric operations.

[6] GeoToolKit has been developed in the groups of A. Cremers and A. Siehl in close cooperation with the Geological Institute and SFB 350 at Bonn University, funded by the German Research Foundation.

A space is a special container class capable of efficient retrieval of its elements according to their location in space specified either exactly by a point or by a spatial interval. A spatial interval, often referred to as a bounding box, is defined in 3D-space as a cuboid with the sides parallel to the axes. Since all operations in the Cartesian coordinate system are considerably faster for cuboids than for other objects, the approximation of spatial objects by their bounding boxes is intensively used as effective pre-check by geometric operations and spatial access methods in GeoToolKit.

A space serves both as a container for spatial objects and as a program interface to the spatial query manager which is realized as an internal GeoToolKit library function linked to a geo-application. The spatial query manager is invoked by member functions of the class `Space`. Practically, a call of any space method means a call of the spatial query manager. A user can add/remove a spatial object to/from a space syntactically in the same way as in the case of usual object collections. However, all changes will go through the spatial query manager. A spatial retrieval is performed through the family of *retrieve* methods. The task of various retrieve methods is to provide a convenient interface for the spatial query manager. In the simplest case a retrieve member function takes a bounding box as a parameter and returns a set of spatial objects contained in or intersected by this bounding box. A user can also formulate his query in the ObjectStore style as a character string. Such interface may be useful for the implementation of interactive retrieval. However, it needs multiple conversions of data and therefore it is not convenient for the internal use within a program. A spatial retrieval involving "indirect" spatial predicates (e.g. intersects) is usually decomposed into two sequential steps. In the first step the query manager retrieves all objects which intersect the bounding box of the given object. On the second step it checks whether the pre-selected objects really intersect the given object.

To provide efficient retrieval, a space must have a spatial index. In order to make an arbitrary user-defined index known to GeoToolKit, it must fit the interface defined within the abstract `AccessMethod` class. GeoToolKit supports two indexing methods: the R-Tree [39] for a pure spatial retrieval and the LSD-Tree [42] for a mixed spatial/thematic retrieval.

Temporal Extensions. The ability to follow a topological evolution of geological entities is of special interest for geo-scientists. Geological entities are characterized by relative large and irregular time intervals between time states available. On the contrary, for the smooth animation small regular time intervals are required. Therefore missing time states need to be interpolated. An interpolation between primitive simplex objects (simplexes) is straightforward. An interpolation of complexes can be reduced to the simplex-to-simplex interpolation only if complexes have the same cardinality. However, an object may change its size and/or shape with respect to time in such a way that it will need more simplexes for the adequate representation. For example, in the result of deformations a flat platform (for the representation of which two triangles are quite enough) may transform into a spherical surface which will need a much larger number of triangles for the qualitative representation. GEODEFORM [1] is a geo-scientific application for calculating geological deformations which have been developed at the Geological Institute of the Bonn University. For the visualization of spatial objects changing in time GEODEFORM uses a model proposed in the graphical library GRAPE [62]. According to this model, each time state contains two representations of the same object with different number of simplexes (a discretization factor). The first representation (post-discretization) corresponds to the approximation of the current state of the object with the discretization required by its current size and shape. The second one (pre-discretization) corresponds to the approximation of the current state of the object but with the discretization used in the previous state. Due to this extension an interpolation can always be performed between representations with the same discretization factor: the post-discretization of the previous state and the pre-discretization of the current state of the object.

To provide an appropriate maintenance of a large number of spatio-temporal objects, GEODEFORM was extended with a database component developed on the GEOTOOLKIT basis. The task was to integrate into GEOTOOLKIT's pure spatial classes a concept of time so that a spatial functionality already available could be re-utilized and a maximal level of compatibility with GRAPE was provided. To represent different time states of the same spatial object, we introduced a class `TimeStep` (see Figure 7.23), which contains a time tag and two references (pre and post) to spatial objects. If the pre- and post-discretization factors are equal, pre and post links simply refer to the same spatial object. The model proposed is general enough since an arbitrary spatial object can be chosen as a representation of a time state. In the case of GEODEFORM there are geological strata and faults modeled through GeoStore's Stratum and Fault classes which in turn are defined as a specialization of GEOTOOLKIT's `Surface` class. A sequence of `TimeStep` instances characterizing different states of the same spatial object are gathered into a spatio-temporal object (class `TimeSequence`). Being a specialization of the abstract class `SpatialObject`, an instance of `TimeScene` can be treated in the same way as any other spatial object, i.e. it can be inserted into a space as well as participate in all geometric operations. The spatial

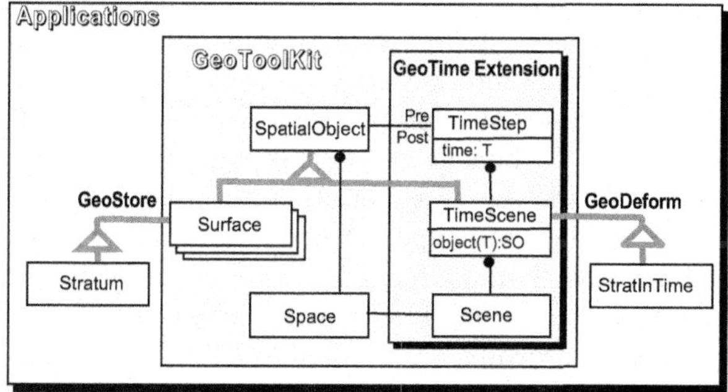

Fig. 7.23. Temporal extensions for GeoToolKit

functionality is delegated by default to the spatial object referred to in the latest `TimeStep` instance. A selection for the interpolation differs from a common selection with a specified key. If there is no object in the *time sequence* that hits exactly the time stamp t specified in the retrieve function, instead of NULL it should return a pair of neighbor time steps with time tags t1 and t2, so that $t1 < t < t2$. The same is valid for the time interval. If the interval's margins do not exactly hit the time step instances in the sequence, the resulting set includes all time steps fitting the interval completely extended with the nearest ancestor of the time step with the lowest time tag and the nearest successor of the time step with the highest time tag. Any `TimeSequence` instance can be inserted in and spatially retrieved from the GeoToolKit's space as any other spatial object. However, to perform a temporal retrieval, a special container class `Scene` is introduced. This class is capable of both spatial and temporal retrieval.

7.7.4 Implementation Details

GeoToolKit is open for new spatial and temporal indexing methods. A general technique is developed which allows the integration of arbitrary spatial indexing methods within the object-oriented DBMS ObjectStore. A new container type for the maintenance of spatial objects (space) has to be defined and for this class a spatial query manager has to be implemented. The spatial query manager overloads the OODBMS native query manager. It parses a query, extracts spatial predicates and checks whether a spatial index is available. If no spatial index is associated with the space, it simply forwards the query to the native query manager. If a spatial index is found, the spatial query performs index-based retrieval. The results (if not empty) together with the rest of the query are forwarded to the native query manager. Since the spatial query manager performs the role of a preprocessor, the syntax of the native query language can be preserved or even extended.

To enable the cooperation between the spatial query manager and spatial indexing methods, they must have a common interface. This requirement is not as restrictive as it seems to be at the first glance because the majority of the spatial indexes exclusively deals with the bounding box approximation of spatial objects. Index developers do not even need to modify the sources to fit the required interface. A usual technique is to develop an adapter class which carries out all necessary conversions and then it simply calls corresponding functions according to the pre-defined interface.

The situation is more complicated in the case of so-called combined indexing. Combined indexes may be beneficial when separately neither a spatial nor a thematic component of the search criterion is selective enough. Distinct to the bounding box, the number and types of non-spatial attributes cannot be pre-defined. To combine spatial and heterogeneous non-spatial subkeys, GeoToolKit offers a special construction. The multikey class provides a uniform access to an arbitrary subkey according to its number (dimension). The retrieval with the multikey is similar to the spatial retrieval with the only difference that a generic multikey substitutes the bounding box.

A query can be formulated and called either in the same way as a usual ObjectStore query or by means of the additional retrieve function. The retrieve function demands from the user to separate explicitly the spatial and non-spatial parts of the query. The spatial part is represented as a bounding box or as a multidimensional key. If a space has more than one spatial index and a user wants a particular index to be used for the retrieval, he can explicitly specify it in the retrieve function.

To maintain the indexes consistently within GeoToolKit, every member function (dealing with spatial or temporal updates) incorporates additional checks before and after performing updates. The pre-check function tests whether an object is contained in at least one indexed space. If yes, the bounding box is stored. The post-check function activates re-indexing only when the bounding box has been changed. To eliminate re-indexing by serial updates, GeoToolKit provides an update block.

7.7.5 Case Studies

GeoToolKit has been tested with different geological applications like well management, 3D geological modeling and geological restoration based on time-dependent geometries [1]. Figure 7.24 shows three time steps of a basin evolution modeled with GeoDeform coupled with GeoToolKit. The figure shows a perspective view of the southern part of the Lower Rhine Basin, Germany, towards the Northeast with the base of the Oligocene and synthetic as well as antithetic faults. Black arrows indicate direction and cumulative amount of extension, whereas negative numbers in *italics* show subsidence at selected points.

In GeoDeform's concept of geological time-space modeling [62], a 4D model consists of a set of states of modeled 3D objects at different times. The time steps between the object states can be different for each object chain and need not to be equidistant. Interpolation (e.g., linear) can be done between known object states. A "time frame" representation of the 4D model is the set of object states at a distinct system time. Time-dependent 3D objects can be modified, added to, or deleted from the 4D model. The possibility to handle varying discretization and topology for each object at each time step makes the concept very flexible for interactive geological modeling and visualization. From the 4D model, a time schedule of subsidence and deformation can be deduced and calibrated against the geological observations, providing new insights into the origin of the basin (Figure 7.24).

Examples for temporal and spatio-temporal queries are:

- os_Set<TimeStep>tmp=scene.retrieve(time>-20.0 && time<0);
- os_Set<TimeStep>tmp=scene.retrieve(bb,time>-20.0 && time<0);

The first query yields all states of geological objects in a scene which existed 20 million years ago until the recent state. The second one additionally selects the geological objects which are located within a specified bounding box.

The development of GeoDeform, a database component coupled with GeoToolKit, proved the advantages of the tool kit approach. GeoDeform classes

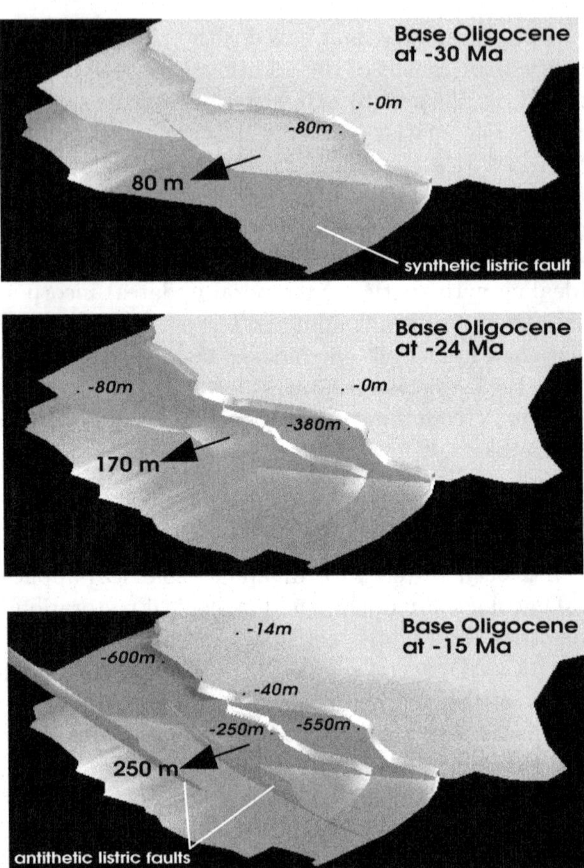

Fig. 7.24. Three time steps of basin evolution in the Lower Rhine Basin, Germany, modeled by GeoDeform coupled with GeoToolKit [1]

contain now only geology-specific members. Geometric relationships between geometric objects are hidden within GeoToolKit. However, the GeoToolKit functionality (spatial retrieval, indexing, etc.) is still fully available for the extended objects.

The developers can focus on application semantics instead of optimally assembling spatial objects from multiple relational tables or the re-implementation of routine geometric algorithms. Classes designed for particular applications can either be used directly or can be refined for other applications.

7.8 Conclusions

We conclude this chapter with a brief summary of the architectures of the prototype systems and their contributions to spatio-temporal data management.

System	Architecture	Main Contribution
CONCERT	extensible	generic index support
SECONDO	extensible	framework for spatio-temporal data types
DEDALE	extensible	constraint model with handling of interpolated spatial data
TIGER	layered RDBMS	(spatio-)/temporal query language
GEOTOOLKIT	layered OODBMS	geometric 3D/4D data types and indexes

Fig. 7.25. Comparison of the prototype systems

System Architectures. As depicted in Figure 7.25, CONCERT, SECONDO, and DEDALE are representatives of the extensible system architecture. These systems allow a more or less seamless extension towards spatio-temporal database systems. CONCERT's extensibility aims at a flexible index support for arbitrary data stores, whereas SECONDO supports the design of spatio-temporal data types on top of a data store. DEDALE's data storage and query engine consist of an extensible C++-implementation. TIGER is a representative of the layered system architecture. Its temporal database query language is based on relational database technology. GEOTOOLKIT is closely coupled with the OODBMS ObjectStore, which basically serves as an ordinary data store. GEOTOOLKIT allows to extend ObjectStore with spatio-temporal data types and access methods. In this sense, GEOTOOLKIT is a representative of the extensible system architecture. However, one can also argue that GEOTOOLKIT is build on top of ObjectStore and thus is a representative of the layered system architecture.

Specific Contributions to Spatio-temporal Data Management. CONCERT provides a framework of a generic index, which can be directly used for the refinement of specific spatio-temporal data types and access methods. This index has several properties:

- *It is generic in the sense of data types.* It is valid and usable for all data types following the `SPATIAL` concept of CONCERT independent of their location or size.
- *It is generic in the sense of tree behavior.* Changing the tree heuristics leads to different tree derivations.
- *It is derived from the main issue of each data management system — the data.* In contrast, other approaches usually derive their genericity by generalizing algorithms or methods.

Throughout the whole code of CONCERT's generic spatio-temporal index, no explicit assumption is made about the data types and storage formats of the spatio-temporal data. Only the concept typical operations are used as an interface. The objects themselves are simply treated as abstract objects, i.e., as uninterpreted byte sequences with a few operations defined on them. Therefore, it is irrelevant to the kernel system where the real data objects reside — as long as they can be accessed via the concept typical operations. Instead of objects themselves, it is possible to store only place holders (e.g. a URL or any sort of a pointer to the actual object) and access the real objects only when processing the concept typical operations, for example, via remote procedure call. This fact allows the kernel system to cope with the interoperability issue. The actual data can reside in heterogeneous repositories. The kernel only needs to know the operations and handles to access it and provide physical design and query capabilities over the external data.

SECONDO has been introduced as a generic development environment for non-standard multi-user database systems. At the bottom architecture level, SECONDO offers tools for efficient and comfortable handling of nested lists and catalogs, a simplified interface to the underlying SHORE storage manager, a tool for efficient management of tuples with embedded large objects, and an SOS compiler. Algebra modules for standard and relational data types and operators as well as simple user interface clients have been implemented. The core part of SECONDO, its extensible query processor, is characterized by the following highlights:

- Its formal basis to describe a generic query plan algebra giving a clear algorithm for translating a query plan into an operator tree.
- Functional abstraction is a well-defined concept in SOS. This leads to a very clean, simple, and general treatment of parameter expressions of operators.
- `stream` is a built-in type constructor in SECONDO. Simply writing the keyword `stream` in the type mapping of an operator lets the query processor automatically set up calls of the evaluation function for this operator in stream mode. For this reason, SECONDO can uniformly handle streams of anything, not just tuples. Also, a query plan can freely mix stream and non-stream operators.
- SECONDO includes complete type checking, type mapping, and resolution of operator overloading.

DEDALE contributes to spatio-temporal databases by its extension to handle interpolated spatial data. Its model covers moving objects like trajectories and interpolated spatial data like Triangulated Irregular Networks. The constraint data model sees interpolated pointsets as infinite relations and expresses queries on these pointsets with the standard SQL language. It also provides an accurate finite representation for interpolated data. Furthermore, an algorithm to evaluate queries at a low cost is supplied.

TIGER is a temporal database system that demonstrates the use of ATSQL's statement modifiers to manage temporal information. It can also be extended to spatio-temporal database access. It features an enhanced support for the time dimension through ATSQL. The highlights concerning the TIGER implementation are:

- the intelligent reuse of existing database technology,
- a seamless extension of database systems with external modules, and
- an applet interface that supports distance learning as application of TIGER.

Emphasizing the temporal management of data, TIGER counterbalances the other prototype systems discussed in this chapter, which primarily focus on the spatial aspects of data management.

GEOTOOLKIT is a spatio-temporal extension of the OODBMS ObjectStore. It has been designed to support especially geological database applications which are intrinsically space (3D) and time-dependent (4D). The highlights of the GEOTOOLKIT implementation are:

- 3D geometric data types based on simplicial complexes,
- advanced 3D geometric algorithms,
- extensions for temporal data handling,
- coupling with a 3D visualization tool, and
- an open CORBA-based system architecture.

GEOTOOLKIT's geometric 3D data types, especially the triangle and tetrahedron networks, allow advanced spatial database queries like the intersection of a 3D-object with a set of other 3D-objects. The temporal extensions of GEOTOOLKIT provides the selection of snapshots of 3D objects and an adequate visualization.

GEOTOOLKIT has been tested during the last years by different geological groups at Bonn University. The most advanced application on top of GEOTOOLKIT is GEOSTORE, an information system for geologically defined 3D geometries in the Lower Rhine Basin, Germany. Furthermore, a data management tool and a spatio-temporal data browser have been developed.

In summary, although none of the prototype systems presented in this chapter are complete spatio-temporal database management system, they implement important issues of spatial and temporal database management systems and thereby provide the platform for future systems with full spatio-temporal support. There is also the legitimate hope that these prototype systems could be predecessors of components and services to be integrated in an overall spatio-temporal system architecture.

References

1. R. Alms, O. Balovnev, M. Breunig, A.B. Cremers, T. Jentzsch, and A. Siehl. Space-Time Modelling of the Lower Rhine Basin Supported by an Object-Oriented Database. *Physics and Chemistry of the Earth*, 23(3):251–260, 1998.
2. ARC/INFO. ArcInfo 8, a New Standard in Professional GIS, ESRI. Brochure `http://www.esri.com/library/brochures/pdfs/arcinfo8ad.pdf`, ESRI, 2000.
3. O. Balovnev, M. Breunig, and A.B. Cremers. From GeoStore to GeoToolKit: The Second Step. In M. Scholl and A. Voisard, eds., *Advances in Spatial Databases, Proc. 5th Int. Symposium, SSD'97*, LNCS, Vol. 1262, pp. 223–237, Springer-Verlag, 1997.
4. O. Balovnev, M. Breunig, A.B. Cremers, and M. Pant. Building Geo-Scientific Applications on Top of GeoToolKit: a case study of Data Integration. In *Proc. 10th Int. Conf. on Scientific and Statistical Database Management*, pp. 260–269, IEEE Computer Science Press, 1998.

5. D.S. Batory, J.R. Barnett, J.F. Garza, K.P. Smith, K. Tsukuda, B.C. Twichell, and T. E. Wise. GENESIS: An Extensible Database Management System. *IEEE Transactions on Software Engineering*, 14(11):1711–1730, 1988.
6. L. Becker and R.H. Güting. The GraphDB Algebra: Specification of Advanced Data Models with Second-Order Signature. Informatik-Report 183, FernUniversität Hagen, Germany, 1995.
7. N. Beckmann, H.-P. Kriegel, R. Schneider, and B. Seeger. The R*-tree: An Efficient and Robust Access Method for Points and Rectangles. In H. Garcia-Molina and H. Jagadish, eds., *Proc. 1990 ACM SIGMOD Int. Conf. on Management of Data*, ACM SIGMOD Record, Vol. 19, No. 2, pp. 322–331, ACM Press, 1990.
8. S. Blott, H. Kaufmann, L. Relly, and H.-J. Schek. Buffering Long Externally-Defined Objects. In *Persistent Object Systems, Proc. 6th Int. Workshop*, Workshops in Computing, pp. 40–53, Springer-Verlag, 1995.
9. S. Blott, L. Relly, and H.-J. Schek. An Abstract-Object Storage Manager. In H. V. Jagadish and I. S. Mumick, eds., *Proc. 1996 ACM SIGMOD Int. Conf. on Management of Data*, ACM SIGMOD Record, Vol. 25, No. 2, pp. 330–340, ACM Press, 1996.
10. T. Bode, A.B. Cremers, and J. Freitag. OMS – An Extensible Object Management System. In R. Bayer, Härder, and P. C. Lockemann, eds., *Objektbanken für Experten*, pp. 29–54, Informatik aktuell, Springer-Verlag, 1992.
11. M.H. Böhlen. *Managing Temporal Knowledge in Deductive Databases*. PhD thesis, Eidgenössisch Technische Hochschule (ETH) Zürich, Switzerland, 1994.
12. M.H. Böhlen. Temporal Database System Implementations. *ACM SIGMOD Record*, 24(4):53–60, 1995.
13. M.H. Böhlen, R. Busatto, and C.S. Jensen. Point-Versus Interval-Based Temporal Data Models. In *Proc. 14th IEEE Int. Conf. on Data Engineering, ICDE'98*, pp. 192–200, IEEE Computer Society Press, 1998.
14. M.H. Böhlen and C.S. Jensen. Seamless Integration of Time into SQL. Technical Report R-96-2049, Department of Computer Science, Aalborg University, Denmark, 1996.
15. M.H. Böhlen, C.S. Jensen, and B. Skjellaug. Spatio-Temporal Database Support for Legacy Applications. In L. Haas and A. Tiwary, eds., *SIGMOD'98, Proc. 1998 ACM SIGMOD Int. Conf. on Management of Data*, ACM SIGMOD Record, Vol. 25, No. 2, pp. 226–234, ACM Press, 1998.
16. M.H. Böhlen, R.T. Snodgrass, and M. D. Soo. Coalescing in Temporal Databases. In T.M. Vijayaraman, A.P. Buchmann, C. Mohan, and N.L. Sarda, eds., *Proc. 22nd Int. Conf. on Very Large Data Bases, VLDB'96*, pp. 180–191, Morgan Kaufmann, 1996.
17. M. Breunig, A.B. Cremers, H.-J. Götze, S. Schmidt, R. Seidemann, S. Shumilov, and A. Siehl. First Steps Towards an Interoperable GIS - An Example from Southern Lower Saxony. *Physics and Chemistry of the Earth*, 24(3):179–189, 1999.
18. M.J. Carey, D.J. DeWitt, D. Frank, G. Graefe, M. Muralikrishna, J.E. Richardson, and E. J. Shekita. The Architecture of the EXODUS Extensible DBMS. In K.R. Dittrich and U. Dayal, eds., *Proc. 1st Int. Workshop on Object-Oriented Database Systems*, pp. 52–65, IEEE Computer Society Press, 1986.
19. M.J. Carey, D.J. DeWitt, M.J. Franklin, N.E. Hall, M.L. McAuliffe, J.F. Naughton, D.T. Schuh, M.H. Solomon, C.K. Tan, O.G. Tsatalos, S.J. White, and M.J. Zwilling. Shoring Up Persistent Applications. In *Proc. 1994 ACM SIGMOD Int. Conf. on Management of Data*, ACM SIGMOD Record, Vol. 23, No. 2, pp. 383–394, ACM Press, 1994.

20. W.F. Clocksin and C.S. Mellish. *Programming in Prolog*. Springer-Verlag, 3 edition, 1987.
21. L. Comet. The Ubiquitous B-tree. *ACM Computing Surveys*, 11(2):121–137, 1979.
22. S. Dieker and R.H. Güting. Plug and Play with Query Algebras: SECONDO – A Generic DBMS Development Environment. Informatik-Report 249, FernUniversität Hagen, Germany, 1999. In *Proceedings of International Database Engineering and Applications Symposium (IDEAS 2000), September 2000.*
23. S. Dieker and R.H. Güting. Efficient Handling of Tuples with Embedded Large Objects. *Data & Knowledge Engineering*, 32(3):247–269, 2000.
24. S. Dieker, R.H. Güting, and M. Rodríguez Luaces. A Tool for Nesting and Clustering Large Objects. Informatik-Report 265, FernUniversität Hagen, Germany, 2000. In *Proceedings of the 12th International Conference on Scientific and Statistical Database Management, July 2000.*
25. EcoWin, Hanson & Partners, Gothenburg, Sweden. *EcoWin Time Series Extender*, 1999. `http://www.ecowin.com`.
26. R.A. Finkel and J.L. Bently. Quad Trees: A Data Structure for Retrieval on Composite Keys. *Acta Informatica*, 4(1):1–9, 1974.
27. GOCAD Techn. Documentation, 2000. `http://www.ensg.u-nancy.fr/GOCAD`.
28. G. Graefe. Query Evaluation Techniques For Large Databases. *ACM Computing Surveys*, 25(2):73–170, 1993.
29. G. Graefe. Volcano — An Extensible and Parallel Query Evaluation System. *IEEE Transactions on Knowledge and Data Engineering*, 6(1):120–135, 1994.
30. G. Graefe and W.J. McLenna. The Volcano Optimizer Generator: Extensibility and Efficient Search. In A. Elmagarmid and E. Neuhold, eds., *Proc. 9th IEEE Int. Conf. on Data Engineering, ICDE'93*, pp. 209–218, IEEE Computer Society Press, 1993.
31. S. Grumbach, P. Rigaux, M. Scholl, and L. Segoufin. The DEDALE/ Prototype. In G. Kuper, L. Libkin, and J. Paradaens, eds., *Constraint Database Systems*, pp. 365–382, Springer-Verlag, 2000.
32. S. Grumbach, P. Rigaux, and L. Segoufin. The DEDALE/ System for Complex Spatial Queries. In L. Haas and A. Tiwary, eds., *SIGMOD'98, Proc. of the 1998 ACM SIGMOD Int. Conf. on Management of Data*, ACM SIGMOD Record, Vol. 25, No. 2, pp. 213–224, ACM Press, 1998.
33. S. Grumbach, P. Rigaux, and L. Segoufin. Manipulating Interpolated Data is Easier than you Thought. *Proceedings of VLDB 2000*, pp. 156-165, Cairo, Egypt, September 2000.
34. C. Gurret, Y. Manolopoulos, A. Papadopoulos, and P. Rigaux. The BASIS System: A Benchmarking Approach for Spatial Index Structures. In M.H. Böhlen, C.S. Jensen, and M. Scholl, eds., *Proc. of the Int. Workshop on Spatio-Temporal Database Management*, LNCS, Vol. 1678, pp. 152–170, Springer-Verlag, 1999.
35. R.H. Güting. Gral: An Extensible Relational Database System for Geometric Applications. In P.M.G. Apers and G. Wiederhold, eds., *Proc. 15th Int. Conf. on Very Large Data Bases, VLDB'89*, pp. 33–44, Morgan Kaufmann, 1989.
36. R.H. Güting. Second-Order Signature: A Tool for Specifying Data Models, Query Processing, and Optimization. In P. Buneman and S. Jajodia, eds., *Proc. 1993 ACM SIGMOD Int. Conf. on Management of Data*, ACM SIGMOD Record, Vol. 22, No. 2, pp. 277–286, ACM Press, 1993.
37. R.H. Güting. GraphDB: Modeling and Querying Graphs in Databases. In J. B. Bocca, Matthias Jarke, and C. Zaniolo, eds., *Proc. 20th Int. Conf. on Very Large Data Bases, VLDB'94*, pp. 297–308, Morgan Kaufmann, 1994.

38. R.H. Güting, S. Dieker, C. Freundorfer, L. Becker, and H. Schenk. SECONDO/QP: Implementation of a Generic Query Processor. In T.J.M. Bench-Capon, G. Soda, and A.M. Tjoa, eds., *Database and Expert Systems Applications, Proc. 10th Int. Conf., DEXA'99*, LNCS, Vol. 1677, pp. 66–87, Springer-Verlag, 1999.
39. A. Guttman. R-trees: A Dynamic Index Structure for Spatial Searching. In B. Yormark, ed., *Proc. 1984 ACM SIGMOD Int. Conf. on Management of Data*, ACM SIGMOD Record, Vol. 14, No. 2, pp. 47–57, ACM Press, 1984.
40. L.M. Haas, W. Chang, G.M. Lohman, J. McPherson, P.F. Wilms, G. Lapis, B. Lindsay, H. Pirahesh, M. Carey, and E. Shekita. Starburst Mid-flight: As the dust clears. *IEEE Transactions on Knowledge and Data Engineering*, 2(1):143–160, 1990.
41. J. M. Hellerstein, J.F. Naughton, and A. Pfeffer. Generalized Search Trees for Database Systems. In U. Dayal, P.M.D. Gray, and S. Nishio, eds., *Proc. 21st Int. Conf. on Very Large Data Bases, VLDB'95*, pp. 562–573, Morgan Kaufmann, 1995.
42. A. Henrich, H.-W. Six, and P. Widmayer. The LSD Tree: Spatial Access to Multidimensional Point and Non-Point Objects. In P.M.G. Apers and G. Wiederhold, eds., *Proc. 15th Int. Conf. on Very Large Data Bases, VLDB'89*, pp. 45–54, Morgan Kaufmann, 1989.
43. Informix Software, Inc., Menlo Park, CA. *Excalibur Text Search DataBlade Module: User's Guide, Version 1.1*, 1997.
44. Informix Software, Inc., Menlo Park, CA. *INFORMIX Geodetic DataBlade Module: User's Guide, Version 2.1*, 1997.
45. Informix Software, Inc., Menlo Park, CA. *INFORMIX Spatial DataBlade Module: User's Guide, Version 2.2*, 1997.
46. Informix Software, Inc., Menlo Park, CA. *INFORMIX TimeSeries DataBlade Module: User's Guide, Version 3.1*, 1997.
47. Informix Software, Inc., Menlo Park, CA. *Extending INFORMIX-Universal Server: Data Types, Version 9.1*, 1998.
48. International Organization for Standardization & American National Standards Institute, ANSI/ISO/IEC 9075-2:99. *ISO International Standard: Database Language SQL - Part 2: Foundation)*, September 1999.
49. IONA Technologies Ltd. Orbix Programmers's Guide, Version 2.3, 1997.
50. H. Luttermann and A. Blobel. Chronos: A Spatiotemporal Data Server for a GIS. In *Proc. 9th. Int. Symposium on Computer Science in Environmental Protection*, pp. 135–142, Metropolis, 1995.
51. J.L. Mallet. GOCAD: A Computer Aided Design Program for Geological Applications. In A. K. Turner, ed., *Three-Dimensional Modeling with Geoscientific Information Systems*, pp. 123–142, Kluwer Academic Publishers, 1992.
52. N. Mamoulis and D. Papadias. Integration of Spatial Join Algorithms for Joining Multiple Inputs. In A. Delis, C. Faloutsos, and S. Ghandeharizadeh, eds., *Proc. 1999 ACM SIGMOD Int. Conf. on Management of Data*, ACM SIGMOD Record, Vol. 28, No. 2, pp. 1–12, ACM Press, 1999.
53. J. Melton and A.R. Simon. *Understanding the New SQL — A Complete Guide*. Morgan Kaufmann, 1993.
54. S. Morehouse. A Geo-Relational Model for Spatial Information. In *Proceedings of Auto Carto 7*, pp. 338–357, 1985.
55. J. Nievergelt, H. Hinterberger, and K.C. Sevcik. The Grid File: An Adaptable, Symmetric Multikey File Structure. *ACM Transactions on Database Systems*, 9(1):38–71, 1984.
56. ObjectStore – Online Product Documentation. `http://www.odi.com`.

57. J. Ong, D. Fogg, and M. Stonebraker. Implementation of Data Abstraction in the Relational Database System Ingres. *SIGMOD Record*, 14(1):1–14, 1984.
58. Oracle Corporation. *Oracle8i Spatial: User's Guide and Reference, Release 8.1.6*, 1999.
59. Oracle Corporation. *Oracle8i TimeSeries: User's Guide, Release 8.1.6*, 1999.
60. Oracle Corporation – Product Documentation. `http://www.oracle.com`.
61. A. Papadopoulos, P. Rigaux, and M. Scholl. A Performance Evaluation of Spatial Join Processing Strategies. In R.H. Güting, D. Papadias, and F.H. Lochovsky, eds., *Advances in Spatial Databases, Proc. 6th Int. Symposium, SSD'99*, LNCS, Vol. 1651, pp. 286–307, Springer-Verlag, 1999.
62. K. Polthier and M. Rumpf. A Concept for Time-Dependent Processes. In M. Göbel, H. Müller, and B. Urban, eds., *Visualization in Scientific Computing*, pp. 137–153, Springer-Verlag, 1995.
63. L. Relly. *Open Storage Systems: Pysical Database Design for External Objects.* PhD thesis, Eidgenössisch Technische Hochschule (ETH) Zürich, ETH-Zentrum, CH-8092 Zürich, Switzerland, 1999. (In German).
64. L. Relly and U. Röhm. Plug and Play: Interoperability in CONCERT. In A. Vckovski, K.E. Brassel, and H.-J. Schek, eds., *Interoperating Geographic Information Systems, Proc. 2nd Int. Conf., INTEROP'99*, LNCS, Vol. 1580, pp. 277–291, Springer-Verlag, 1999.
65. L. Relly, H.-J. Schek, O. Henricsson, and S. Nebiker. Physical Database Design for Raster Images in Concert. In M. Scholl and A. Voisard, eds., *Advances in Spatial Databases, Proc. 5th Int. Symposium, SSD'97*, LNCS, Vol. 1262, pp. 259–279, Springer-Verlag, 1997.
66. L. Relly, H. Schuldt, and H.-J. Schek. Exporting Database Functionality — The Concert Way. *Bulletin of the IEEE Technical Committee on Data Engineering*, 21(3):43–51, 1998.
67. B. Salzberg and V.J. Tsotras. A Comparison of Access Methods for Temporal Data. TimeCenter Technical Report TR-18, TimeCenter, 1997.
68. H.-J. Schek, H.-B. Paul, and M.H. Scholl. The DASDBS Project: Objectives, Experiences, and Future Prospects. *IEEE Transactions on Knowledge and Data Engineering*, 2(1):25–43, 1990.
69. H.-J. Schek and W. Waterfeld. A Database Kernel System for Geoscientific Applications. In D. Marble, ed., *Proc. of the 2nd Symposium on Spatial Data Handling*, pp. 273–288, 1986.
70. P. Seshadri, M. Livny, and R. Ramakrishnan. The Design and Implementation of a Sequence Database System. In T.M. Vijayaraman, A.P. Buchmann, C. Mohan, and N.L. Sarda, eds., *Proc. 22nd Int. Conf. on Very Large Data Bases, VLDB'96*, pp. 99–110, Morgan Kaufmann, 1996.
71. P. Seshadri, M. Livny, and R. Ramakrishnan. The Case for Enhanced Abstract Datatypes. In M. Jarke, M.J. Carey, K.R. Dittrich, F.H. Lochovsky, P. Loucopoulos, and M. A. Jeusfeld, eds., *Proc. 23rd Int. Conf. on Very Large Data Bases, VLDB'97*, pp. 66–75, Morgan Kaufmann, 1997.
72. Smallworld. SMALLWORLD, the Geographical Information System SMALLWORLD GIS. SMALLWORLD Report, SMALLWORLD Systems GmbH, Ratingen, Germany, 2000.
73. M. Stonebraker. Inclusion of New Types in Relational Database Systems. In G. Wiederhold, ed., *Proc. 2nd IEEE Int. Conf. on Data Engineering, ICDE'86*, pp. 262–269, IEEE Computer Society Press, 1986.

74. M. Stonebraker and L. A. Rowe. The Design of POSTGRES. In C. Zaniolo, ed., *Proc. 1986 ACM SIGMOD Int. Conf. on Management of Data, Washington, D.C.*, ACM SIGMOD Record, Vol. 15, No. 2, pp. 340–355, ACM Press, 1986.
75. M. Stonebraker, L.A. Rowe, and M. Hirohama. The Implementation of POSTGRES. *IEEE Transactions on Knowledge and Data Engineering*, 2(1):125–142, 1990.
76. Y. Theodoridis, J.R.O. Silva, and M.A. Nascimento. On the Generation of Spatiotemporal Datasets. In R.H. Güting, D. Papadias, and F.H. Lochovsky, eds., *Advances in Spatial Databases, Proc. 6th Int. Symposium, SSD'99*, LNCS, Vol. 1651, pp. 147–164, Springer-Verlag, 1999.
77. P.F. Wilms, P.M. Schwarz, H.-J. Schek, and L.M. Haas. Incorporating Data Types in an Extensible Database Architecture. In C. Beeri, J W. Schmidt, and U. Dayal, eds., *Proc. 3rd Int. Conf. on Data and Knowledge Bases: Improving Usability and Responsiveness*, pp. 180–192, Morgan Kaufmann, 1988.

8 Advanced Uses: Composing Interactive Spatio-temporal Documents

Isabelle Mirbel[1], Barbara Pernici[2],
Babis Theodoulidis[3], Alex Vakaloudis[2], and Michalis Vazirgiannis[4]

[1] Laboratoire I3S, Sophia-Antipolis, France
[2] Politecnico di Milano, Italy
[3] University of Manchester - Institute of Science and Technology, United Kingdom
[4] Athens University of Economics and Business, Greece

8.1 Introduction

Lately a new generation of application domains is emerging. Such applications, heavily dynamic and interactive, include interactive multimedia applications, virtual reality worlds, digital movies and 3D animations. They deal with intensive spatio-temporal dependencies between the participating objects with motion becoming a central issue. Indeed in the aforementioned application domains, much of the information conveyed and manipulated has spatial and/or temporal aspects. Synthetic worlds (like VRML worlds, digital movies, interactive multimedia scenarios etc.) can be started at any point in time, and there is a multitude of events that may occur (e.g., 3D object collisions) which may trigger other actions (e.g., change of the motion of the objects that collided) provided that some constraints hold. The session concept provides a new spatio-temporal context that has a different temporal origin and perhaps a different evolution according to the internal/external interaction that takes place. In this case the concept of scenario is an important entity representing the spatio-temporal course of the context (i.e., spatio-temporal actions possibly related) in conjunction with the occurring interaction (in terms of events) and potentially with the validity of conditions/constraints.

It is therefore evident that there is a strong connection between the topics studied in the modeling of spatio-temporal information and the issues of concern in composing interactive presentations. This is the motivation for the work of this chapter; to apply knowledge earned in the spatio-temporal domain into areas not directly associated with spatio-temporal databases but still uniform with them. The result leads to the integrated study of spatio-temporal information applicable in a range of diverse uses.

In the next section we relate spatio-temporal databases and interactive presentations, formally termed Interactive Spatio-Temporal Documents (ISTDs). We then present an approach for modeling the spatio-temporal components of such presentations. This focuses on the treatment of objects as integrated entities and thus apart from modeling their geometrical and positional characteristics (in an application independent format), it also demonstrates how to relate them to any non-spatial attributes. Next we analyze the method for describing the interactive behavior in these applications following the *Events Conditions Actions*

T. Sellis et al. (Eds.): Spatio-temporal Databases, LNCS 2520, pp. 319–344, 2003.
© Springer-Verlag Berlin Heidelberg 2003

paradigm. The behavior of objects includes geometric transformations of translation, rotation and scaling. Section 8.5 shows how to provide database storage and querying support. Finally we discuss some examples of applying this work and we close with references to related and future work.

8.2 Interactive Presentations and Spatio-temporal Databases

An Interactive Spatio-Temporal Document (ISTD) such as a virtual world is organized in two parts namely, structural and dynamic. The participating objects, called actors or components comprise the structural part. Similarly to traditional multimedia objects that have temporal (e.g. sound), spatial (e.g., image) and spatio-temporal (e.g., video) characteristics, the components of ISTDs are also managed by their spatio-temporal features. The dynamic part of such presentations specifies the behavior of components and also the interaction among them or with the viewer. The viewer is integral part of a presentation as components interact either through I/O devices or through the modification of their properties (e.g., change of position).

Both the structural and the behavioral parts are rich in spatio-temporal concepts. Components are in fact (possibly 3D) spatio-temporal objects. Their behavior takes the form of spatial and temporal movement, variation in the geometrical characteristics and change in the appearance of the objects. As a result of such changes, further modifications may be triggered as the objects interact with each other.

The extensive similarities that appear between spatio-temporal data and ISTDs are reflected on the modeling of these presentations. The requirements for defining a spatio-temporal data model have been analyzed in Chapters 4 and 5. The same ones (e.g. modeling of object continuous evolution, support for orthogonal spatial, temporal and spatio-temporal granularity) apply when defining data models for spatio-temporal scenario components. The following additional requirements originating from the 3D nature of objects and the need for interactive support come also into play:

- Viewer inclusion: Incorporate the viewer of an application as part of the data model and as a query argument.
- Interaction support: Define interaction semantics over 3D-spatio-temporal objects, the viewer and the application in order to add user-defined behavioral modeling.
- Types of elements: Specify the different cases of interaction elements that exist in this domain.
- Composite elements: Show how to construct composite structural and interaction elements in order to capture complex behavior.

Components are objects with complex spatial and temporal features, which make their specification a laborious procedure. The storage of components can facilitate their reusability thus accelerating the process of producing ISTDs. However, the use of a file system manager for this purpose does not provide enough support because components should be retrieved through specialized, context-based criteria. Therefore, employing a DBMS, relational or object-oriented is recommended.

In the following subsections, we define concepts that can be implemented using both relational and object-oriented designs. By providing database support our goal is not only to store components but also to retrieve them with criteria that satisfy the developer's and the application's requirements.

8.3 Modeling the Components of Spatio-temporal Interactive Documents

This section presents 3D-STDM, a data model for the components of ISTDs regarded as 3D-spatio-temporal objects. We specify a mechanism for modeling them and later we demonstrate how to provide database support in terms of storage and querying. First we discuss the particularities of the 3D-spatio-temporal domain.

8.3.1 Particularities of 3D-Spatio-temporal Modeling for Scenario Components

The 3D-spatio-temporal domain derives from the composition of the temporal and the 3D spatial domains inheriting their characteristics. Towards the combination of spatial and temporal concepts many researchers define a single, integrated spatio-temporal framework by simply considering time as the fourth geographical dimension. This simplistic approach is the first and obvious step in spatio-temporal modeling. However, the spatial and temporal composition is not simplistic (e.g., space and time are measured differently; we cannot measure time in feet) and new spatio-temporal-based features which don't exist in neither the spatial nor the temporal domain are introduced.

- First of all, the three geometrical dimensions are orthogonal among themselves. In contrast they are not independent to time as spatial changes happen on a temporal basis. Even more, the temporal distinction between the past and the future is complex and more profound than the analogous between positive and negative of the spatial fields. Finally research work on temporal analysis has proved that time is not a simple one-dimensional linear concept. Modern requirements for temporal functionality are the modeling of future (and possibly branching) events and the description of the same fact under different perspectives (version management).
- 3D-spatio-temporal worlds are utilized for portraying 3D objects and their evolution through time. Since these objects are "real-life" graphical objects, their evolution may only be continuous and never discrete. A real-life graphical object cannot directly just jump from one state into another (as for example the non-graphical "salary" may at some point get a sharp, crisp rise). Instead it can only mutate in a continuous way. Only when there is uncertainty or incomplete knowledge should discrete transitions be recorded.

- 3D-spatio-temporal objects, as they represent actual material, may only change their formation, but never disappear. For example, steel, glass and plastic are at one point in time put together to construct a car and twenty years later some of the parts of this car end up in a scrap yard. The original materials remain the same; what is different is their formation (from amorphous maze, glass becomes part of a car and later part of a scrap yard belongings). A 3D-spatio-temporal data model should be able to follow the different formations that a 3D-spatio-temporal object adopts during its lifetime (i.e., as long as the information system requires its modeling).
- Another interesting feature is the role of the viewer in 3D spatial and 3D spatio-temporal environments. Unlike the case of 2D maps where the viewer remains external to the presentation (a.k.a. map), in 3D worlds he/she fully navigates inside the world thus becoming an integral part of the overall setting. Its position influences the way they conceptualize the relationships among the different objects. Figure 8.1 exemplifies this phenomenon. The two boxes have stable positions. Nevertheless, *viewer 1* believes that *Box 2* is behind *Box 1*. On the contrary, *viewer 2* assumes that *Box 2* is in front of *Box 1*. A third viewer thinks that *Box 1* is on the left of *Box 2*. Due to this occurrence, the viewer's position must be taken into consideration within a 3D-spatio-temporal query language. Under this perspective we should take into consideration Freksa's work on conceptual spatial relationships where qualitative operators (such as left, left-front etc.) are introduced.
- The 3D-spatial domain also introduces the notion of orientation. Some (but not all) 3D objects have a certain "directioning" in the sense that there is a consensus on which their top/bottom or left/right side is defined. No matter where the viewers of a world stand they would all agree for example on the fact that a glass is on the table. That is because the glass and the table have universally agreed top and bottom sides.
- Finally the 3D-spatio-temporal domain is particular from a visualization perspective. Once again, the realistic features of the participating objects forbid the use of their visual properties for visualization purposes. The color, position and geometry in 3D worlds specify the state of an object and their alteration (for visualization) means that the represented object is not the right (i.e., exact) one. Fortunately, as it is explained in Chapter 5, the 3D-spatio-temporal domain has a rich set of properties available for visualization.

From all these particularities it becomes evident that 3D-spatio-temporal modeling, the foundation for specifying ISTDs, cannot be accomplished by neither extending a 2D-spatio-temporal model nor by attaching a simple timestamp to 3D-spatial entities. A 3D-spatio-temporal modeling approach must be autonomous in order to capture these spatio-temporal-sourced characteristics.

8.3.2 Meta-modeling

The design of 3D-spatio-temporal semantics is characterized by their universal and elementary nature. "Universability" in design means that they remain inde-

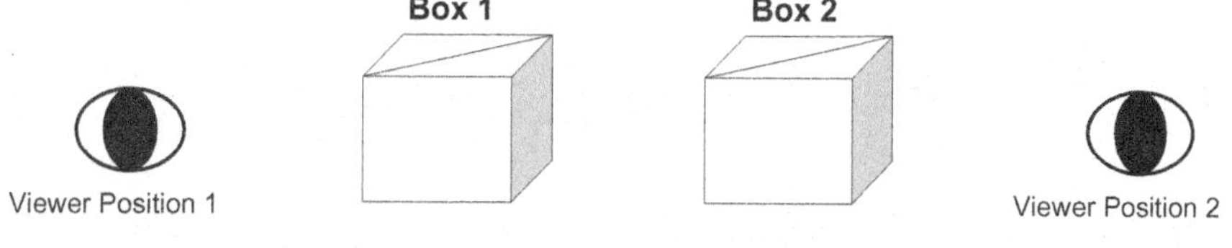

Fig. 8.1. Different viewpoints lead to different relationships

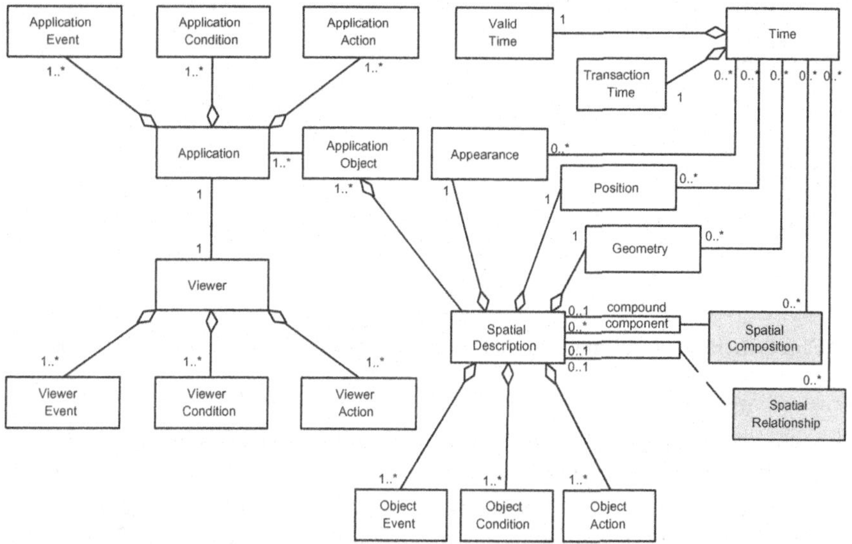

Fig. 8.2. Structural meta-model for spatio-temporal scenario components

pendent from any application and thus can be used in every domain. Elementary nature ensures that they are as minimal as possible. They contain information implied by their types but do not include any context information and do not make any assumptions about how they are visualized. For example, VGeometry describes the geometry and therefore does not include the color or a rendering device.

A relational meta-model is a guide towards creating a framework for the definition of these semantics. A meta-model comes to determine the basic concepts of a domain and to discover knowledge about the data to be managed. It defines

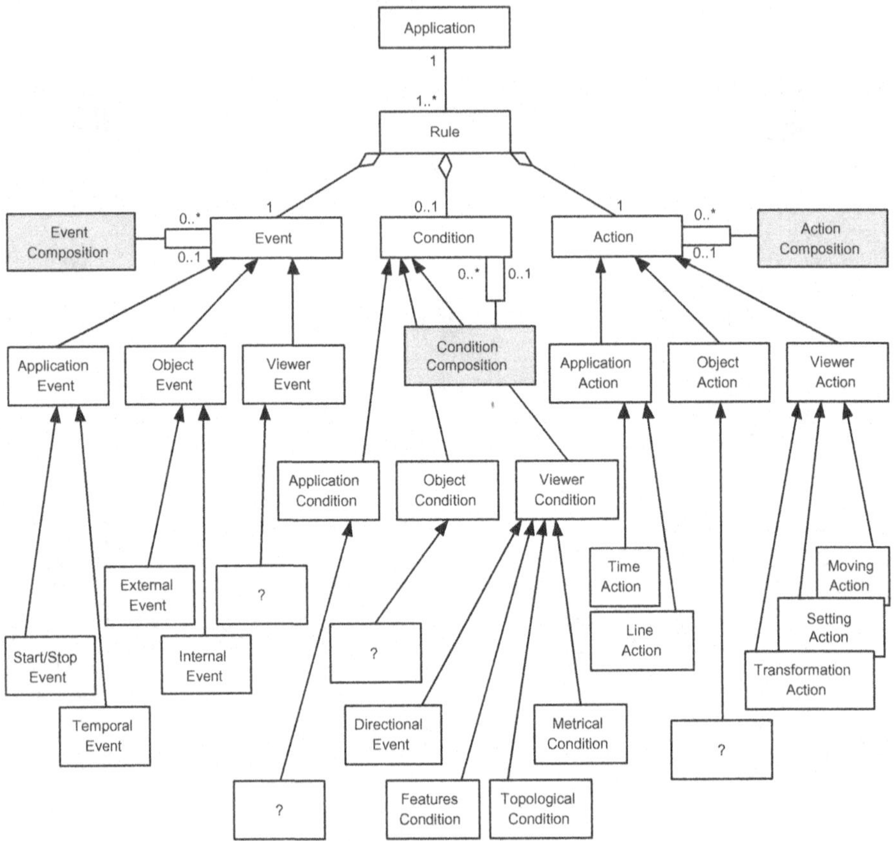

Fig. 8.3. Behavioral meta-model for spatio-temporal scenario components

generic concepts instances of whose are the specific models. Hence, in the design stage it constitutes a reference base for the development of the various models.

We consider that an ISTD consists of two parts, structural and behavioral, presented using UML notation in Figures 8.2 and 8.3 respectively. The entities of the structural part called *Application objects* have two sides, namely attribute and spatial (spatio-temporal). The first one contains the attributes with no spatial context (like the salary of an employee). Because such data may change through time, they acquire an optional bi-temporal timestamp attached on them.

The spatial (spatio-temporal) side describes the viewer and the graphical properties of an object, which can also vary through time. A spatial object is the central concept and it can be either primitive or composite. Primitive objects are the leaf nodes in the object graph and explicitly describe a simple 3D element by specifying its shape (Geometry), its placement inside the world (Position) and its display (Appearance). Primitive spatial objects are grouped together to construct composite ones. The latter consist of primitive and/or complex objects. They need to have only their position specified as their primitive children already describe their geometry and appearance. To generate a spatio-temporal object, a spatial object is combined with temporal constructs in a form of a composite timestamp.

The behavioral part is modeled in terms of *Event Condition Action* (ECA) rules. Events are the fundamental means of interaction and are raised by user actions, by participating objects or by the system. They may be simple or complex, and have attached their spatio-temporal signature (i.e., the space and the time they occurred). Conditions take also as arguments the viewer or objects and finally actions apply to objects. The scenario stands for the integrated behavior of a presentation, i.e. what kind of events it will consume and what actions will be triggered as a result. The different kinds of events, conditions and actions are summarized in Figure 8.3.

In the next subsection, we define the semantics that build the 3D-spatio-temporal Data Model (3D-STDM) for implementing the structural part of this meta-model. These semantics are temporal, spatial and spatio-temporal.

8.3.3 Temporal Semantics

The 3D-STDM temporal semantics provide bi-temporal support to both attribute and spatial data. Even more, they capture the continuous evolution of spatio-temporal objects. First we give formal definitions for the point in time and the trend of object evolution:

Definition 1. A point in time *VTimePoint* is defined as follows:

$$VTimePoint ::= [Year], [Month], [Day], [Hour], [Min], [Sec], [msec], [\mu sec]$$

All *VTimePoint* fields are optional in order to accommodate multiple granularities. Different applications treat time with various details (e.g., year, day, msec). Furthermore, objects in the same application might require different temporal granularities. To switch from one to another we use the following rules:

- When converting from lower to higher granularity (e.g., sec to year), we just delete the parts not required by the higher granularity.
- When converting from higher to lower granularity (e.g., year to sec), we sum all the numbers of lower granularity units that any higher granularity consists of. For instance to convert 5/3/1989 to days we need to calculate the number of days for 1989 years and 3 months and 5 days.

Definition 2. A tuple *VTrend* models the way a 3D-spatio-temporal object changes from an explicitly recorded state into the next one.

$$VTrend ::= mode, [value]$$

Where $mode ::= Integer||String$ and $value ::= Real.$

Tuple *VTrend* applies to spatio-temporal objects only. It is introduced so as to satisfy the requirement for the continuous modeling of the state of an object.

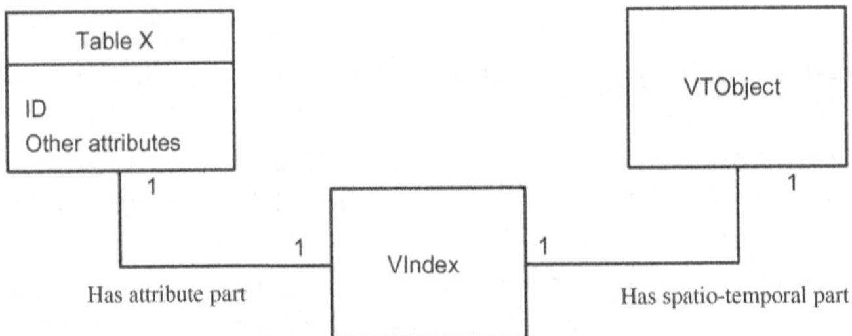

Fig. 8.4. The use of VTrend

VTrend associates two successive recorded states of an object. Field mode encodes the way the object evolved from one state into the next one (e.g., linear, quadratic, cubic etc.). Value is an optional field that comes along with complex modes and specifies the value of the evolution (like the radius in circular movement). Using *VTrend* (Figure 8.4) one can interpolate into the period that two successive object states form and thus extract the state of an object at any point. For instance, if a database has recorded the position of a car at 12:00 and 12:30 and we know that the car is moving with steady speed, we may calculate the car's position at 12:11, 12:20 etc.

An object may evolve in many and application-dependent ways. It is up to the application developer to identify these different ways (modes) and try to simulate them with *VTrend*.

Both point and period timestamps can be formed. *VTimePoint* if used alone is a point timestamp. *VTrend* in conjunction with two successive instances of *VTimePoint* simulates a temporal period. Most important, with the second method, we not only have the state of the object at the ends of the period but through *VTrend* we can calculate the state at any internal point of this period. This combination of *VTimePoint* and *VTrend* leads us to the definition of the integrated timestamp, termed *VTime*:

Definition 3. A Timestamp *VTime* marks the valid and transaction time for a recording. It also provides the *VTrend* link to the previous recording.

$$VTime ::= [valid\_time], [transaction\_time], [mode, value]$$

Where:

$valid\_time ::= VTimePoint,$
$transaction\_time ::= VTimePoint,$
$mode ::= Integer||String$ and
$value ::= Real$

All the attributes of *VTime* are optional. Under this perspective, *VTime* is optional as well. This approach is adopted in order to give the developer the option to apply full bi-temporal timestamp or wherever needed to encompass valid or transaction timestamps only.

VTime will be used to describe internal changes of 3D-spatio-temporal objects. We also need to record alterations in the status of the object (e.g., birth, death) and of the relationships among objects. For this reason we introduce VLife:

Definition 4. A "Life stamp" *VLife* concerns the status of an object. It records the birth and death of the object and points to the object or objects that generated the current object. It is time-stamped.

$$VLife ::= birth, death, life, previds$$

Where:

$previds ::= VTObjectID, VTObjectID$ (explained in the next section)
$birth, death,$
$life ::= VTime.$

8.3.4 3D-Spatial Semantics

The following 3D spatial semantics are also set according to the guidelines of the meta-model of Fig 8.2. First we define the three different parts that comprise a 3D-spatial object namely, geometry, appearance and position.

Definition 5. The geometry *VGeometry* of a primitive 3D-spatial object is either a Box, or a Cylinder or a Cone or a Sphere:

$$VGeometry ::= VGeometryID, (Box||Cylinder||Cone||Sphere)$$

Where:

$VGeometryID ::= Integeridentifier,$
$Box ::= width, depth, height,$
$Cylinder ::= height, radius,$
$Cone ::= height, radius,$
$Sphere ::= radius$
Where $width, depth, height, radius$ are *Real*.

Definition 6. The position *VPosition* of a 3D-spatial object is specified by its translation, rotation, scale and orientation:

$$VPosition ::= VPositionID, Translation, Rotation, Scale, Orientation$$

Where:

$VPositionID ::= Integer,$
$Translation ::= 3DVector,$
$Rotation ::= 3X3Matrix,$
$Scale ::= 3DVector,$
$Orientation ::= 3DVector.$

VPosition is the 3D-Vector abstraction of a 3D-spatial object. It determines the object's placement (translation), size (scale) and direction (rotation, orientation). It is to be used in cases such like distance querying where we need to treat the object as one vector or as one 3D point.

Definition 7. The appearance *VAppearance* of a primitive 3D-spatial object is defined by specifying the following parameters:

$$VAppearance ::= VAppearanceID, (color||texture)$$

Where:

$VAppearanceID ::= Integer,$
$color$ ::=diffuse, specular, emissive, transparency,
$texture$ ::= name of graphics file for the texture diffuse, specular,
$emissive ::= 3DVector,$
$transparency ::= Real.$

By combining these three parts we build the 3D spatial object:

Definition 8. A 3D spatial object *VObject* is defined as:

$$VObject ::= VObjectID, VPosition, [VGeometry][VAppearance]$$

Where:

$VObjectID ::= Integer,$
$VGeometry, VAppearance, VPosition$ as defined above.

The above definition covers both primitive and complex objects. In the case of primitive ones, all three fields of an object must be specified. However, only *VPosition* determines complex objects. The description of their geometry and appearance is delegated to their primitive children.

Finally, the 3D-spatial domain introduces the inclusion of the viewer in the 3D environment. To model the viewer's view, we introduce *VViewpoint.*

Definition 9. The viewer's viewpoint is determined by its position, its direction and its orientation:

$$VViewpoint ::= position, direction, orientation$$

Where $position, direction, orientation ::= 3DVector$

8.3.5 3D-Spatio-temporal Semantics

Up to now, we have listed strictly temporal and 3D-spatial semantics. This section describes the technique for creating firstly spatio-temporal objects and secondly complex objects.

Let us begin with the combination of *VObject* and *VTime*, which leads us to the definition of *VTObject*, capturing the semantics for describing a 3D spatio-temporal object.

Definition 10. A 3D-spatio-temporal object is a 3D-spatial object *VObject* with the particular timestamp *VTime* bound with its attributes.

$$VTObject ::= VObject\ X\ VTime$$

In order to maintain flexible temporal and spatial support, this time stamping is optional. It may occur to those attributes that the system developer considers that they mutate in time. For example the position of the object might need both valid and transaction time recording while its color may remain static (i.e., without any *VTime* support).

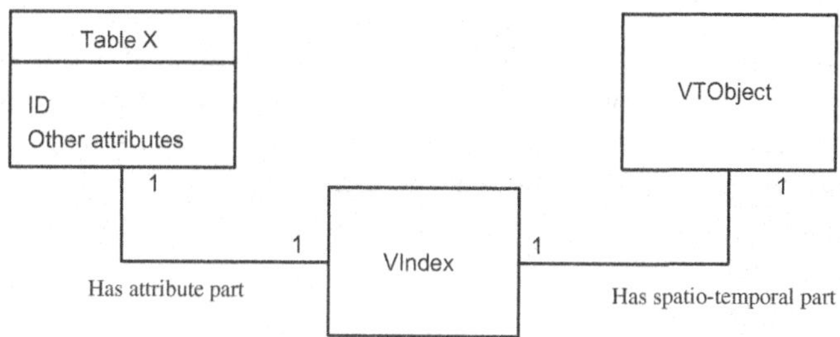

Fig. 8.5. VIndex links the spatio-temporal and the attribute part

Next we introduce the technique for grouping 3D-spatio-temporal Objects together.

Definition 11. The pointer *VIndex* realizes two connections; one between the spatial object and its parent and the other between the spatial and the attribute sides of an object. It also records (through a lifeinst) any changes at the status of the object:

$$VIndex ::= VObjectID, ParentObjectID, TableName,$$
$$KeyID, [timeinst], lifeinst$$

VIndex serves as a bridge between the spatio-temporal and the attribute parts of an object (Figure 8.5). Attributes *VObjectID* and *ParentObjectID* provide the link to the spatial part and its parent respectively. Hence *ParentObjectID* is the notion that groups spatial objects together as it defines a part of relationship. On the other hand, attributes *TableName* and *KeyID* implement the connection to the attribute part. *TableName* is the name that the non-spatial part is classified into while *KeyID* is its primary key value. By using a *VIndex* tuple we can associate the non-spatial attributes of an object (e.g., the constructor of a desk) to the spatial ones (the graphical display of a desk) and vice versa. Because these two associations can change in time, the tuple is time stamped with *timeinst* (wherever it is needed).

8.4 Modeling of Spatio-temporal Behavior

In the previous subsection the static part of the ISTD has been defined in terms of simple and complex objects related to each other with spatial relations. It is important to define ways that the objects in the world behave. We will further use the term scenario to represent the behavioral part. A world involves a variety of individual objects presented according to a set of specifications called Scenario. In our approach a scenario consists of a set of self-standing functional units (scenario tuples) that include: triggering events (for start and stop), presentation actions (in terms of spatio-temporal compositions) to be carried out in the context of the scenario tuple, and related synchronization events. We will define the behavior of objects in terms of rules that represent the behavior of the entities that participate in the scenario.

A rule is a description of the action that an entity should carry out (actions), as a response to a trigger (event), given that a condition is true. This scheme can be modeled using the well known ECA (Event - Condition - Action) scheme from active databases. An ECA rule has the form:

WHEN event G occurs
IF condition S holds
THEN trigger the action A

8.4.1 Modeling Interaction with Events

The concept of event is defined in several research areas. In the area of Active Databases [6] an event is defined as an instantaneous happening of interest. An event is caused by some action that happens at a specific point in time and may be atomic or composite. In the multimedia literature, events are not uniformly defined. In [13] events are defined as a temporal composition of objects, thus they have a temporal duration. In addition to the temporal aspect of an event, which is represented by a temporal instance, there are events in ISTDs that convey spatial information. This is represented by a spatial instance. For example, an event captures the position of visual objects at a certain point in time. Another aspect that is also crucial is that, although the number and multitude of events that are produced both by the user and the system may be huge, we may be interested only in a small subset of them. Thus, events must be treated in a different way in this context, as compared to Active Databases. We define an event in the context of ISTDs as the occurrence of an action, which has attached a spatial and temporal instance. Some interested human or process recognizes the event.

In order to assist the authors in the specification of behavior, we have to provide them with a fundamental repertoire of events. In this framework we further classify the events into categories. The classification of the events is done on the basis of the entity that produces the event. The categories are:

Definition 12. User interaction events are the ones generated explicitly by user interactions. They are mainly input events as the user interacts with the system via input devices such as mouse, keyboard, touch screen, etc.

Definition 13. The category of intra-object events includes events that are related to the internal functionality of an object. This functionality is implemented in object-oriented approaches as method invocation. For instance, the invocation of a method corresponding to temporal access control such as *myObject.start()*, produces an intra-object event. Another source of intra-object events is state changes. The viable states of an object as regards its presentation may be temporal (active, idle, suspended) and/or spatial (show, hidden, layer classification information, etc.) or related to a geometric transformation (translation/rotation/scaling). State events occur when there is a state change of a media object, e.g., image I gets hidden, audio A is started, etc. Intra-object events may indicate a discontinuity in the continuous presentation of a media object.

Definition 14. Inter-object events are events, which occur when two or more objects are involved in the occurrence of an action of interest. These events are raised if spatial and/or temporal relationships between two or more objects hold. In the spatial case, an inter-object event can occur if one object, moving spatially, meets another media object. A temporal inter-object event can occur when the deviation between the synchronized presentations of two continuous media objects exceeds a threshold. Moreover, we may consider spatio-temporal inter-media synchronization events. Such events occur, for instance, when two objects are in relative motion during a specified temporal interval (e.g., object A approaches object B before 2 am).

Definition 15. We classify as complex the events that are defined by the developer. Developers are provided with an initial set of atomic events, namely the ones in the aforementioned categories. In many cases, though, they need to define complex events that are somehow composed of simple ones. Hereafter we present in summary a model for simple and complex events, based on the event concept and classification that has been presented above. A detailed presentation can be found in [15].

Attributes Needed to Represent Generic Events: According to the aforementioned event definition, we need the following attributes to represent a generic event: The subject and object attributes, that are of type *objectList* essentially representing the objects that caused or are affected by the event, respectively. The attribute *spatio_temporal_signature* that takes to the spatial and temporal instances attached to the event when it actually occurs.

As we mentioned before, it is important to provide the tools to the authors for the definition of composite events. The composition of events has two aspects:

i) Algebraic composition is the composition of events according to algebraic operators, adapted to the needs and features of an ISTD,
ii) Spatio-temporal composition reflects the spatial and temporal relationships between events.

In this chapter we will elaborate on the former aspect. The reader may refer to [15] for the latter one.

Algebraic Composition of Events: In many cases the author wants to define specific events that relate to other existing events. We exploited some of the operators presented in other proposals on composite events in Active Databases [6]. We distinguish between the following cases:

Definition 16 - Disjunction.
$e :== OR(e1, ..., en)$: This event occurs when at least one of the events $e1, ..., en$ occurs. For instance we may be interested in the event e occurring when button A ($e1$) or button B ($e2$) was pressed.

Definition 17 - Conjunction.
$e ::= ANY(k, e1, ..., en)$: This event occurs when at least any k of the events $e1, ..., en$ occur. The sequence of occurrence is irrelevant. For example, in an interactive game a user proceeds to the next level when she/he is successful in two out of three tests that generate the corresponding events $e1, e2$, and $e3$.
$e ::= SEQ(e1, ..., en)$: This event occurs when all events $e1, ..., en$ occur in the order appearing in the list. For example, in another interactive game the user proceeds to the next level when she/he succeeds in three tests causing the events $e1, e2$, and $e3$ one after the other.
$e ::= TIMES(n, e1)$: This event occurs when there are n occurrences of event $e1$. This implies that other events may occur in-between the occurrences of $e1$.

In many cases the authors want to apply constraints related to event occurrences in specific temporal intervals. To facilitate this requirement we define a set of operators that are of interest in the context of multimedia applications:

Definition 18 - Inclusion.
$e ::= IN(e1, t\_int)$, event e occurs when event $e1$ occurs during the temporal period t_int with start and end *Vtime* e2,e3 respectively. For example, we might want to detect three mouse clicks in an interval of 1 sec., so that a help window appears. If $t\_int = (e2, e3)$, where $e2$ corresponds to the starting point of a timer while $e3$ corresponds to the end of a timer whose duration is defined as 1 second. The desired event would then be defined as $e = IN(TIMES(3, mouse.click), t\_int)$.

Definition 19 - Negation.
$e ::= NOT(e1, t\_int)$: event e occurs when $e1$ does not occur during the temporal interval t_int.

Definition 20 - Strictly Consecutive Events:
In some cases we are interested in whether a series of events of interest is "pure" or mixed up with other events occurring. The event $e ::= S\_CON(e1, ..., en)$ is raised when all of $e1, ..., en$ have occurred in the order appearing in the list and no other event occurred in between them.

Conditions: A condition is a boolean expression that is evaluated each time a rule is triggered, and if true then the corresponding actions are fired. For instance, in the sentence "If users clicks on object A, *if the object is red,* rotate the object by 90(in the horizontal plane" the words in italics constitute the condition to be checked. The conditions can be classified in the following categories:

- Existence conditions for objects and groups.
- Object State conditions, such as visibility, invisibility, geometric type, dimensions, appearance. As for time dependent media objects, we have the additional temporal states such as "active", "suspended", "idle" etc.
- Group related conditions such as inclusion of an object in a group, number or type of objects in the group etc.
- Spatial relationship conditions, checking the spatial relationship between objects. Such can be: "above", "below", "near", "far", "distance longer than 10 m".
- Relative motion conditions such as: object "moving" "approaching", "goes further", "speed greater than 5m/sec".
- We also consider complex conditions on formed by algebraic conjunction or disjunction operators.

Actions: A rich set of actions is feasible in an ISTD. The interested reader may refer for further details to [16]. The actions are classified in the following broad categories:

Definition 21. Object related actions are related to the spatial features of an on object such as showing/hiding an object or applying a geometric transformation. Here we have to stress that these actions refer to showing/hiding all the constituents (children) of complex objects. We also have the actions representing the set of recognized geometric transformations like: Move, Rotate, Scale. Each of these actions are applied for a specific time interval using a law $f(t)$ defining the evolution of the action in time (t). The actions are formally defined as follows:

$$HideAction ::= HideChildrenID$$
$$ShowAction ::= ParentID, ChildrenID, ChildrenID$$
$$MoveAction ::= MoveID, For, WithLaw$$
$$RotateAction ::= RotateID, For, WithLaw$$
$$ScaleAction ::= ScaleID, For, WithLaw$$
$$ActionParameters ::= ParamID|Params$$

Where:

$For ::= VTimePoint,\ ParentID,\ ChildrenID,\ MoveID,\ RotateID,$
$ScaleID ::= Integer,$
$WithLaw ::= ActionParameters$ and
$Param ::= Float.$

Definition 21. The actions that are viewed from one or more potentially moving viewpoints, are called camera related actions. Cameras that are located in the local coordinate system characterized by its position and orientation handle the viewpoints. The translation and rotation transformations that were mentioned above can be applied to the camera also. Then the related actions are:

MoveCameraAction ::= MoveCameraID, For, WithLaw
RotateCameraAction ::= RotateCameraID, For, WithLaw

Where:

$For ::= VTimePoint,\ MoveCameraID,$
$RotateCameraID ::= Integer,$
$WithLaw ::= ActionParameters$ and
$Param ::= Float.$

These actions imply that the camera moves/rotates for time interval. In many cases we need to impose a time interval between two actions. This is achieved with the TimeGapAction defined as:

$$TimeGapAction ::= Float$$

Also in many cases we need to include a sound in our document, thus we need the predicate: $SoundAction ::= Sound"URL"SourceID$ that assigns a sounds found in a URL to the identifier ID.

Each action can be started, suspended, restarted or stopped (killed) before its completion. Thus we need the control primitive:

$$Control ::= play|pause|kill|rewind$$

The actions can be composed to forms chains of related actions that correspond to sets of grouped actions as perceived by the scenario author. The complex actions are defined as action sequences as follows:

$Sequence ::= SequenceID, SequenceItem, SequenceItem$
$SequenceItem ::= Item, Control, ActionUseAction ::= USEID$

Where:

$SequenceID, UseID ::= Integer,$
$Item ::= Float$

Rule-Based Scenarios: As we have defined the constituents of the rules, we now define the rule as constituents of the scenario, which stands for the behavioral part.

Definition 22. A behavior rule is defined according to the following grammar:

$< Rule >::= TriggeringEvent, Condition, Action$
$TriggeringEvent ::= anysimpleorcomplexevent$
$Condition ::= anysimpleorcomplexevent$
$Action ::= anysimpleorcomplexaction$

8.5 Database Support for Scenario Components

As mentioned above, components of ISTD are regarded as objects varying in 3D-spatial and temporal dimensions. As they are complex objects, their specification is a laborious and time-consuming procedure. By providing database support we offer a gateway to reusability where objects can be stored and retrieved possibly through the use of specialized criteria. In this section we specify how to provide such support.

3D-STDM is defined under the concept of orthogonality among temporal, spatial and spatio-temporal semantics. This is because as it is understood, in an information system there are data that do not have any spatial context (but might vary in time) along with strictly spatial data (that do not mutate) and spatio-temporal data. Having kept these semantics orthogonal, we supply the system designers with control over granularity of support and give them the opportunity to choose where to provide temporal, spatial or spatio-temporal support. This task is delivered by applying the following guidelines:

Temporal Support. Use *VTime* without *VTrend* and attach the *valid_time* and/or *transaction_time* fields to any attributes that require valid and/or transaction time stamping.

Spatial Support. Create a new entry of *VIndex* without using the *timeinst* attribute. Acquire the 3D-spatial description of the object and link it to its attribute part using the rest of the *VIndex* entry attributes. If the 3D-spatial description contains complex objects, use the *ParentObjectID* attribute to provide links inside the spatial description graph.

Spatio-temporal Support. The change in the state of a single object is handled by connecting the new state to its preceding one using *VTrend.* In detail, the types of different changes (from an ontological perspective [9]) are dealt with as:

- Birth: Similarly to spatial support, create a new entry of *VIndex*, acquire the initial spatial description and link it to the attribute part. This time use *timeinst*. *VLife* should have birth value equal to as *timeinst*, death value null and no previds. Even more, attach *VTime* (with *VTrend*) to all spatial attributes that vary in time. Depending on this variation, it may be either a full bi-temporal or a valid time or a transaction time timestamp.
- Death: For the object's entry in *VIndex* update its "life". The death field should acquire the appropriate *VTime* value.

- Reincarnation: Follow the same procedure as in Birth but this time, previd should be the *ID* of the object(s) that the current object reincarnates.
- Evolution: Mark the new object's birth, the old object's death and update the previd of the new object with the *ID* of the old one.
- Aggregation: Create a new entry in *VIndex* including *timeinst*. Set life's birth to the appropriate value and populate the previds field with the identifiers of the aggregating objects. For each of the latter ones, mark their death.
- Separation: Create as many entries in *VIndex* as the produced objects. Set their birth and the death of the producing object.
- Spawning: Follow the same procedure as in aggregation but without marking the death of the spawned object.
- Mixture: Mark the death of the objects that are to be mixed (except of the object that incorporates the mixture) and update the previds of the resulting object.
- Fusion: Follow the same procedure as in separation but without marking the death of the fused object.

SQL (OQL) statements can handle temporal, spatial and spatio-temporal support. Finally we briefly introduce special statements for the definition and the insertion of VObject and VTObject structures (see Figure 8.6).

Operator	Signature
CREATE VTObject	CREATE VTObject (vtobj, id, parid,table,key, time, [attName, temSup])
CREATE VObject	CREATE VObject (vobj, id, parid, table, key,)
INSERT VNEWVALUE	INSERT VNEWVALUE (name, value, time, id)
INSERT ANEWVALUE	INSERT ANEWVALUE (name, key, value, time)
INSERT VNEWVINDEX	INSERT VNEWINDEX (objid, parid, table, key, time, life)

Fig. 8.6. Creation and manipulation operators

8.5.1 Querying and Accessing Stored Components

Querying stored objects is delivered with 3D-STQL, which is developed as a set of functions and predicates that can be incorporated into query statements. This approach was preferred to that of extending SQL (or OQL) because:

- 3D-STQL can be directly implemented in a number of popular platforms such as Oracle and O2. When providing query language extensions, the use of such platforms is not straightforward.
- It is not bound to either SQL or OQL but instead it can be used as a supplement to both of them. The operators are formally defined hence there is not any question of lacking a formalized foundation.

- Finally 3D-STQL, as it deals with graphical information, should be accessed through a graphical user interface. The details of implementation should be made transparent to the user who would ask queries in a visual, not a textual way. Therefore, from the user's perspective it is insignificant how the operators relate to the query language.

Table of Figure 8.7 lists the operators of 3D-STQL organized as temporal, spatial and spatio-temporal. Apart from the conventional spatial predicates we introduce 3D-spatial operators whose result depends on the viewer's position. For the evaluation of spatio-temporal topological operators we followed the principle that a relationship is true (or not) in a period if it is true (or not) for each time point in that period. Otherwise the result is indefinite.

Temporal Operators	VObject At(VTObject, VtimePoint) VTObject Retrieve (VTObject, VTimePoint, VTimePoint) Boolean Before (VtimePoint, VTimePoint, VTimePoint, VTimePoint) Boolean After (VtimePoint, VTimePoint, VTimePoint, VTimePoint) Boolean Begin (VtimePoint, VTimePoint, VTimePoint, VTimePoint) Boolean Finish (VtimePoint, VTimePoint, VTimePoint, VTimePoint) Boolean Overlap (VtimePoint, VTimePoint, VTimePoint, VTimePoint) Boolean Contain (VtimePoint, VTimePoint ,VTimePoint, VTimePoint) Boolean Equal (VtimePoint, VTimePoint ,VTimePoint, VTimePoint)
Spatial Operators	
Topological	real Distance((VObject, Vobject) realDistance((VObject, plane) real Volume(VObject) Boolean Equal(VObject, Vobject) Boolean Disjoint(VObject, VObject) Boolean Meet(VObject, Vobject) Boolean Overlap(VObject, VObject) Boolean In(VObject, VObject)
Viewpoint Dependent	Boolean IsVisible(VObject, plane) VBoolean Infront(VObject, VObject) VBoolean Behind(VObject, VObject) VBoolean Left(VObject, VObject) VBoolean Right(Vobject, VObject) VBoolean Above(VObject, VObject) VBoolean Below(VObject, VObject)
Viewpoint Independent	VBoolean Upon(Vobject, VObject) VBoolean Beneath(VObject, VObject)
3D-spatio-temporal Operators	
Topological	Boolean Equal(VTObject, VTObject, VTimePoint, VTimePoint) Boolean Disjoint(VTObject, VTObject, VTimePoint, VTimePoint) Boolean Meet(VTObject, VTObject, VTimePoint, VTimePoint) Boolean Overlap(VTObject, VTObject, VTimePoint, VTimePoint) Boolean In(VTObject, VTObject, VTimePoint, VTimePoint)
Viewpoint Dependent	VBoolean Infront(VTObject, VTObject, VTimePoint, VTimePoint) VBoolean Behind(VTObject, VTObject, VTimePoint, VTimePoint) VBoolean Left(VTObject, VTObject, VTimePoint, VTimePoint) VBoolean Right(VTObject, VTObject, VTimePoint, VTimePoint) VBoolean Above(VTObject, VTObject, VTimePoint, VtimePoint) VBoolean Below(VTObject, VTObject, VTimePoint, VTimePoint)
Viewpoint Independent	VBoolean Upon(VTObject, VTObject, VTimePoint, VTimePoint) VBoolean Beneath(VTObject, VTObject, VTimePoint, VTimePoint)
Behavioral Operators	Boolean Grow(VTObject, VTObject) Boolean Shrink(VTObject, VTimePoint, VTimePoint) VBoolean Approach(VTObject, VTimePoint, VTimePoint, VViewpoint) VBoolean Leave(VTObject, VTimePoint, VTimePoint) Real Differentiation(VTObject.(Attribute), VTimePoint, VTimePoint)

Fig. 8.7. 3D-STQL full set of operators

As our objective was to produce a universal and application-independent set, these operators are kept generic. Nevertheless, they can be employed for the definition of complex functions. For example given the *VObject v1,v2* and *v3* and the *VViewpoint vp1*, spatio-temporal predicates *between* and *upon* are defined as:

- Between (v1, v2, v3, vp1): Infront(v1,v2,vp1) AND Infront(v2,v3,vp1)
- Upon (v1,v2): Over(v1,v2) AND Meet(v1,v2)

Composite spatio-temporal behavior can be expressed in a similar way as a combination of spatio-temporal and temporal operators. For example:

- Overtake (v1, v2, vp1, t1, t2, t3, t4): Behind (v1, v2, vp1, t1, t2) AND Infront (v1, v2, vp1, t3, t4)

8.5.2 A Global Architecture

Figure 8.8 reviews how to provide database support to the construction of ISTD. A presentation is populated by components retrieved from a database repository with the use of specialized criteria. This way we increase reusability (a component can be used in more than one presentations) while ensuring that the components retrieved meet the user's requirements. After a customization stage which arranges components in a presentation and thus produces the scenario (consisting of structural and behavioral elements), the latter is mapped to a spatial or spatio-temporal data format. Conversely such data formats are used for the population of the repository.

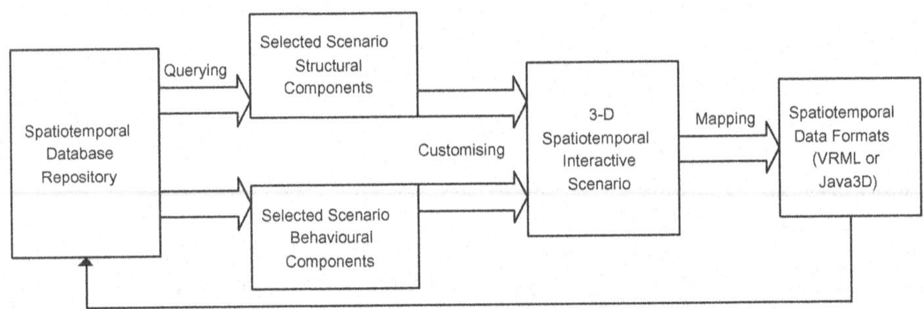

Fig. 8.8. Database support for Interactive spatio-temporal documents

8.6 Examples of Applications

Hereafter we present a very simple example leading with a table and a lamp. In this example we use a relational database and we show how data related to the table and the lamp are deployed through the different relations. An object may be built as a composition of basic ones and the table is described as a composition between four legs (Cylinders with IDs 3, 4, 5, 6) and a plane (Box with ID 2).

For the modeling of spatio-temporal data the following tuples of *VTIndex* are used (Figure 8.9)

VTObjectID	VTime	TTime	Parent ObjectID	Table Name	KeyID	LifeInst
1		2000	-1	-	-	-
2		2000	1	-	-	-
3		2000	1	-	-	-
4		2000	1	-	-	-
5		2000	1	-	-	-
6		2000	1	-	-	-

Fig. 8.9. *VTIndex* Tuples referring to the table

After that, one tuple in *VGeometry* is used to store the geometry of the plane (Figure 8.10)

VGeometryID	BoxID	ConeID	CylinderID	SphereID	VTObjectID
1	1	-1	-1	-1	3

Fig. 8.10. *VGeometry* tuple referring to the pane of the table

The geometry of the table is specified with tuples in VBox (Figure 8.11) and in detail with tuples in *VPoint* table (e.g., height in Figure 8.12). All values have a transaction timestamp.

BoxID	WidthID	HeightID	DepthID	VPointID
1	1	2	1	1

Fig. 8.11. VPoint tuples referring to the height of the table

VTime	TTime	Value
-	1	.3

Fig. 8.12. Timestamps for the plane of the table

The lamp is modeled with a bottom plane, a cylinder and a cone. Two switches relate to the lamp; when the first is pressed it turns the lamp on and rotates the lamp around the vertical axis with the rotational speed of 30 degrees/sec. The other switch turns off the lamp and stops the rotation.

```
ON CLICK (onSwitch)
IF lamp.SwitchLight = FALSE
DO
Lap.SwitchLight = TRUE
Lamp.rotate(vertical_axis,30).start()
END DO

ON CLICK (offSwitch)
IF lamp.SwitchLight = TRUE
DO
Lap.SwitchLight = FALSE
Lamp.rotate(current()).stop()
END DO
```

8.7 Related Work

There are several tools, which produce enormous code for simple geometry (using more complex geometries -IndexedFaceSet- to define it with thousands of 3Dcoordinates; like i.e., AC3D). In [8], SCORE, a distributed object oriented system for 3D real time visualization is presented. It achieves a reusable design that can be used to develop other virtual 3D buildings where spatial navigation is required. In this work, emphasis is given to hypermedia and spatial navigations. Though the system lacks a methodology and a clear model for spatial navigation.

The specification of 3D-spatio-temporal models can derive form works that model animation in a 3D environment. MAM/VRS is an object-oriented approach [5] that defines object ADT for both the spatial (geometrical) and the temporal attributes of a 3D-spatio-temporal object. Using these ADT the user can specify interactive animations but though the flow of time is controlled there is no relationship between the spatial and the temporal constructs.

A continuation of this work is Virtual Reality Modeling Language (VRML), an attempt to introduce a common and standard language for advanced graphics and virtual reality. VRML aims to be efficient enough to be used in standalone applications and in the World Wide Web. Theoretically it can accommodate anything from 3D geometry and MIDI data to JPEG images, links and simple text. It was created in 1995 as a consortium of academic and commercial partners. Its first version (VRML 1.0) addressed the modeling of 3D objects in an open and extensible format. VRML consists (in its latest version) of 55 commands or nodes. Each node specifies a geometry, an appearance, the background, the behavior etc. VRML caters for the representation of composite objects organized in a scene graph. The elements of a scene graph have individual positional properties and appearance. As a whole however, they yield to the transition or the behavior of the scene graph. If for example an object is placed at point (2, 1, 1) and the scene graph where it belongs is placed at (-2, -1, -1) then the absolute position of the object is at (0, 0, 0). VRML follows an orthonormal 3D system of coordinates. It supports time thus providing a gateway for spatio-temporal presentations. Time in VRML starts at January 1st, 1970 (although this absolute date is rarely important) flows continuously and cannot be stopped.

Applications of VRML vary from scientific visualizations to 3D pictures for CAD [2] to virtual presentations of stadium, city centers [12] and airports [11]. The advantages from using VRML are interoperability, Internet gateway and easy 3D navigation.

VRML is a data format and although it provides an interface to programming languages, the use of programming for advanced interaction it is not always the best solution in terms of performance and security. Sun Microsystems has developed Java3D [7], a Java API with classes to model and display 3D data. Java3D copies the philosophy of VRML with classes that duplicate VRML nodes and the scene graph technique. The main difference is that 3D modeling is done with programming and the final result is a Java application or an applet (for browsers which in the future will support Java3D.

The need to extend VRML to the direction of database support has been acknowledged by VAG (VRML Architecture Group), the governing body for VRML [17] which established the VRML Database Working Group [18]. Its goals are to define a standard set of API for persistence, scalability and security. Their fulfillment will enable VRML worlds to be stored in multi-user relational and object databases and be the foundation for building reusable complex and robust worlds.

Two proposals have been submitted in this direction. Adding regions defines region node, which has an identifier and fields containing built-in topological information (distance and visible region). VRML Data Repository [19] defines C and Java mappings, which enable VRML to be mapped to procedural and object databases. Although these two works are still in the development stage, they clearly demonstrate the interest in providing a database interface to VRML.

Another proposal, which relates VRML and databases, is specified in [14]. It focusses on the issues of reusability and sharing of VRML data. They facilitate these objectives by specifying semantic associations of VRML files to non-spatial information and by storing both of them in an object database. The user can retrieve VRML worlds by querying the associated information. While this effort provides an answer, it does not constitute a complete solution; firstly, retrieval is delivered with non-spatial criteria while VRML objects are primarily graphical objects and secondly its reusability is limited because it is based on object databases and excludes relational models.

In [4], the concept of virtual environment (VE) is introduced. The main feature of this system is its navigation means, which is spatial, and the interactivity facilities. VEs are described through three fundamental kinds of element: context(s) (for structuring the environment), object(s) (which can be passive, reactive or active and movable or non-movable), and users that can activate objects, collect or move them. On the contrary of our approach we consider only one context and we only distinguish the avatar from the other objects. The dynamic features of a VE are based on the dynamic features of its elements. All elements of a VE are described in terms of their behavior, interaction with the other components and their dependencies on them. Collaboration between components (user-object, object-context, user-context, user-user, context-context) is defined through constellation's behavior. Constellations define object instances at runtime and the relationships among them. In our approach relationships between objects are described through reactions to events under conditions. Based on the constellation analysis, the authors present two behavioral patterns: the compound pattern and the collector pattern. These patterns focus on the way of propagating the compound's behavior to the group.

In [3] methods to create complex scenarios for the IOWA-driving simulator are presented. A scenario specifies the behavior of synthetic vehicles, pedestrians, traffic control devices and variations in weather and lighting. To model basic behaviors of vehicles and other active entities, hierarchical concurrent state machines are used. The hierarchy of the state machine allows abstraction in behavior modeling; the concurrency allows attending to multiple constraints and goals (through constraints resolution methods). Each state machine encodes a behavior, each instance of a state machine deals with a vehicle. Diversity is achieved by varying the attributes of the vehicle (preferred driving speed, desired following distance, etc.). To create complex scenarios (i.e., scenarios that meet experimental needs while maintaining diversity, reactivity and realism in object behaviors), the generated sequences of events blend smoothly. In addition, behavior controller objects are provided (and implemented also through hierarchical concurrent state machines) to allow interactions among the object of the virtual worlds through sending of messages. A scene editor and a scenario editor have been developed.

In [5] an interesting approach is presented for uniform modeling of geometry and behavior of a 3D graphics and animation framework. More specifically the authors specify separately geometry and behavior using corresponding nodes organized in two separate directed acyclic graphs, the geometry and the behavior graph. Moreover they provide rendering independent graphics objects that can be visualized by geometry nodes constrained by time and event-rated behavior nodes and finally provide high-level 3D widget nodes that combine geometry and behavior nodes.

In [10] an integrated effort for building 3D animations is presented. The system provides a small set of modeling primitives, namely the graphical objects, their properties and the callbacks. The graphical objects are used for constructing scenes, the properties are time-variant and are used for animating various aspects of a scene while the call backs are the primitives used to handle interaction. As for the graphical objects, they basically define the geometric content of the animation also defining groups that are essentially the scene graphs. The properties are of several kinds and define the behavior of the objects. The interaction in this system is limited to user input (such as selection) and does not cover the internal animation interaction, which may be very rich.

In [1] a language is presented for specification of virtual agents behavior. The language presents a higher level behavioral model putting emphasis in temporal logic background enriched with qualitative spatial and temporal predicates. The system uses Petri-nets for rendering the animation.

8.8 Conclusions

We described mechanisms for managing the structural and behavioral parts of ISTDs. Such applications are rich in spatio-temporal features both in their structure and behavior and therefore this task was made easier by studying the work done in the spatio-temporal databases area. We also showed how to store the components of ISTDs in database repositories in order to reuse them for future, customized composition of worlds. The need for 3D interactive content in the areas of cultural, tourism and transportation application is profound. Yet current languages do not provide database support and are not high-level enough and easy to use for specifications.

Particular requirements for modeling the structural part of this domain are:

i) the continuous treatment of time,
ii) the inclusion of the viewer in the data model semantics,
iii) the linking to the descriptive (attribute) data.

Given these requirements we defined 3D-STDM, the metric, object-based data model of 3D-spatio-temporal data. It follows an open format and hence can be employed in order to store spatio-temporal components in either object-oriented or relational databases. Indeed there is a discussion over the suitability of the relational model to represent specialized information (such as spatio-temporal). Nevertheless we chose not to discard the relational model because it is still a popular database model and is supported by well built and test commercial database servers (e.g., Oracle, Sybase etc.). The main semantics of 3D-STDM namely, *VTime*, *VObject* and *VTObject* are orthogonal among themselves thus giving the developer the opportunity to provide grades of temporal, spatial and spatio-temporal support.

Towards modeling of the behavioral part, we have presented a model that enables definition of interactive spatio-temporal scenarios. The fundamental features of the language are: (i) complex 3D objects (any complex object can be defined through a set of elementary ones connected via a set of spatial relationships); (ii) ECA-like rules where E is the triggering event, C the spatio-temporal conditions that must be fulfilled and A a set of actions connected by temporal intervals.

The main advantages of our approach are the following:

- It provides a high level generic declarative definition of complex 3D scenarios - therefore, suitable for data base storage and retrieval (issue of further research);
- It covers internal and external interaction;
- It handles complex objects specification based on spatial relationships, making thus definitions natural and easy;

- A complete coverage of all three spatial transformations (translation, rotation, scaling) in 3D is provided in addition to a composition scheme (i.e., a translation simultaneously with a rotation).

The visualization of ISTDs is currently implemented using VRML but it can also be implemented in any visualization language.

References

1. A. Del Bimbo and E. Vicario. Specification by example of virtual agents behavior *IEEE Transactions on Visualization and Computer Graphics*, 1(2), December 1995.
2. VRML Venus pictures, `http://www-venus.cern.ch/VENUS/LHC_pict.htm`.
3. J.F. Cremer and J.K. Kearney. Scenario authoring for virtual environment. In *Proc. Of IMAGE VII Conf.*, Tucson, AZ, June 12-17, 1994.
4. A. Diaz and R. Melster. Patterns for modeling behavior in virtual environment applications. *Second Workshop on Hypermedia Development: Design patterns in hypermedia* (in conjunction with Hypertext 99). Darmstadt, Germany, February 1999.
5. J. Dollner and K. Hinrichs, Object-Oriented 3D Modelling, Animation and Interaction. *J. of Visualisation and Computer Animation*, Vol. 8, p33-64, 1997.
6. N.H. Gehani, H.V. Jagadish, and O. Shmueli. Composite Event Specification in an Active Database: Model and Implementation. In *Proceedings of VLDB Conference*, 1992, pp. 327–338.
7. Java3D Specification. `http://java.sun.com/products/java-media/3D/`.
8. R. Melster, A. Diaz, and B. Groth. SCORE - The virtual museum, development of a distributed, object-oriented system for 3D real-time visualization. Technical report 1998-15, TU Berlin, Germany. October 1998.
9. D. Medak. Lifestyles - An Algebraic Approach to Change in Identity. In *STDBM'99*, Edinburgh, UK, September 1999.
10. M. Najork and M. Brown. Obliq-3d: A high Level, Fast Turnaround 3D Animation System. *IEEE Transactions on Visualization and Computer Graphics*, 1(2), June 1995.
11. Schiphol Airport in 3D, `http://www.schiphol.nl/maps/3d.htm`.
12. VRML world of piazza di Siena `http://www.planetitaly.com/Spazio/VRML/siena.wrl.gz`.
13. G. Schloss and M. Wynblatt. Providing definition and temporal structure from multimedia data. *Multimedia Systems Journal*, 3:264-277, 1995.
14. M.Kamiura, H. Oiso, K. Tajima, and K. Tanaka. Spatial Views and LOD-Based Access Control in VRML-object Databases. Springer Lecture Notes in Computer Science Vol. 1274.
15. M. Vazirgiannis and S. Boll. Events in Interactive Multimedia Applications: Modeling and Implementation Design. In *Proceedings of IEEE International Conference on Multimedia Computing and Systems (ICMCS'97)*, June 1997, Ottawa, Canada.
16. M. Vazirgiannis and S. Lazaridis. STEDEL: A language for interactive spatio-temporal contexts. Technical report, Athens Univerity of Economics & Business.
17. The VRML Architecture Group `http://vag.vrml.org`.
18. The VRML Database Working Group `http://www.vrml.org/WorkingGroups/dbwork/`.
19. VRML Data Repository API, Oracle proposal, 1997 `http://www.olab.com/vrml/overview.html`.

9 Spatio-temporal Databases in the Years Ahead

Manolis Koubarakis[1], Yannis Theodoridis[2], and Timos Sellis[3]

[1] Technical University of Crete, Greece
[2] University of Piraeus, Greece
[3] National Technical University of Athens, Greece

9.1 Introduction

CHOROCHRONOS has been a fruitful and enjoyable project. It contributed many innovative ideas in the areas of ontology and data modeling, query evaluation and prototype systems for spatio-temporal databases. Our ideas have already found uses in various application domains such as moving object databases (see Chapter 4), environmental information systems (see the Dedale application in Chapter 5), interactive multimedia applications and virtual worlds (see Chapter 8).

CHOROCHRONOS has opened many avenues for research in spatio-temporal databases, but it also left us with lots of challenging research problems awaiting solution. Many of these problems have already been emphasized in the concluding sections of each chapter, and there is no reason to repeat them here.[1] As an epilogue to this book, we would like to challenge the reader by discussing three important application areas and the role spatio-temporal databases can play in these.

9.2 Mobile and Wireless Computing

The main concept of interest here is the concept of *location* (location of mobile clients, moving application objects and so on) and how it changes over time (a nice recent survey of this area is [10]). This application area has motivated a lot of spatio-temporal research recently (for research carried out outside CHOROCHRONOS see [13,14,4] and [11,6]) but there are many aspects of the problem that have not been looked at in detail. In particular, all approaches seem to adopt a centralized database view of the problem while the problem is clearly distributed [10]. Towards this direction, a recently launched European project [3], where CHOROCHRONOS researchers participate, considers both moving and stationery objects as agents that play the roles of data servers, producers and clients interchangeably. Open issues in all aspects of mobile databases arise under this consideration.

[1] We are sure the seasoned database researcher can easily imagine many others!

T. Sellis et al. (Eds.): Spatio-temporal Databases, LNCS 2520, pp. 345–347, 2003.
© Springer-Verlag Berlin Heidelberg 2003

9.3 Data Warehousing and Mining

There is currently a huge amount of spatio-temporal data that has been collected over the years. The concept of a data warehouse has naturally been extended from alphanumeric data to temporal, spatial and spatio-temporal data and efficient implementation of OLAP (on-line analytical processing) operations have been studied [7,5,12,8,9]. Of particular interest here is the mining of spatio-temporal patterns, since it can lead to important observations in many applications (e.g., environmental monitoring and fleet management). There is very little work in this area and much remains to be done [15].

9.4 The Semantic Web

According to the Web's inventor Tim Berners-Lee, Jim Hendler and Ora Lassila "the Semantic Web will bring structure to the meaningful content of Web pages, creating an environment where software agents roaming from page to page can readily carry out sophisticated tasks for users." [2,1]. The main technologies underlying this vision are *knowledge representation* and *ontologies* for formalizing meaning, *software agents* as the most useful abstraction for a computational entity and *XML* as the universal language for encoding and sharing information.

There is currently a lot of excitement in this area with much research done under the auspices of W3C[2] and DARPA (e.g., see the effort to develop DAML - DARPA's Agent Markup Language[3]). Ideas from spatio-temporal ontologies, models, languages and query processing algorithms (as discussed in various chapters of this book) cannot really be absent from the Semantic Web vision and there is much interesting research to be carried out by spatio-temporal database researchers.

A related effort is the Geography Markup Language (GML) proposed by the OpenGIS consortium.[4] This is an XML-based language for the storage and transfer of geographic information, including both the spatial and non-spatial properties of geographic objects. This is a good step towards achieving interoperability among existing geographical applications.

9.5 Conclusions

We believe that new applications like the ones sketched above will be the ultimate test for the arsenal of spatio-temporal concepts, models and algorithms developed in CHOROCHRONOS. We invite the reader to join us in taking up the challenge!

---

[2] `http://www.w3c.org`.

[3] `http://www.darpa.mil`.

[4] `http://www.opengis.org`.

References

1. Tim Berners-Lee and Mark Fischetti. *Weaving the Web: The original design and ultimate destiny of the World Wide Web, by its inventor*. Harper, 1999.
2. Tim Berners-Lee, James Hendler, and Ora Lassila. The Semantic Web. Scientific American, May 2001.
3. DBGlobe: A Data-centric Approach to Global Computing. See `http://softsys.cs.uoi.gr/dbglobe/index.htm`.
4. M. Hadjieleftheriou, G. Kollios, V. Tsotras, and D. Gunopulos. Efficient Indexing of Spatiotemporal Objects. In *Proceedings of the 8th International Conference on Extending Database Technology (EDBT 2002)*, Prague, March 2002.
5. J. Han, N. Stefanovic, and K. Koperski. Selective Materialisation: An Efficient Method for Spatial Datacube Construction. In *Proceedings of the Pacific-Asia Conference on Knowledge Discovery and Data Mining (PAKDD'98)*, pages 144–158, Melbourne, Australia, April 1998.
6. I. Lazaridis, K. Porkaew, and S. Mehrotra. Dynamic Queries over Mobile Objects. In *Proceedings of the 8th International Conference on Extending Database Technology (EDBT 2002)*, Prague, March 2002.
7. A.O. Mendelzon and A.A. Vaisman. Temporal Queries in OLAP. In *Proceedings of the 26th International Conference on Very Large Databases (VLDB 2000)*, pages 242–253, Cairo, Egypt, September 2000.
8. D. Papadias, P. Kalnis, J. Zhang, and Y. Tao. Efficient OLAP Operations in Spatial Data Warehouses. In *Proceedings of the 7th International Symposium on Spatial and Temporal Databases (SSTD 2001)*, pages 59–78, Redondo Beach, California, USA, July 2001.
9. D. Papadias, Y. Tao, P. Kalnis, and J. Zhang. Indexing Spatiotemporal Data Warehouses. In *Proceedings of the 18th International Conference on Data Engineering (ICDE 2002)*, San Jose, California, February 2002.
10. E. Pitoura and G. Samaras. Locating Objects in Mobile Computing. *IEEE Transactions on Knowledge and Data Engineering*, 13(4):571–592, 2001.
11. K. Porkaew, I. Lazaridis, and S. Mehrotra. Querying Mobile Objects in Spatiotemporal Databases. In *Proceedings of the 7th International Symposium on Spatial and Temporal Databases (SSTD 2001)*, pages 59–78, Redondo Beach, California, USA, July 2001.
12. N. Stefanovic, J. Han, and K. Koperski. Object-based Selective Materialisation for Efficient Implementation of Spatial Data Cubes. *IEEE Transactions on Knowledge and Data Engineering*, 12(6):938–958, 2000.
13. J. Su, H. Xu, and O.H. Ibarra. Moving Objects: Logical Relationships and Queries. In *Proceedings of the 7th International Symposium on Spatial and Temporal Databases (SSTD 2001)*, pages 3–19, Redondo Beach, California, USA, July 2001.
14. G. Trajcevski, O. Wolfson, F. Zhang, and S. Chamberlain. The Geometry of Uncertainty in Moving Objects Databases. In *Proceedings of the 8th International Conference on Extending Database Technology (EDBT 2002)*, Prague, March 2002.
15. I. Tsoukatos and D. Gunopulos. Efficient Mining of Spatiotemporal Patterns. In *Proceedings of the 7th International Symposium on Spatial and Temporal Databases (SSTD 2001)*, pages 425–442, Redondo Beach, California, USA, July 2001.

List of Contributors

Michael H. Böhlen
Department of Computer Science
Aalborg University
DK-9220 Aalborg Øst
Denmark
boehlen@cs.auc.dk

Martin Breunig
Institute of Environmental Sciences
University of Vechta
P.O. Box 1553
D-49364 Vechta
Germany
mbreunig@iuw.uni-vechta.de

Stefan Dieker
Praktische Informatik IV
FernUniversität Hagen
58084 Hagen
Germany
Stefan.Dieker@FernUni-Hagen.de

Martin Erwig
Praktische Informatik IV
FernUniversität Hagen
58084 Hagen
Germany
Martin.Erwig@FernUni-Hagen.de

Luca Forlizzi
Dipartimento di Informatica
Università degli Studi di L'Aquila
I-67010 L'Aquila, Italy
forlizzi@univaq.it

Andrew U. Frank
Dept. of Geoinformation
Technical University of Vienna
A-1040 Vienna
Austria
frank@geoinfo.tuwien.ac.at

Stéphane Grumbach
INRIA
Rocquencourt BP 105
78153 Le Chesnay Cedex
France
stephane.grumbach@inria.fr

Ralf Harmut Güting
Praktische Informatik IV
FernUniversität Hagen
58084 Hagen, Germany
Ralf-Hartmut.Gueting
@FernUni-Hagen.de

Christian S. Jensen
Department of Computer Science
Aalborg University
DK-9220 Aalborg Øst
Denmark
csj@cs.auc.dk

Manolis Koubarakis
Dept. of Electronic
and Computer Engineering
Technical University of Crete
University Campus - Kounoupidiana
GR-73100 Chania, Crete, Greece
manolis@intelligence.tuc.gr

Nikos Lorentzos
Informatics Laboratory
Agricultural University of Athens
Iera Odos 75
GR-11855 Athens, Greece
lorentzos@aua.gr

Yannis Manolopoulos
Department of Informatics
Aristotle University
GR-54006 Thessaloniki, Greece
manolopo@csd.auth.gr

Isabelle Mirbel
Laboratoire I3S
Les algorithmes
2000 Route des Lucioles
BP 121
06903 Sophia Antipolis Cedex
France
Isabelle.Mirbel@unice.fr

Enrico Nardelli
Dipartimento di Informatica
Università degli Studi di L'Aquila
I-67010 L'Aquila
Italy
Istituto di Analisi dei Sistemi
ed Informatica "Antonio Ruberti"
Consiglio Nazionale delle Ricerche
I-00185 Roma
Italy
nardelli@univaq.it

Adriano Di Pasquale
Dipartimento di Informatica
Università degli Studi di L'Aquila
I-67010 L'Aquila
Italy
dipasqua@univaq.it

Barbara Pernici
Dip. Elettronica e Informazione
Politecnico di Milano
Piazza Leonardo da Vinci 32
I-20133 Milano
Italy
pernici@elet.polimi.it

Dieter Pfoser
Department of Computer Science
Aalborg University
DK-9220 Aalborg Øst
Denmark
pfoser@cs.auc.dk

Rosanne Price
School of Computer Science
and Software Engineering
Monash University
Caulfield East
VIC 3145, Australia
rosanne@cs.rmit.edu.au

Guido Proietti
Dipartimento di Informatica
Università degli Studi di L'Aquila
I-67010 L'Aquila, Italy
Istituto di Analisi dei Sistemi
ed Informatica "Antonio Ruberti"
Consiglio Nazionale delle Ricerche
I-00185 Roma, Italy
proietti@univaq.it

Lukas Relly
Kusenstr. 15
8700 Küsnacht
Switzerland
lukas@relly.ch

Philippe Rigaux
CNAM-Paris
292, rue St. Martin
75141 Paris Cedex 03
France
rigaux@cnam.fr

Simonas Šaltenis
Department of Computer Science
Aalborg University
DK-9220 Aalborg Øst
Denmark
simas@cs.auc.dk

Hans-Jörg Schek
ETH Zürich
Institute of Information Systems
ETH Zentrum, IFW C49.1
CH-8092 Zürich
Switzerland
schek@inf.ethz.ch

Markus Schneider
Praktische Informatik IV
FernUniversität Hagen
58084 Hagen
Germany
Markus.Schneider
@FernUni-Hagen.de

Michel Scholl
INRIA
Rocquencourt BP 105
78153 Le Chesnay Cedex
France
michel.scholl@inria.fr

Timos Sellis
Dept. of Electrical
and Computer Engineering
National Technical
University of Athens
GR-15773 Zographou
Athens
Greece
timos@dblab.ece.ntua.gr

Spiros Skiadopoulos
Dept. of Electrical
and Computer Engineering
National Technical
University of Athens
GR-15773 Zographou
Athens
Greece
spiros@dblab.ece.ntua.gr

Yannis Theodoridis
Department of Informatics
University of Piraeus
80 Karaoli & Dimitriou Str.
GR-18534 Piraeus
Greece
ytheod@unipi.gr

Babis Theodoulidis
Dept. of Computation
UMIST, P.O. Box 88
Manchester
M60 1QD
United Kingdom
babis@co.umist.ac.uk

Nectaria Tryfona
Department of Computer Science
Aalborg University
DK-9220 Aalborg Øst
Denmark
tryfona@cs.auc.dk

Can Türker
ETH Zürich
Institute of Information Systems
ETH Zentrum, IFW C47.2
CH-8092 Zürich
Switzerland
tuerker@inf.ethz.ch

Theodoros Tzouramanis
Data Engineering Laboratory
Department of Informatics
Aristotle University
GR-54006 Thessaloniki
Greece
theo@csd.auth.gr

Alex Vakaloudis
Dip. Elettronica e Informazione
Politecnico di Milano
Piazza Leonardo da Vinci 32
I-20133 Milano, Italy
vakaloudisa@cpw.co.uk

Michael Vassilakopoulos
Data Engineering Laboratory
Department of Informatics
Aristotle University
GR-54006 Thessaloniki
Greece
mvass@computer.org

Michalis Vazirgiannis
Dept. of Informatics
Athens University
of Economics and Business
76 Patision Str.
10434 Athens
Greece
mvazirg@aueb.gr

Jose R.R. Viqueira
Databases Laboratory
Computer Science Department
University of A Coruna
Campus de Elvina S/N
15071 A Coruna
Spain
joserios@udc.es

Lecture Notes in Computer Science

For information about Vols. 1–2662
please contact your bookseller or Springer-Verlag

Vol. 2520: M. Koubarakis, T. Sellis, A.U. Frank, S. Grumbach, R.H. Güting, C.S. Jensen, N. Lorentzos, Y. Manolopoulos, E. Nardelli, B. Pernici, H.-J. Schek, M. Scholl, B. Theodoulidis, N. Tryfona (Eds.), Spatio-Temporal Databases. XIV, 352 pages. 2003.

Vol. 2663: E. Menasalvas, J. Segovia, P.S. Szczepaniak (Eds.), Advances in Web Intelligence. Proceedings, 2003. XII, 350 pages. 2003. (Subseries LNAI).

Vol. 2664: M. Leuschel (Ed.), Logic Based Program Synthesis and Transformation. Proceedings, 2002. X, 281 pages. 2003.

Vol. 2665: H. Chen, R. Miranda, D.D. Zeng, C. Demchak, J. Schroeder, T. Madhusudan (Eds.), Intelligence and Security Informatics. Proceedings, 2003. XIV, 392 pages. 2003.

Vol. 2666: C. Guerra, S. Istrail (Eds.), Mathematical Methods for Protein Structure Analysis and Design. Proceedings, 2000. XI, 157 pages. 2003. (Subseries LNBI).

Vol. 2667: V. Kumar, M.L. Gavrilova, C.J.K. Tan, P. L'Ecuyer (Eds.), Computational Science and Its Applications – ICCSA 2003. Proceedings, Part I. 2003. XXXIV, 1060 pages. 2003.

Vol. 2668: V. Kumar, M.L. Gavrilova, C.J.K. Tan, P. L'Ecuyer (Eds.), Computational Science and Its Applications – ICCSA 2003. Proceedings, Part II. 2003. XXXIV, 942 pages. 2003.

Vol. 2669: V. Kumar, M.L. Gavrilova, C.J.K. Tan, P. L'Ecuyer (Eds.), Computational Science and Its Applications – ICCSA 2003. Proceedings, Part III. 2003. XXXIV, 948 pages. 2003.

Vol. 2670: R. Peña, T. Arts (Eds.), Implementation of Functional Languages. Proceedings, 2002. X, 249 pages. 2003.

Vol. 2671: Y. Xiang, B. Chaib-draa (Eds.), Advances in Artificial Intelligence. Proceedings, 2003. XIV, 642 pages. 2003. (Subseries LNAI).

Vol. 2672: M. Endler, D. Schmidt (Eds.), Middleware 2003. Proceedings, 2003. XIII, 513 pages. 2003.

Vol. 2673: N. Ayache, H. Delingette (Eds.), Surgery Simulation and Soft Tissue Modeling. Proceedings, 2003. XII, 386 pages. 2003.

Vol. 2674: I.E. Magnin, J. Montagnat, P. Clarysse, J. Nenonen, T. Katila (Eds.), Functional Imaging and Modeling of the Heart. Proceedings, 2003. XI, 308 pages. 2003.

Vol. 2675: M. Marchesi, G. Succi (Eds.), Extreme Programming and Agile Processes in Software Engineering. Proceedings, 2003. XV, 464 pages. 2003.

Vol. 2676: R. Baeza-Yates, E. Chávez, M. Crochemore (Eds.), Combinatorial Pattern Matching. Proceedings, 2003. XI, 403 pages. 2003.

Vol. 2678: W. van der Aalst, A. ter Hofstede, M. Weske (Eds.), Business Process Management. Proceedings, 2003. XI, 391 pages. 2003.

Vol. 2679: W. van der Aalst, E. Best (Eds.), Applications and Theory of Petri Nets 2003. Proceedings, 2003. XI, 508 pages. 2003.

Vol. 2680: P. Blackburn, C. Ghidini, R.M. Turner, F. Giunchiglia (Eds.), Modeling and Using Context. Proceedings, 2003. XII, 525 pages. 2003. (Subseries LNAI).

Vol. 2681: J. Eder, M. Missikoff (Eds.), Advanced Information Systems Engineering. Proceedings, 2003. XV, 740 pages. 2003.

Vol. 2683: A. Rangarajan, M. Figueiredo, J. Zerubia (Eds.), Energy Minimization Methods in Computer Vision and Pattern Recognition. Proceedings, 2003. XI, 534 pages. 2003.

Vol. 2684: M.V. Butz, O. Sigaud, P. Gérard (Eds.), Anticipatory Behavior in Adaptive Learning Systems. X, 303 pages. 2003. (Subseries LNAI).

Vol. 2685: C. Freksa, W. Brauer, C. Habel, K.F. Wender (Eds.), Spatial Cognition III. X, 415 pages. 2003. (Subseries LNAI).

Vol. 2686: J. Mira, J.R. Álvarez (Eds.), Computational Methods in Neural Modeling. Proceedings, Part I. 2003. XXVII, 764 pages. 2003.

Vol. 2687: J. Mira, J.R. Álvarez (Eds.), Artificial Neural Nets Problem Solving Methods. Proceedings, Part II. 2003. XXVII, 820 pages. 2003.

Vol. 2688: J. Kittler, M.S. Nixon (Eds.), Audio- and Video-Based Biometric Person Authentication. Proceedings, 2003. XVII, 978 pages. 2003.

Vol. 2689: K.D. Ashley, D.G. Bridge (Eds.), Case-Based Reasoning Research and Development. Proceedings, 2003. XV, 734 pages. 2003. (Subseries LNAI).

Vol. 2691: V. Mařík, J. Müller, M. Pěchouček (Eds.), Multi-Agent Systems and Applications III. Proceedings, 2003. XIV, 660 pages. 2003. (Subseries LNAI).

Vol. 2692: P. Nixon, S. Terzis (Eds.), Trust Management. Proceedings, 2003. X, 349 pages. 2003.

Vol. 2693: A. Cechich, M. Piattini, A. Vallecillo (Eds.), Component-Based Software Quality. X, 403 pages. 2003.

Vol. 2694: R. Cousot (Ed.), Static Analysis. Proceedings, 2003. XIV, 505 pages. 2003.

Vol. 2695: L.D. Griffin, M. Lillholm (Eds.), Scale Space Methods in Computer Vision. Proceedings, 2003. XII, 816 pages. 2003.

Vol. 2696: J. Feigenbaum (Ed.), Digital Rights Management. Proceedings, 2002. X, 221 pages. 2003.

Vol. 2697: T. Warnow, B. Zhu (Eds.), Computing and Combinatorics. Proceedings, 2003. XIII, 560 pages. 2003.

Vol. 2698: W. Burakowski, B. Koch, A. Bęben (Eds.), Architectures for Quality of Service in the Internet. Proceedings, 2003. XI, 305 pages. 2003.

Vol. 2701: M. Hofmann (Ed.), Typed Lambda Calculi and Applications. Proceedings, 2003. VIII, 317 pages. 2003.

Vol. 2702: P. Brusilovsky, A. Corbett, F. de Rosis (Eds.), User Modeling 2003. Proceedings, 2003. XIV, 436 pages. 2003. (Subseries LNAI).

Vol. 2704: S.-T. Huang, T. Herman (Eds.), Self-Stabilizing Systems. Proceedings, 2003. X, 215 pages. 2003.

Vol. 2706: R. Nieuwenhuis (Ed.), Rewriting Techniques and Applications. Proceedings, 2003. XI, 515 pages. 2003.

Vol. 2707: K. Jeffay, I. Stoica, K. Wehrle (Eds.), Quality of Service – IWQoS 2003. Proceedings, 2003. XI, 517 pages. 2003.

Vol. 2708: R. Reed, J. Reed (Eds.), SDL 2003: System Design. Proceedings, 2003. XI, 405 pages. 2003.

Vol. 2709: T. Windeatt, F. Roli (Eds.), Multiple Classifier Systems. Proceedings, 2003. X, 406 pages. 2003.

Vol. 2710: Z. Ésik, Z, Fülöp (Eds.), Developments in Language Theory. Proceedings, 2003. XI, 437 pages. 2003.

Vol. 2711: T.D. Nielsen, N.L. Zhang (Eds.), Symbolic and Quantitative Approaches to Reasoning with Uncertainty. Proceedings, 2003. XII, 608 pages. 2003. (Subseries LNAI).

Vol. 2712: A. James, B. Lings, M. Younas (Eds.), New Horizons in Information Management. Proceedings, 2003. XII, 281 pages. 2003.

Vol. 2713: C.-W. Chung, C.-K. Kim, W. Kim, T.-W. Ling, K.-H. Song (Eds.), Web and Communication Technologies and Internet-Related Social Issues – HSI 2003. Proceedings, 2003. XXII, 773 pages. 2003.

Vol. 2714: O. Kaynak, E. Alpaydin, E. Oja, L. Xu (Eds.), Artificial Neural Networks and Neural Information Processing – ICANN/ICONIP 2003. Proceedings, 2003. XXII, 1188 pages. 2003.

Vol. 2715: T. Bilgiç, B. De Baets, O. Kaynak (Eds.), Fuzzy Sets and Systems – IFSA 2003. Proceedings, 2003. XV, 735 pages. 2003. (Subseries LNAI).

Vol. 2716: M.J. Voss (Ed.), OpenMP Shared Memory Parallel Programming. Proceedings, 2003. VIII, 271 pages. 2003.

Vol. 2718: P. W. H. Chung, C. Hinde, M. Ali (Eds.), Developments in Applied Artificial Intelligence. Proceedings, 2003. XIV, 817 pages. 2003. (Subseries LNAI).

Vol. 2719: J.C.M. Baeten, J.K. Lenstra, J. Parrow, G.J. Woeginger (Eds.), Automata, Languages and Programming. Proceedings, 2003. XVIII, 1199 pages. 2003.

Vol. 2720: M. Marques Freire, P. Lorenz, M.M.-O. Lee (Eds.), High-Speed Networks and Multimedia Communications. Proceedings, 2003. XIII, 582 pages. 2003.

Vol. 2721: N.J. Mamede, J. Baptista, I. Trancoso, M. das Graças Volpe Nunes (Eds.), Computational Processing of the Portuguese Language. Proceedings, 2003. XIV, 268 pages. 2003. (Subseries LNAI).

Vol. 2722: J.M. Cueva Lovelle, B.M. González Rodríguez, L. Joyanes Aguilar, J.E. Labra Gayo, M. del Puerto Paule Ruiz (Eds.), Web Engineering. Proceedings, 2003. XIX, 554 pages. 2003.

Vol. 2723: E. Cantú-Paz, J.A. Foster, K. Deb, L.D. Davis, R. Roy, U.-M. O'Reilly, H.-G. Beyer, R. Standish, G. Kendall, S. Wilson, M. Harman, J. Wegener, D. Dasgupta, M.A. Potter, A.C. Schultz, K.A. Dowsland, N. Jonoska, J. Miller (Eds.), Genetic and Evolutionary Computation – GECCO 2003. Proceedings, Part I. 2003. XLVII, 1252 pages. 2003.

Vol. 2724: E. Cantú-Paz, J.A. Foster, K. Deb, L.D. Davis, R. Roy, U.-M. O'Reilly, H.-G. Beyer, R. Standish, G. Kendall, S. Wilson, M. Harman, J. Wegener, D. Dasgupta, M.A. Potter, A.C. Schultz, K.A. Dowsland, N. Jonoska, J. Miller (Eds.), Genetic and Evolutionary Computation – GECCO 2003. Proceedings, Part II. 2003. XLVII, 1274 pages. 2003.

Vol. 2725: W.A. Hunt, Jr., F. Somenzi (Eds.), Computer Aided Verification. Proceedings, 2003. XII, 462 pages. 2003.

Vol. 2726: E. Hancock, M. Vento (Eds.), Graph Based Representations in Pattern Recognition. Proceedings, 2003. VIII, 271 pages. 2003.

Vol. 2727: R. Safavi-Naini, J. Seberry (Eds.), Information Security and Privacy. Proceedings, 2003. XII, 534 pages. 2003.

Vol. 2728: E.M. Bakker, T.S. Huang, M.S. Lew, N. Sebe, X.S. Zhou (Eds.), Image and Video Retrieval. Proceedings, 2003. XIII, 512 pages. 2003.

Vol. 2731: C.S. Calude, M.J. Dinneen, V. Vajnovszki (Eds.), Discrete Mathematics and Theoretical Computer Science. Proceedings, 2003. VIII, 301 pages. 2003.

Vol. 2732: C. Taylor, J.A. Noble (Eds.), Information Processing in Medical Imaging. Proceedings, 2003. XVI, 698 pages. 2003.

Vol. 2733: A. Butz, A. Krüger, P. Olivier (Eds.), Smart Graphics. Proceedings, 2003. XI, 261 pages. 2003.

Vol. 2734: P. Perner, A. Rosenfeld (Eds.), Machine Learning and Data Mining in Pattern Recognition. Proceedings, 2003. XII, 440 pages. 2003. (Subseries LNAI).

Vol. 2741: F. Baader (Ed.), Automated Deduction – CADE-19. Proceedings, 2003. XII, 503 pages. 2003. (Subseries LNAI).

Vol. 2743: L. Cardelli (Ed.), ECOOP 2003 – Object-Oriented Programming. Proceedings, 2003. X, 501 pages. 2003.

Vol. 2745: M. Guo, L.T. Yang (Eds.), Parallel and Distributed Processing and Applications. Proceedings, 2003. XII, 450 pages. 2003.

Vol. 2746: A. de Moor, W. Lex, B. Ganter (Eds.), Conceptual Structures for Knowledge Creation and Communication. Proceedings, 2003. XI, 405 pages. 2003. (Subseries LNAI).

Vol. 2748: F. Dehne, J.-R. Sack, M. Smid (Eds.), Algorithms and Data Structures. Proceedings, 2003. XII, 522 pages. 2003.

Vol. 2749: J. Bigun, T. Gustavsson (Eds.), Image Analysis. Proceedings, 2003. XXII, 1174 pages. 2003.

Vol. 2750: T. Hadzilacos, Y. Manolopoulos, J.F. Roddick, Y. Theodoridis (Eds.), Advances in Spatial and Temporal Databases. Proceedings, 2003. XIII, 525 pages. 2003.

Vol. 2751: A. Lingas, B.J. Nilsson (Eds.), Fundamentals of Computation Theory. Proceedings, 2003. XII, 433 pages. 2003.

Vol. 2753: F. Maurer, D. Wells (Eds.), Extreme Programming and Agile Methods – XP/Agile Universe 2003. Proceedings, 2003. XI, 215 pages. 2003.

Vol. 2759: O.H. Ibarra, Z. Dang (Eds.), Implementation and Application of Automata. Proceedings, 2003. XI, 312 pages. 2003.